A. Sakai · W. Larcher

Frost Survival of Plants

Responses and Adaptation to Freezing Stress

With 200 Figures

Springer-Verlag
Berlin Heidelberg New York
London Paris Tokyo

Professor em. Dr. AKIRA SAKAI
The Institute of Low Temperature Science
Hokkaido University
Sapporo, Japan

Professor Dr. WALTER LARCHER
Institut für Botanik der Universität Innsbruck
Sternwartestraße 15
A-6020 Innsbruck, Austria

ISBN-13:978-3-642-71747-5 e-ISBN-13:978-3-642-71745-1
DOI: 10.1007/978-3-642-71745-1

Library of Congress Cataloging in Publication Data. Sakai, A. (Akira), 1920– . Frost survival of plants. (Ecological studies; v. 62). Bibliography: p. Includes indexes. 1. Plants, Effect of cold on. 2. Plants–Frost resistance. 3. Acclimatization (Plants) 4. Plants–Wounds and injuries. I. Larcher, W. (Walter), 1929— . II. Title. III. Series. QK756.S235 1987 581.19′165 86–31544.

Softcover reprint of the hardcover 1st edition 1987

2131/3130-543210

Ecological Studies, Vol. 62

Analysis and Synthesis

Ecological Studies

Volume 49
Forest Ecosystems in Industrial Regions Studies on the Cycling of Energy, Nutrients and Pollutants in the Niepolomice Forest, Southern Poland
Edited by W. Grodziński, J. Weiner, and P.F. Maycock
1984. XVIII. 277p., 116 figures. cloth
ISBN 3-540-13498-0

Volume 50
The Gulf of Agaba (Elat) Ecological Micropaleontology
By Z. Reiss and L. Hottinger
1984. VIII, 360p., 207 figures. cloth
ISBN 3-540-13486-7

Volume 51
Soil Salinity under Irrigation Processes and Management
Edited by I. Shainberg and J. Shalhevet
1984. X, 358p., 133 figures. cloth
ISBN 3-540-13565-0

Volume 52
Air Pollution by Photochemical Oxidants Formation, Transport, Control and Effects on Plants
Edited by Robert Guderian
1985. XI, 346p., 54 figures. cloth
ISBN 3-540-13966-4

Volume 53
The Gavish Sabkha A Model of a Hypersaline Ecosystem
Edited by G.M. Friedman and W.E. Krumbein
1985. X, 484p., 246 figures. cloth
ISBN 3-540-15245-8

Volume 54
Tidal Flat Ecology An Experimental Approach to Species Interactions
By Karsten Reise
1985. X, 198p., 69 figures. cloth
ISBN 3-540-15447-7

Volume 55
A Eutrophic Lake Lake Mendota, Wisconsin
By Thomas D. Brock
1985. XII, 308p., 82 figures. cloth
ISBN 3-540-96184-4

Volume 56
Resources and Society A Systems Ecology Study of the Island of Gotland, Sweden
By James J. Zucchetto and Ann-Mari Jansson
1985. X, 248p., 70 figures. cloth
ISBN 3-540-96151-8

Volume 57
Forest Ecosystems in the Alaskan Taiga A Synthesis of Structure and Function
Edited by K. Van Cleve, F.S. Chapin III, L.A. Viereck, C.T. Dyrness and P.W. Flanagan
1986. X, 240p., 81 figures. cloth
ISBN 3-540-96251-4

Volume 58
Ecology of Biological Invasions of North America and Hawaii
Edited by H.A. Mooney and J.A. Drake
1986. X, 320p., 25 figures. cloth
ISBN 3-540-96289-1

Volume 59
Acid Deposition and the Acidification of Soils and Waters
By J.O Reuss and D.W. Johnson
1986. VIII, 120p., 37 figures. cloth
ISBN 3-540-96290-5

Volume 60
Amazonian Rain Forests
Edited by C.F. Jordan
1987. X, 133p., 55 figures. cloth
ISBN 3-540-96397-9

Volume 61
Potentials and Limitations of Ecosystem Analysis
Edited by E.-D. Schulze and H. Zwölfer
1987. XII, 435p., 141 figures. cloth
ISBN 3-540-17138-X

Volume 62
Frost Survival of Plants
By A. Sakai and W. Larcher
1987. XI, 321 p., 200 figures. cloth
ISBN 3-540-17332-3

Preface

Low temperature represents, together with drought and salt stress, one of the most important environmental constraints limiting the productivity and the distribution of plants on the Earth. Winter survival, in particular, is a highly complex phenomenon, with regards to both stress factors and stress responses. The danger from winter cold is the result not only of its primary effect, i.e. the formation of ice in plant tissues; additional threats are presented by the freezing of water in and on the ground and by the load and duration of the snow cover. In recent years, a number of books and reviews on the subject of chilling and frost resistance in plants have appeared: all of these publications, however, concentrate principally on the mechanisms of injury and resistance to freezing at the cellular or molecular level. We are convinced that analysis of the ultrastructural and biochemical alterations in the cell and particularly in the plasma membrane during freezing is the key to understanding the limits of frost resistance and the mechanisms of cold acclimation. This is undoubtedly the immediate task facing those of us engaged in resistance research. It is nevertheless our opinion that, in addition to understanding the basic physiological events, we should be careful not to overlook the importance of the comparative aspects of the freezing processes, the components of stress avoidance and tolerance and the specific levels of resistance. Frost resistance is a genetically determined ecophysiological trait which is expressed under environmental constraints. The ability to harden is the basis for the differences between individuals, ecotypes, varieties and species with respect to their potential frost resistance. The gene pool of every population exhibits a range of variations both in hardening potential and seasonality timing, enabling it to survive a certain degree of change in the environment. The variability of frost resistance within a population is essential for the survival of a species following frost of unusual severity or of untimely occurrence, as well as for its adaptation to long-term fluctuations in climate. The degree of scatter of resistance within the progeny is characteristic for a species and provides a measure of its scope and selection. Thus, the quantitative analysis of phenomena over the full range of their variability among plant groups, species and genotypes provides important information from the viewpoint of ecology, phylogenetics and applied science. Climatic stress promotes evolutionary adaptation and accelerates the differentiation of ecotypes and species.

Progression in the cold hardiness of plants along latitudinal and altitudinal gradients, and transitions between the various categories of cold hardiness have been hypothesized as being steps of evolutionary adaptation to low temperatures during the geological periods in which colder climates prevailed, as well as being connected with the spread of plants to colder regions.

The book has been designed to include both the cellular and the comparative aspects of freezing stress and plant survival. With this special problem in mind, an attempt has been made to cover the entire range of ecophysiological research, from the biochemical to an ecological viewpoint. We have not attempted to produce a comprehensive compendium of plant cryobiology, but rather to present data on responses of plants to freezing stress, drawing to a large extent upon our personal experience gained in more than three decades of experimental research. Let us hope that the information presented here will make a significant contribution to the efforts of investigators engaged in planning research strategies for the study of plant hardiness in order to reduce crop losses resulting from frost and related stresses. Furthermore, it is our hope that more attention than hitherto will be given to the ecological interpretation of cold stress and cold resistance.

Many colleagues have helped us by means of discussion, by providing us with unpublished data or by placing at our disposal original photographs and graphs: we gratefully acknowledge the help of members of the Institute of Low Temperature Science at the Hokkaido University in Sapporo: S. Yoshida, T. Niki, T. Sato, M. Ishikawa, M. Uemura, F. Yoshie; of the Hokkaido Forest Tree Breeding Institute: S. Eiga; and of the Institute of Botany at the University of Innsbruck: M. Bodner and U. Tappeiner. Further, we should like to express our thanks to C. J. Weiser and K. Tanino (Oregon State University), E. L. Proebsting (Washington State University), P. Wardle (DSIR, New Zealand), A. Kacperska (Warsaw University), M. J. Earnshaw (University of Manchester), M. Senser (University of Munich) and E. Beck (University of Bayreuth) for valuable suggestions and stimulating discussions. We should also like to take this opportunity to extend our gratitude to all persons and institutions who have helped us over the years by sending us their publications. We are most grateful to J. Wagner, A. Deutsch, R. Gapp and S. Hirn for their assistance in preparing the manuscript and with the compilation of indices. Our special thanks go to J. Wieser who has translated part of the text and carefully checked the remainder.

It is a great pleasure to express our sincere thanks to Dr. K. F. Springer and the Editor responsible for the *Ecological Studies,* Prof. Dr. O. L. Lange, for including our book in their program, also to Dr. D. Czeschlik and his colleagues at Springer-Verlag for their cooperation and care in its preparation.

Sapporo and Innsbruck, January 1987

Akira Sakai
Walter Larcher

Contents

Abbreviations

°C	Degree Celsius (centigrade); temperature level
DR	Directly recorded temperature (°C)
DTA	Differential thermal analysis (K)
HTE	High temperature exotherm
K	Kelvin; unit of temperature, temperature difference
LN_2	Liquid nitrogen (−196 °C)
LT	Lowest temperature at a defined degree of lethality; quantification of frost resistance
LT_i	Initial injury temperature; incipient killing temperature
LT_o	Lowest temperature without injury
LT_{50}	Temperature at 50% lethality
LTE	Low temperature exotherm
LTE_{50}	Median temperature of LTEs
R_{act}	Actual level of frost resistance corresponding to the state of hardening at a given time
R_{pot}	Maximum attainable level of frost resistance
T_f	Threshold freezing temperature
T_{sc}	Threshold supercooling temperature

For *Terminology* and *Definitions* see pp. 305–306.

1. Low Temperature and Frost as Environmental Factors

1.1 Low Temperature Hazards

Cold and frost are important environmental factors limiting the productivity and distribution of plants (Dilley et al. 1975; Larcher 1981a; Larcher and Bauer 1981). In tropical mountains air temperatures below 8° to 10 °C and soil temperatures below +15 °C set the limits for the altitudinal range of chilling-sensitive wild plants and plantation crops. Recurrent frost plays a decisive role in determining the distribution area of tropical and the majority of subtropical plants. It is very probable that low winter temperatures are responsible for the latitudinal and altitudinal limits of broad-leaved evergreen woody plants of regions with mild winters. Freezes with minimum temperatures of -5° to -10 °C frequently cause severe damage to citrus and olive plantations in regions with Mediterranean climate. In the cool temperate and the boreal zones spring frost has been considered as an important factor limiting the distribution of various woody plant species. Spring frosts also represent a regularly occurring risk for horticulture and sylviculture in the temperate zone. The chief problem involved in establishing tree plantations for energy farming in tundra regions is summer frost. In arctic and subarctic regions the low summer temperatures are the main obstacle to successful agriculture and horticulture, thus rendering permanent settlement impracticable.

1.2 Cold, Frost and Snow

Low temperatures are the result of a negative heat balance of a body, microsite or larger area of the earth's surface. Globally, the energy balance of the earth is the difference between energy input as short-wave irradiation, and energy losses due to the emission of long-wave radiation. During the phase of insolation, from shortly after sunrise until shortly before sunset, energy flows to the earth's surface and the radiation balance is positive. In the nocturnal phase, the amount of energy lost due to "radiational cooling" is the greater, the longer the night and the less the long-wave radiation from the earth's surface is prevented by clouds, fog or other turbidity. Long, clear nights accompanied by low air humidity especially favour loss of energy. The radiation balance is thus principally dependent on:

1. the angle of incidence of the irradiation, which depends upon the geographical latitude of the site concerned, the season and the slope of the terrain;

2. the ratio of day length to night length; and on
3. topographical variations in net radiation.

Cold is the thermodynamic state characterized by low kinetic energy of the molecules. The biological significance of cold lies in the retardation of chemical processes (Van't Hoff's rule), displacement of equilibrium reactions in the direction of energy liberation (Le Chatelier principle) and in alterations in biologically important structures (e.g. phase transitions in the lipid components of biomembrane).

Frost is a condition in which temperatures fall below 0 °C. The process of solidification of water and the formation of ice in aqueous solutions is termed *freezing*. In plant tissues, freezing of water occurs at temperatures lower than 0 °C (see Chap. 2); therefore, climatological data on frost (as meteorologically defined) do not necessarily give the appropriate information for judging the jeopardization of plants. For this reason, additional data concerning temperature drops below thresholds which are harmful for life functions and viability should also be given (Table 1.1). Of great significance for the plant is the level to which the temperature sinks during the frost, the duration of the frost, the time of onset, whether it is confined to the area surrounding the shoot or if it also penetrates the ground.

Ground frost is an important phenomenon in stress ecology: not only does it limit the survival of underground organs of the plant, it also prevents the uptake of water and mineral nutrients from the soil. In addition, by loosening the soil structure, especially by geliturbation and solifluction in arctic (Sigafoos 1952) and high-mountain soils (Franz 1979), and by needle ice formation, ground frost can also expose roots and uproot young plants (frost heaving).

Snow is a low-temperature event which, although providing certain advantages for the vegetation, can also become a stress factor. Although plants are protected by the snow from very low temperatures, wind, and winter desiccation, they are deprived of light: prolonged snow cover weakens the plants by impeding metabolic processes and

Table 1.1. Occurrence of freezing events at the alpine timberline with respect to their cryobiological relevance. (From Tranquillini and Holzer 1958)

Air temperature	Frequency (percent observations)				
	Oct.	Nov.	Dec.	Feb.	March
Daily maximum					
below 0 °C[a]	3	30	43	26	4
below −8 °C[b]	0	0	4	4	0
Daily minimum					
above 0 °C	35	0	0	0	0
above −4 °C[c]	61	14	17	15	11
Freeze-thaw cycles					
min < 0 °C, max > 0 °C	61	70	56	74	96
min < −8 °C, max > −4 °C	16	43	35	52	43

[a] Meteorological standard.
[b] Supercooling threshold of conifer needles.
[c] Freezing threshold temperature of conifer needles.

by reducing the growing season. Heavy snowfall also means that the plants have to support large loads. The deleterious effect of snow on the vegetation thus depends on the depth of the snow cover and its duration.

1.3 The Occurrence of Cold, Frost and Snow

Cold is a stress factor of widespread occurrence. Low temperatures and frost increase zonally with rising geographical latitude, regionally with distance from the coast and altitudinally in mountains and highlands.

At intermediate and high geographical *latitudes* the radiation balance becomes increasingly negative from summer towards winter due to the lower solar elevation and the increasing night length. This results in cooling and the accumulation of cold air masses which are transported by the atmospheric circulation to lower latitudes where they bring about a drop in temperature. A typical advective frost occurs when a body of cold air of several cubic kilometers is brought in from a distant area of origin. The cooling affects all altitudes and is not confined to a particular time of day. Cold air can also approach from smaller areas of origin in the more immediate vicinity, in which case the body of cold air is smaller (1000 m^3); it flows slowly along the valleys and across slopes into the valley basin and hollows in the terrain, where it accumulates and becomes covered by a layer of warmer air (temperature inversion). The greatest decrease in temperature is attained when the intrusion of cold air masses coincide with strong radiational cooling on clear calm nights.

Table 1.2. Average annual minimum air temperatures of the Earth (Hoffmann 1963)

Temperature range (°C)	Land area (%)
> 15°	6.6
15 ... 10	9.4
10 ... 5	9.5
5 ... 0	10.5
0 ... − 5	11.0
− 5 ... −10	4.4
−10 ... −15	3.5
−15 ... −20	3.5
−20 ... −25	4.2
−25 ... −30	3.9
−30 ... −35	3.9
−35 ... −40	4.6
−40 ... −45	6.3
−45 ... −50	5.4
−50 ... −55	3.9
−55 ... −60	2.1
−60 ... −70	3.5
−70 ... −80	2.0
−80 ... −90	1.8

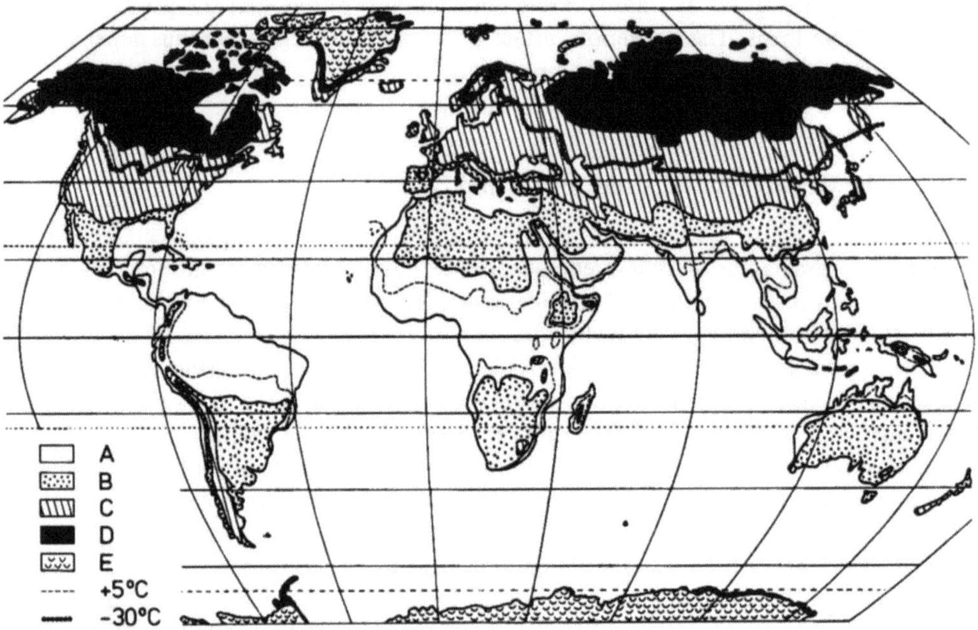

Fig. 1.1. Global occurrence of low temperatures and frost. *A* frost-free; *B* episodic frost to −10 °C; *C* average annual minimum between −10 °C and −40 °C; *D* average annual minimum below −40 °C; *E* polar ice. (From Larcher and Bauer 1981). For detailed frost zone maps see Heinze and Schreiber (1984; Europe, W. Asia), Sakai (1978a; Japan), Ouellet and Sherk (1968; Canada) and USDA-ARS Publ. No. 814 (USA)

The only places on the earth where temperatures do not fall below +15 °C are the continuously warm, humid lowland regions of the Amazon Basin, the Congo Basin and parts of southeast Asia (6.6% of the total continental area of the earth). Only 25% of the entire continental area can be considered absolutely safe from frost (Fig. 1.1; Table 1.2). For 64% of the earth's land mass the mean minimum air temperature is below 0 °C, for 48% below − 10 °C (i.e. the approximate survival limit for freezing-sensitive plants; cf. Sect. 3.1), for 35% it is below − 20 °C (the approximate survival limit for broad-leaved evergreen trees of maritime temperate regions; cf. Sect. 7.2), and for 25% it is below − 40 °C (distribution limit for woody plants with deep supercooling xylem; cf. Sect. 4.2). In eastern Siberia, in inland Alaska and northwestern Canada, where the trees still form stands, the temperature regularly decreases to − 50 °C. Trees wintering in these severe climates remain frozen for 3 or 4 months or more. An absolute temperature minimum of − 68 °C has been recorded for Verkhoyansk, the coldest eastern Siberian lowland station, and one of − 71 °C for Oimyakon, also in eastern Siberia, but situated in a high-lying valley (Borisov 1965; Lydolph 1977). Temperature minima of − 45 °C to − 55 °C have been reported from coastal stations in the Antarctic and values of about − 91 °C from the interior of the continent (Orvig 1970; Kuhn et al. 1973). In those polar regions and in peak regions of high mountains where the temperature never exceeds 0 °C (8.5% of the total land area of the earth) active life is only possible in habitats that offer some special kind of advantage, such as rocks protruding above the ice, which provide a better microclimate (Siegel et al. 1969). Here, apart from cryo-

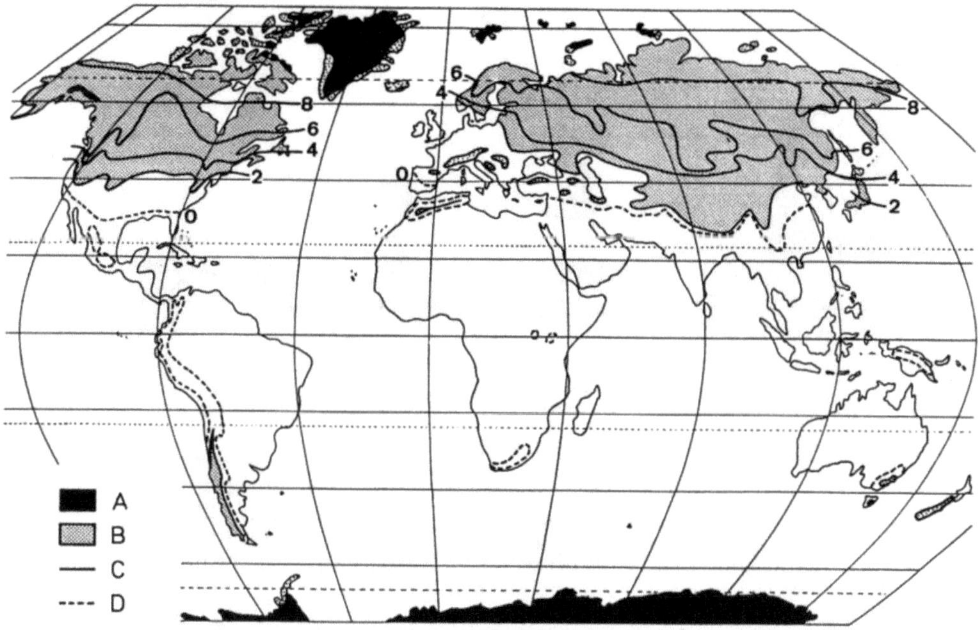

Fig. 1.2. Global distribution of snow cover. *A* Permanent snow and ice; *B* stable snow cover every year; *C* duration of snow cover (*numerals* indicate months with stable snow); *D* snow cover forms almost every year, but is unstable. (From Mellor 1964)

plankton, psychrophytic algae and bacteria, are to be found poikilohydric lichens and mosses that can survive even the lowest temperatures of the earth in a desiccated or frozen state.

Approximately 90% of the oceanic waters of the world is colder than 5 °C (Baross and Morita 1978). In northern regions of the Arctic Ocean where macroalgae still occur the water temperatures remain between 0° and −1.8 °C (the freezing point of seawater) throughout the year (Lüning 1985). A covering of ice forms on the northern polar seas at latitudes higher than 60° N (except the Norwegian and Barents Seas where the Gulf Stream advances). The upper decimeters of the ice sheet of inland waters in the Arctic may cool down to −10° to −20 °C in winter, although even here the temperature below a depth of about 1 m does not drop below 0 °C (Brewer 1958).

As the temperature drops, the proportion of *snow* in the total precipitation increases, with the result that at latitudes above 70° N and 60° S, half of the precipitation falls as snow (Lauscher 1976). In the regions of central Asia and N. America with a continental climate and where cold air penetrates into lower latitudes, the limits of frequent snowfall (25% precipitation as snow) are displaced by 10° latitude to the south. Figure 1.2 shows the distribution and the duration of a snow cover on the earth.

The permafrost zone extends to latitudes of 65°–60° N; patchy permafrost is encountered as far south as 50° N, and even further in central Asia (Ives 1974; Lockwood 1974). Where the ground remains frozen for a large part of the year, particularly in permafrost regions, the forest gradually thins out and gives way to tundra vegetation (Tikhomirov 1971; Sakai et al. 1979; Oechel and Lawrence 1985; Fig. 1.3).

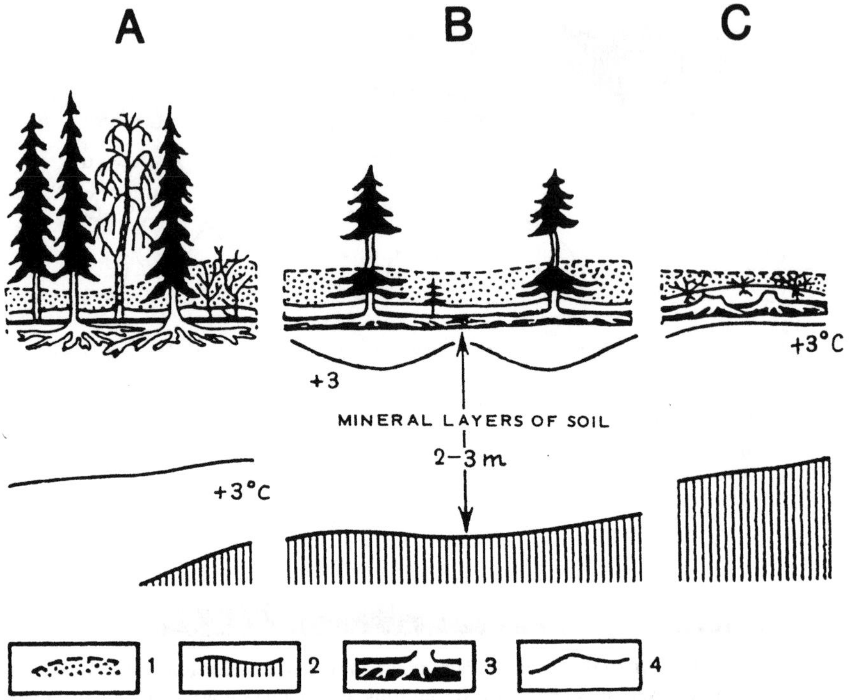

Fig. 1.3A–C. Environmental conditions in the subarctic taiga-tundra ecotone. **A** Northern taiga; **B** forest tundra; **C** tundra. *1* Snow cover in winter; *2* permafrost; *3* peat; *4* position of soil isotherms at end of August. (After Govorukhin from Tikhomirov 1971)

The zonal and regional patterns in the occurrence of low temperatures and frost vary with altitude and topography.

In the *mountainous regions* of the tropics and subtropics, the mean and the absolute temperature minima drop by 0.45–0.65 K per 100 m altitude, which is well within the normal range of temperature lapse (Lauscher 1976/77). The same applies to the mountains of the temperate zone in summer (Franz 1979; Fig. 1.4). At the upper distribution limit of vascular plants, lowest temperatures of −10° to −13 °C were recorded in July and August at elevations between 3000–3700 m in the Alps (Lauscher 1960; Swiss Institute for Meteorology 1955–1975) and in the Caucasus (Nakhutsrishvili and Gamtsemlidze 1984). In winter, however, as a result of temperature inversion, the temperature lapse is smaller. With increasing elevation ground frost and snow extend over longer periods, and days with minimum and maximum temperatures at or below 0 °C become more frequent (Tables 1.3 and 1.4). On an average, at 2000 m in the European Alps the daily mean air temperature remains at or below 0 °C for nearly half the year; at the same elevation on 39% of all days of the year the soil is frozen at a depth of 10 cm, and from the end of October to the end of April more than 75% of the precipitation falls as snow (Lauscher 1976), the snow cover usually persisting for 60% of the year.

Topography may be responsible for low temperature environments arising from cold air currents at the valley floors and in depressions. The meteorological explanation of air movements of this kind can be found in textbooks of atmospheric physics

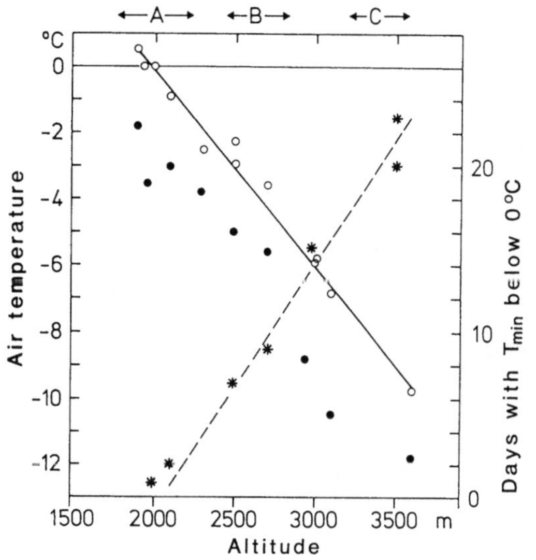

Fig. 1.4. Frost in July at various altitudes in the Alps: (•) absolute minimum air temperature recorded during the observation period (at least 20 years): (○) mean of the lowest temperatures; (*) mean number of days with minimum below 0 °C. Altitudinal ranges of vegetation boundaries: *A* timberline; *B* zone of fragmentation of herbaceous communities; *C* upper limit of occurrence of vascular plants. [Data from Lauscher (1960), Schüepp (1967, 1968), Fliri (1975) and Annals of the Swiss Institute for Meteorology (1955–1975]

Table 1.3. Winter temperatures (°C) in the European Alps. [From Larcher (1985b), extended by data from Annals of the Swiss Institute of Meteorology (1930)]

Altitude (m)	Mean of the coldest month (°C)	Mean daily minimum of coldest month (°C)	Mean daily maximum of coldest month (°C)	Absolute lowest air temperature[a] (°C)
500	– 3	– 5 ... –7	+1 ... +2	–27 (a)
1000	– 4	– 6 ... –7	0 ... +1	–28 (b)
1500	– 4.5	– 7 ... –8	0 ... –1	–27 (c)
2000	– 7	– 8 ... –10	–2 ... –4	–29 (d)
2500	–10	–11 ... –12	–4 ... –6	–30 (e)
3000	–12.5	–15	–10	–37 (f)
3500	–15			–37 (g)

[a] Absolute temperature minima from (a) Innsbruck 580 m; (b) Rinn near Innsbruck 900 m; (c) Davos 1560 m; (d) Mt. Patscherkofel near Innsbruck 2045 m; (e) Grand St. Bernhard 2479 m and Weißfluhjoch 2667 m; (f) Sonnblick 3106 m; (g) Jungfraujoch 3576 m.

Table 1.4. Average duration of periods with frost and snow at various altitudes in the Alps. (Data from Lauscher 1946; Schüepp 1968; Fliri 1975; Lauscher and Lauscher 1980)

Altitude (m)	Annual number of days with minimum air temperature ⩽ 0 °C	Annual number of days with maximum air temperature ⩽ 0 °C	Annual number of days with soil temperature at 10 cm ⩽ 0 °C	Number of days with snow cover
500	120	30	58	70
1000	140	40	72	127
1500	170	60	96	167
2000	200	95	145	214
2500	250	150		280
3000	300	230		354
3500	350	290		365

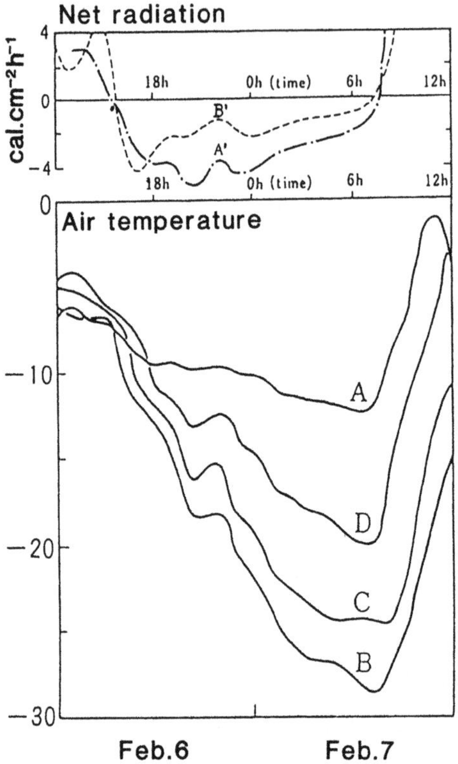

Fig. 1.5

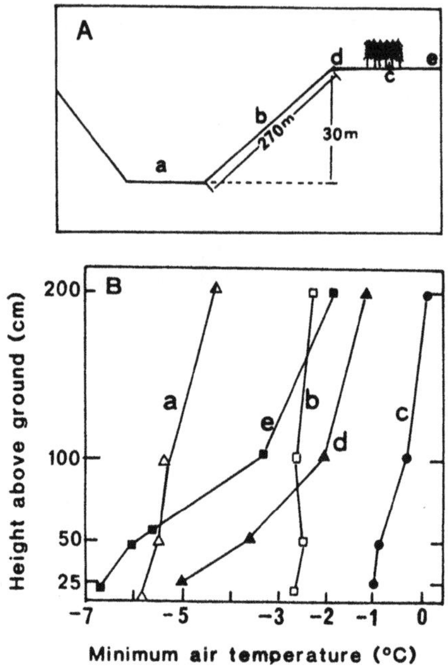

Fig. 1.6

Fig. 1.5. Radiation cooling in a depression during a clear night (Hokkaido University Forest Toikanbetsu). Net radiation at the top of the slope (*A'*) and the bottom of the depression (*B'*). Air temperatures at the bottom (*B*), at 35 m above the bottom (*C*), at 105 m (*D*) and at 230 m, on the plateau (*A*). (From Ishikawa 1977)

Fig. 1.6A,B. Minimum air temperature profiles the morning after a late frost in accordance to topography (from Konda and Sasaki 1959). Four-year-old seedlings of *Abies sachalinensis* planted on the bottom of the depression (*a*) and on the plateau (*e*) were killed by frost, whereas plants under a canopy (*c*) and on the slope (*b*) remained undamaged. (From Konda and Sasaki 1959)

and microclimatology; examples are shown in Figs. 1.5–1.7. The differences in temperature between the valley floors and the slopes depend upon the nature of the terrain and on the possibility for the cold air to escape from the depressions. Cold air also accumulates in emission areas, such as bogs and mires, where heat is poorly conducted from ground to surface, resulting in frosts well into the summer. Snow deposition and the duration of the snow cover, too, are strongly influenced by the topoclimate. This applies especially in high mountains where microsite differentiation during winter is primarily determined by wind-dependent distribution of snow (see Fig. 8.1).

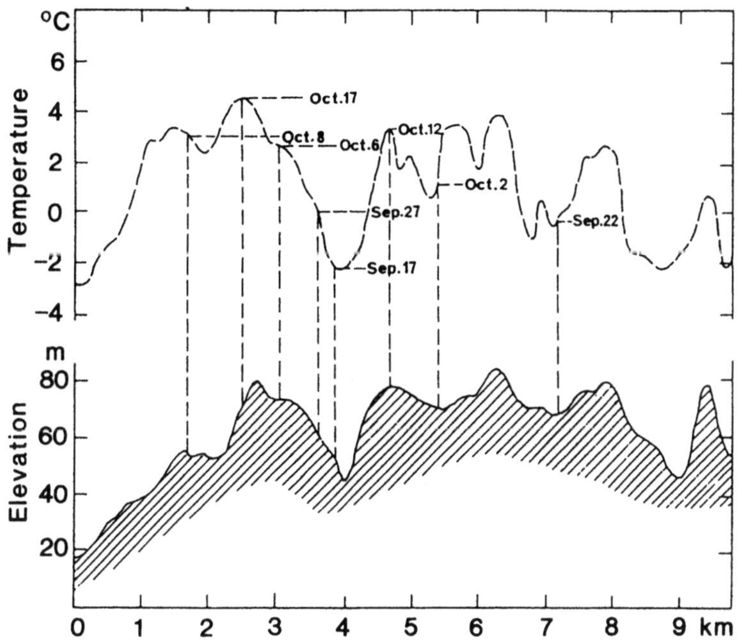

Fig. 1.7. Dependence of night frost temperatures and average early frost dates on topography. (From Bootsma 1976)

1.4 Frost and Snow in the Plant's Environment

Following convection of cold air, the temperature of plants usually equalizes to that of the environment within a very short time (e.g. apple flower buds within 10–20 min; Pisek 1958). Only massive organs containing large quantities of water lose heat more slowly.

The lowest temperature in a *stand of plants* is immediately beneath the upper surface (Fig. 1.8). A forest canopy, on the other hand, reduces the thermal radiant loss and thus the freezing risk of the understorey layer (cf. Table 1.5). In open plant communities the effective surfaces for radiational cooling are provided by the peripheral leaves and the soil surface, above which a boundary layer of cold air develops. During

Table 1.5. Lowest air and soil temperatures at the Hokkaido University Forest, Tomakomai, during winter 1979/80. (From Yoshie and Sakai 1982)

Habitat	Minimum air temperature (°C) at 1 cm above the litter	Minimum soil temperature (°C)		
		0 cm	10 cm	20 cm
Deciduous broad-leaved forest	-11.9	- 4.5	-1.3	-0.3
Larch forest	-12.5	- 6.0	-1.5	-0.3
Fir forest	-13.0	- 8.5	-4.5	-2.3
Bare ground	-24.4	-10.9	-9.5	-6.9

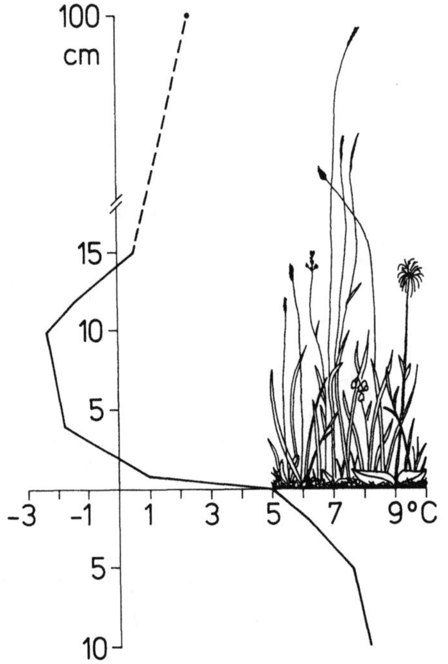

Fig. 1.8. Vertical temperature profile in an alpine pasture (Aveno-Nardetum, 1960 m a.s.l.) at 04.00 h in late August. The lowest temperature appears at the surface of the dense vegetation layer. (From Tappeiner 1985)

a cold night, the air layers at the surface of the stand and near the ground are 2 to 5 K cooler than the air at a height of 2 m, which is the reference level for meteorological measurements (Fig. 1.9). Therefore, climatic data, particularly temperature extremes, as recorded at 2 m above the ground in the weather shelters of meteorological stations are, at best, applicable to solitary trees, and may be very different from the temperatures nearer the ground or in a dense vegetation. An especially large drop in temperature can be observed in the lowest air layer if heat transfer from the ground is prevented by an insulating layer of litter or peat. In such situations, frost events are more frequent and the frost-free period becomes shorter. Marcellos and Single (1975) and Bootsma (1976) present regression constants and coefficients for the estimation of the probability of frost near the ground, based on differences between screen minimum and grass minimum.

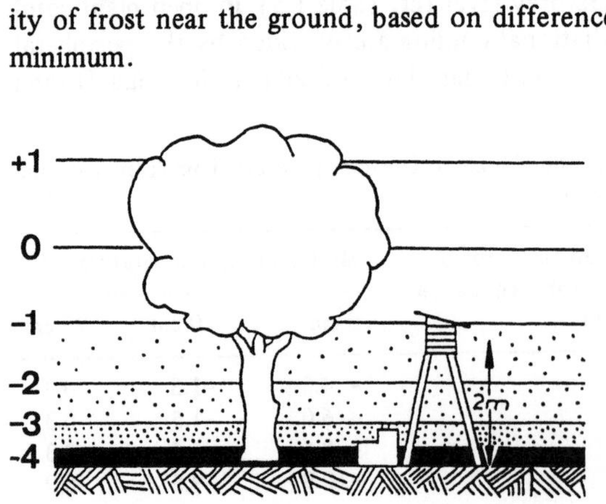

Fig. 1.9. Stratification of cold air during a clear, calm night. (After Hofmann, from Burckhardt 1963)

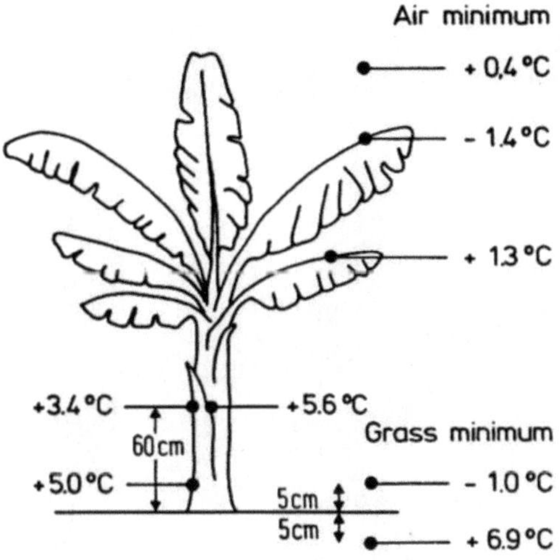

Fig. 1.10

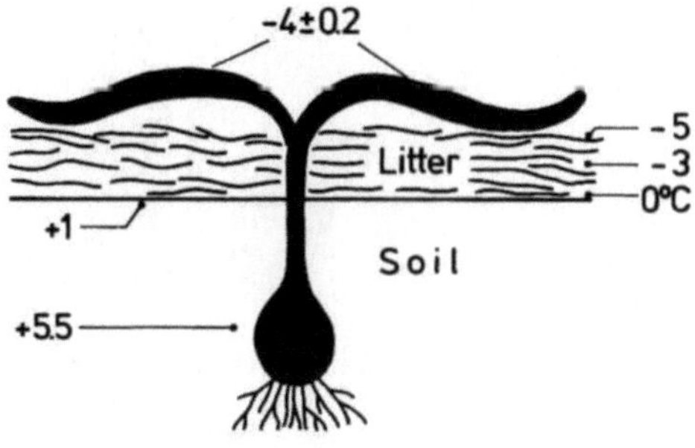

Fig. 1.11

Fig. 1.10. Minimum temperatures of various parts of a banana plant, and air and soil temperatures during radiation cooling. (From Shmueli 1960)

Fig. 1.11. Temperature distribution on a geophyte *(Scilla sibirica)* and in its microenvironment during a spring frost. (From Goryshina 1969)

Under conditions of radiational cooling, *solitary plants* cool to about 1 to 3 K below the minimum air temperature (Fig. 1.10), depending upon growth habit, degree of exposure and the incoming heat from the surroundings. Even in a small area on the same topograph plants are exposed to different low temperatures depending on their life forms and microhabitats (Fig. 1.11). Temperature differences may also arise across the individual leaf, due to the faster heat transfer at the tip and the edge (Fig. 1.12). This may lead to characteristic injury patterns.

Frost penetrates the *soil* very slowly, i.e. only if low temperatures persist for a considerable length of time, in which case the speed of cooling and the depth which the frost finally reaches depend upon the ground cover, the type of soil and its water content. Dry soil freezes sooner than wet soil, and bare ground sooner than covered soil. Freezing of soil is delayed by the supercooling of its water: with a water content of 30% (w/v) a soil may temporarily supercool to −1.4 °C, and with 10% (w/v) down to −3.0 °C (Eckel 1960). When, after a period of supercooling, which may last several hours, the soil water begins to freeze, heat of crystallization is set free as the ice forms, and the soil temperature remains at about 0 °C whilst freezing is in progress. The speed and depth of freezing depend on the nature of the soil. Comparative studies by Kreutz (1942) revealed that in a coarse, dry soil, frost penetrated at a mean speed of 2 cm day^{-1} and reached a depth of 67 cm. In wet humus, on the other hand, the speed of frost penetration was only 0.6 cm day^{-1} and it only reached a depth of 32 cm, but the ice did not disappear in spring from beneath its thawed surface until 25 days later than in the dry soil. Mean minimum air and soil temperatures for bare ground and various forest floors are compared in Table 1.5. The annual minimum soil temperatures below the forest were 6.4 to 8.2 K higher than below bare ground.

Fig. 1.12a,b. Cooling patterns in plants. **a** Marginal location of frost injury on a leaf of *Nerium oleander* due to facilitated heat transfer at the tip and the edge. (Photo: W. Larcher). **b** Temperature distribution on a spruce seedling visualized by thermovision technique; the needle tips are more than 1 K colder than other parts of the plant. (From Hagner 1969)

At high altitudes, frost penetrates deeply into the rock. At 3700 m in the Jungfrau massif, Switzerland, at air temperatures of −28 °C in winter, Mathys (1974) recorded temperatures as low as −27 °C at a depth of 5 cm in rocks facing north, and −25 °C at the same depth in rocks facing south. At depths of 40 cm the absolute temperature minima were about −24 °C. Even in the warmer seasons, ground frost can occur in high mountains; at an altitude of 3000 m, temperatures as low as −2 °C can be recorded in the uppermost soil layers, and may last for several days (Moser et al. 1977; Fig. 1.13). At an altitude of 3700 m the rock temperature on northern slopes drops each night to below 0 °C and even down to −5 °C (Mathys 1974).

A *covering of snow* protects the soil from invasion by frost. The clear correlation existing between snow depth and ground frost can be expressed as a negative regression (Fig. 1.14). Depending on the distribution of snow there may be large topographic differences in winter temperature minima in the soil as well as in the depth of penetration of frost (Aulitzky 1961; Fig. 1.15). Due to the air held in the pores, snow is a very poor conductor of heat, which means that temperature changes at the snow surface are only slowly transmitted to the interior of the snow cover (Fig. 1.16). Only the surface layer is subjected to daily temperature fluctuations, the minimum temperature being observed at 0 to 10 cm below the surface (Fig. 1.17). If snow falls before the soil

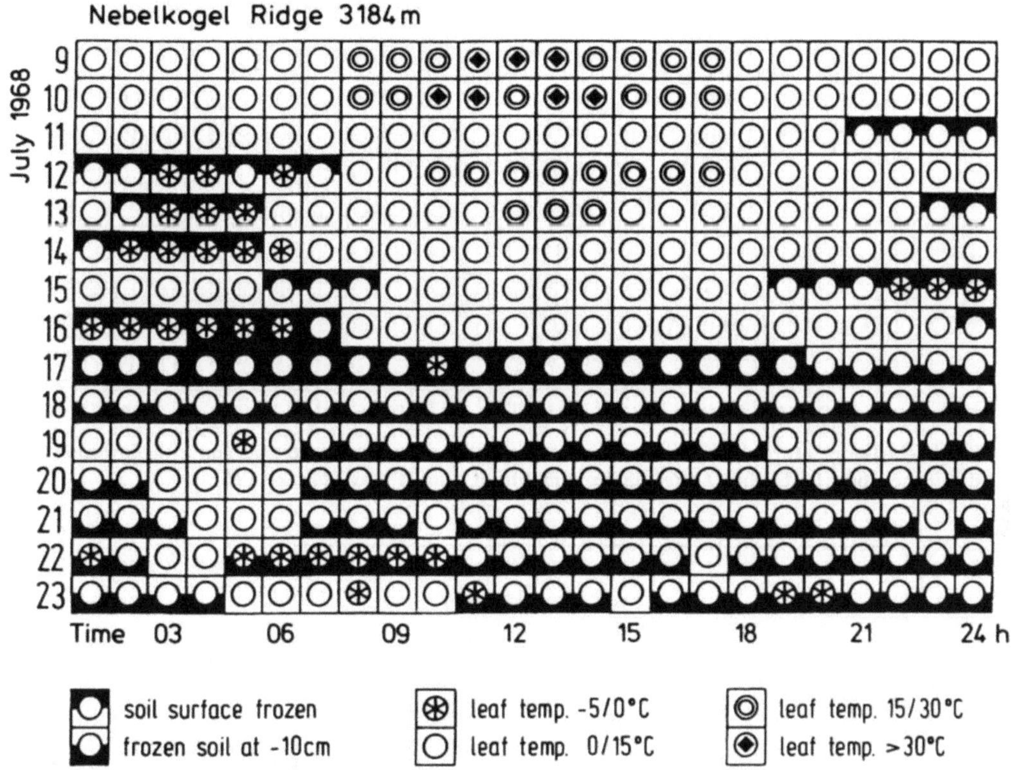

Fig. 1.13. Leaf temperatures of *Ranunculus glacialis* and frequency of ground frost in July at a high altitude site in the Alps. (From Moser et al. 1977)

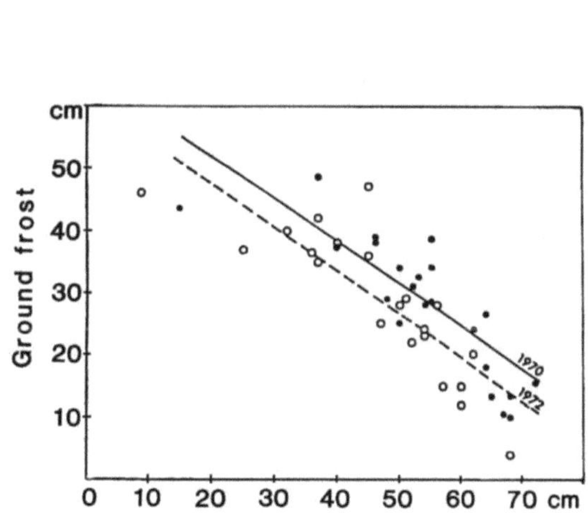

Fig. 1.14. Relationship between snow cover and depth of ground frost in a Finnish mire. (From Eurola 1975)

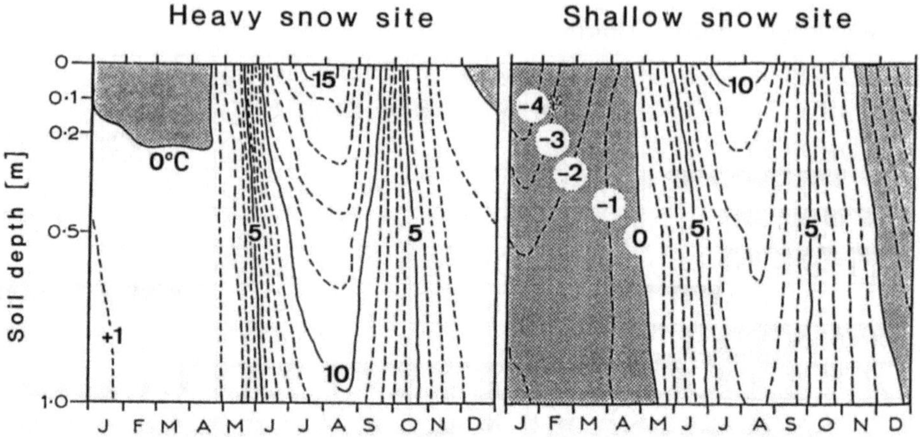

Fig. 1.15. Temperature distribution at different depths on a snow-protected and a shallow-snow site at the alpine timberline. *Letters* indicate month of year. (From Aulitzky 1961)

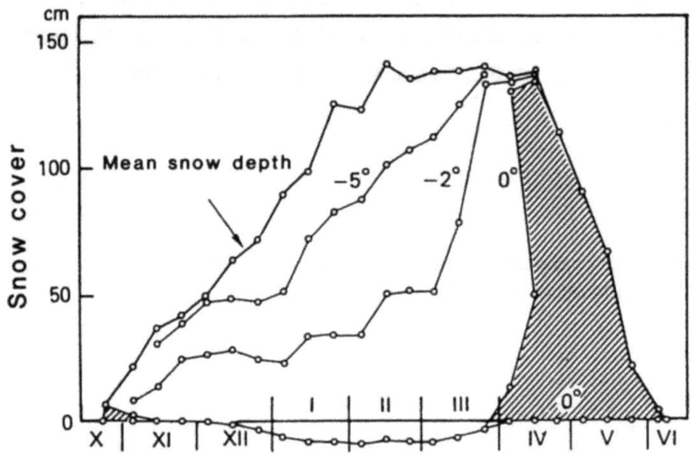

Fig. 1.16. Penetration of frost in a snow cover during winter. (From Aulitzky et al. 1982)

is frozen, a covering of 2 m keeps the underlying ground at a temperature between 0° and −2 °C (Michaelis 1934a).

Figure 1.18 shows the soil temperatures at depths of 5 cm and 30 cm beneath an alpine dwarf shrub heath on Mt. Patscherkofel near Innsbruck (1920 m a.s.l.). As long as the ground is free of snow, the temperature at a soil depth of 5 cm follows every fluctuation in air temperature, with a fairly large daily amplitude, even during overcast weather. At a depth of 30 cm the soil still stores heat. Following an extended period of cold in January, with air temperature minima of −17 °C and maxima consistently below −5 °C, the soil temperature in snow-free ground, even at a depth of 30 cm where the daily fluctuations amount to only tenths of a degree, declines to −5 °C. Immediately thereafter, 50 cm snow fell and the soil temperatures at depths of 5 and

Fig. 1.17. Midday temperatures at various depths in a snow cover (1969-02-26). Mean air temperatures in February are –5 °C in Sapporo and –18 °C in Moshiri. (From Kojima et al. 1970)

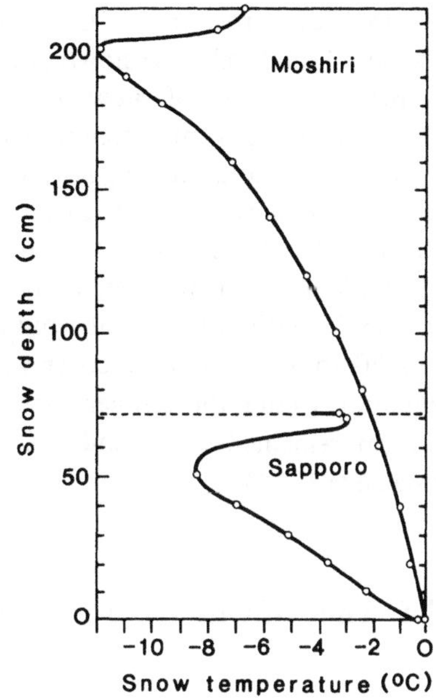

Fig. 1.18. Daily fluctuation of air temperatures (2 m above ground) and soil temperatures (at 5 and 30 cm depth) in relation to accumulation and melting of snow at the alpine timberline. (From Larcher 1957)

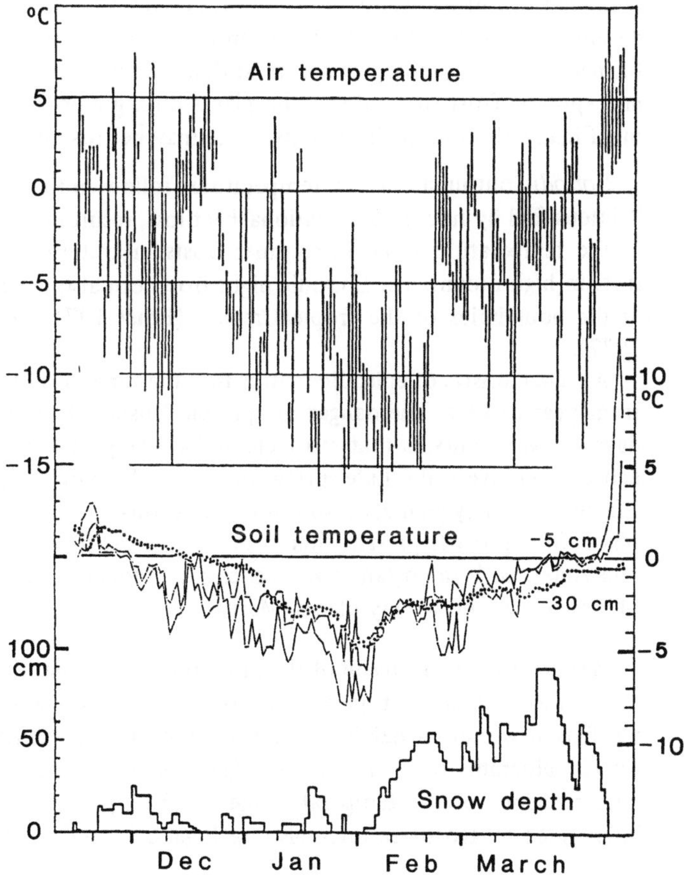

30 cm at once rose to -3 °C. In March, despite another brief cold wave, the temperature at all soil levels rose slowly, but constantly.

Beneath a layer of snow thicker than 20 cm, the temperature seldom falls below -5 °C under the winter conditions of intermediate latitudes. In regions where winters are extremely cold and long-lasting, the soil temperature decreases to much lower levels, even under snow. In the High Arctic the soil temperature at a depth of 1 m decreases to -20 °C and even to -30 °C in the upper layers (Corbet 1972; Lewis and Callaghan 1976). Mid-winter mean air temperatures in Point Barrow, Alaska, ranged from -30° to -35 °C with extremes of around -50 °C; the temperature of the ground surface below 50 to 100 cm snow cover was -20° to -25 °C (Scholander et al. 1953). In the antarctic winter the temperature of moss communities below a snow cover of about 50 cm remained at around -20 °C, while the air temperature approached -40 °C (Matsuda 1964).

1.5 Time of Onset, Severity and Duration of Frost: The Freezing Risk

The extent to which a plant is endangered by frost depends not only on the magnitude of the drop in temperature, but also on the time of onset and duration of the negative temperatures. Frost events should always be regarded in combination with the stage of development of the vegetation. According to its time of occurrence, a distinction can be made between periodic winter frost that occurs when the plants are in a resting state and episodic frost that catches the plants during a phase when they are actively growing. Finally, in some environments, there may be frost at any time of year.

Episodic outbursts of low temperatures follow in the wake of polar air masses and, if intensified by nocturnal radiational heat loss, result in spring and early autumn frosts in the temperature zone, in summer frosts in arctic regions, in sudden snowfall and frost in high mountains. Episodic night frosts can also occur in subtropical deserts and at the boundaries of the tropical region (Ernst 1971; Silberbauer-Gottsberger et al. 1977).

A characteristic of episodic frost is that the time of outbreak may vary considerably from year to year. The danger of episodic frosts, which can take effect within a few days or even hours, is that most plants have very little resistance to cold when in an active vegetative state. Late spring frost coincides with the most sensitive phases of the life cycle of many plants, i.e. sprouting, flowering and the beginning of the main growth phase. Spring frosts present therefore the greatest freeze threat to plants in the temperate zones, with important significance for regional and microtopographic distribution of herbaceous and woody plants (Matuszkiewicz 1977).

Periodic frost is the result of the variation of the radiation balance at higher latitudes due to the seasonal photoperiodism. An extreme seasonal climate with severe, continuous frost in winter prevails in arctic, subarctic and continental boreal regions (Fig. 1.19; for climatograms of N. American taiga sites see Oechel and Lawrence 1985). In the milder regions of the temperate zone the frost periods are of varying length and are interrupted by frost-free intervals (see Figs. 6.13 and 6.18). In such regions the dura-

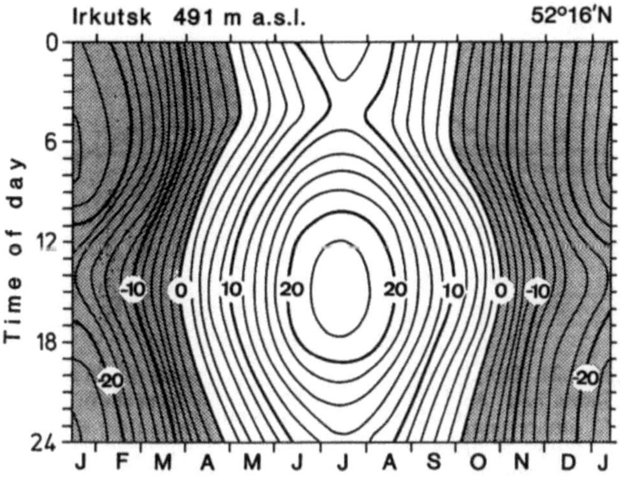

Fig. 1.19. Average hourly air temperatures at Irkutsk as a function of time of day (*ordinate*) and time of year (*abscissa*). *Letters* indicate month of year. (From Troll 1964)

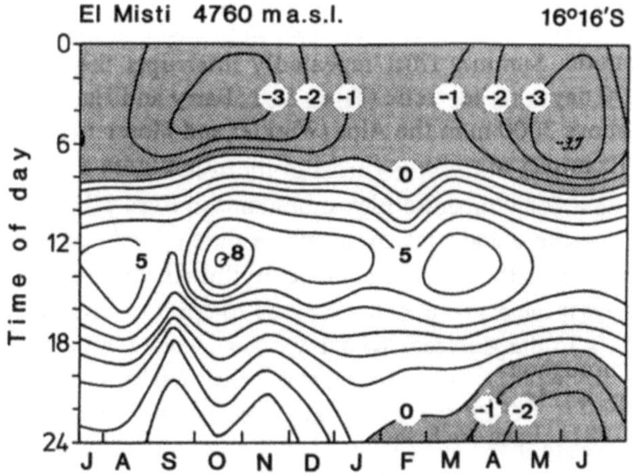

Fig. 1.20. Average hourly air temperatures characteristic of a tropical high mountain station. (From Troll 1961)

tion of periods of consecutive days with subfreezing temperatures (daily maximum below 0 °C) and with freeze-thaw cycle (only the daily minimum temperature below 0 °C) is of primary importance for winter survival of plants.

When winter frost sets in, particularly sensitive or exposed organs have already been discarded. The perennating organs are in a dormant state, a state which is causally associated with enhanced frost hardiness (see Chap. 5). Therefore, only in the form of

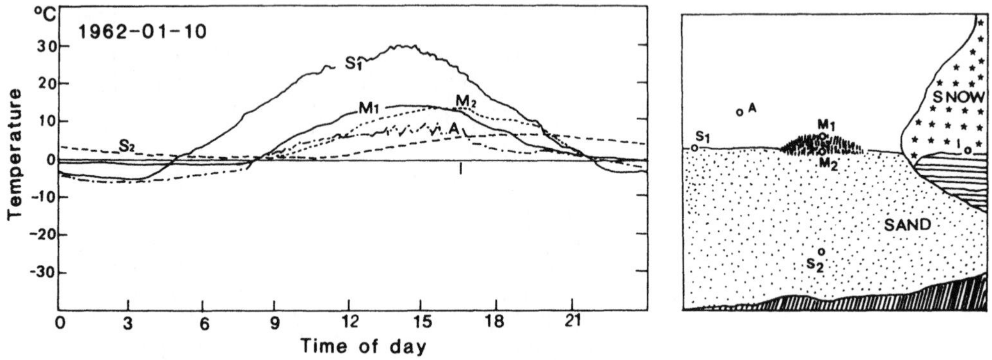

Fig. 1.21. Temperature fluctuations on a summer day in a moss cushion and its microenvironment (Showa Base, Antarctica). *A* Air temperature 3 cm above ground; *I* temperature of the snow layer; M_1 surface temperature of the moss; M_2 temperature at the center (1 cm below surface) of the moss cushion; S_1, S_2 temperature on the soil surface and 10 cm below the moss community, respectively. (From Matsuda 1964)

exceptionally long-lasting and extremely low temperatures does winter frost represent a danger to acclimatized plants. Injurious winters can be expected to occur on an average 10–12 times each century in the temperate zone (Quamme 1976; Lauscher 1985) and in Mediterranean regions (Larcher 1954; Morettini 1961).

At any time of year frost may occur in high altitudes, in the Arctic and the Antarctic. Summer frost repeatedly interrupts the growing season, which is brief, 40 to 70 days, in the Arctic (Bliss 1971; Barry and Hare 1974; Tieszen 1978) and in altitudes above 3000 m in the Alps (Winkler and Moser 1967; Bliss 1971, 1985; Larcher 1980a). Examples of summer frost in the arctic tundra are shown by Chapin and Shaver (1985) and for temperate high mountains in Figs. 1.13 and 7.30. On mountains and highlands of the equatorial zone, where no well-defined thermic seasons appear, the day/night cycle is the decisive climatic rhythm; there, at altitudes above 4000 m at any time of the year the plants may be exposed to episodic night frosts (Fig. 1.20; see also Sect. 7.5.2).

In the Antarctic cold deserts, where air temperatures remain between $-5°$ and -15 °C throughout the polar summer, the surface temperatures on rocks facing north increase to +7 °C during clear days, but remain below zero on overcast days (Kappen et al. 1981). Even such short thawing periods allow cryptoendolithic lichens to colonize on sandstones and epilithic lichens and moss cushions in rock gaps (Kappen 1985). The microclimatic conditions of a moss community growing on a southwestern slope at the sandy area near the Japanese Syowa Base at East Ongul Island, Antarctica, were studied by Matsuda (1964). The moss community appeared in summer, mainly from January to February, and disappeared under snow for about 10 months from March to mid-December. In summer, the temperatures of the soil surface and the surface of the moss community rose during the day to as much as 19° and 30 °C respectively, although the air temperature 3 cm above the ground ranged from 5° to 0 °C; at night the temperature in the cushion core remained near 0 °C, while the moss surface sometimes cooled to -5 °C or below (Fig. 1.21).

1.6 Temperature Fluctuations in Wintering Trees

Frequent alternation of freezing and thawing in winter lessens the state of hardening of wintering plants which, particularly in trees, represents an additional hazard (see Sect. 3.3.2).

The daily temperature fluctuation in tree stems was measured during winter in Sapporo (Fig. 1.22). The temperature in the centre of the xylem (C) remained subzero,

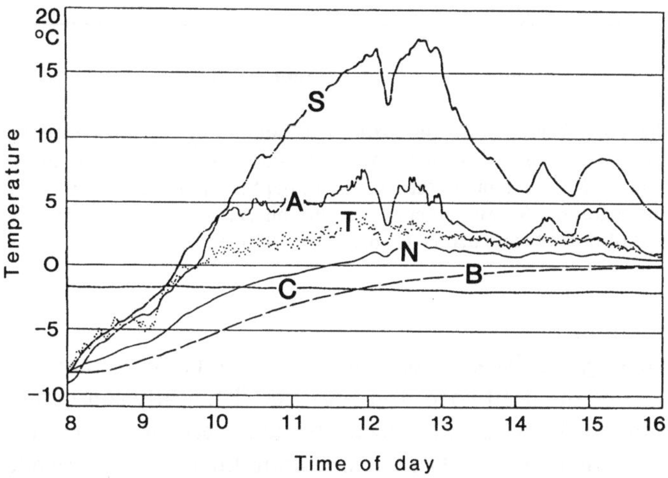

Fig. 1.22. Temperature fluctuation in wintering trees. Temperatures in the center of woody stems of different diameter: *A* 1.5 cm (*Kalopanax*); *B* 13.5 cm (*Kalopanax*); *C* 86 cm (*Ulmus*). Bark temperatures of *Kalopanax* (13.5 cm stem diameter) at 2.5 cm below the surface: *S* facing south; *N* facing north; *T* air temperature. (From Sakai 1966c)

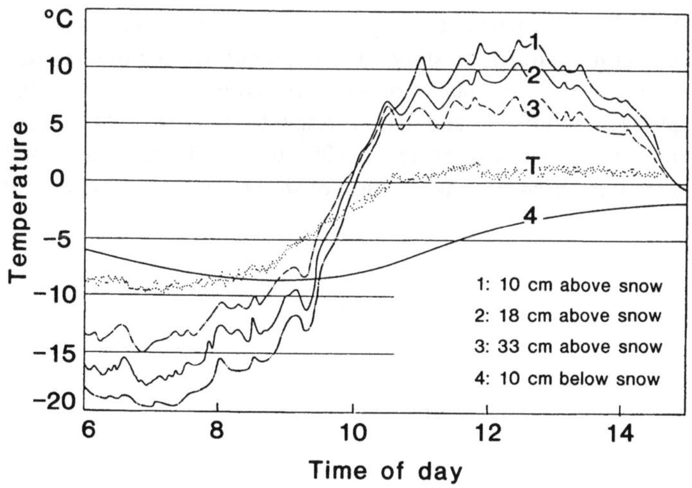

Fig. 1.23. Bark temperature of a slender mulberry stem (1 cm diameter) in mid-winter at various heights above (*1*, *2*, *3*) and below (*4*) the snow surface; *T* air temperature. (From Sakai 1966c)

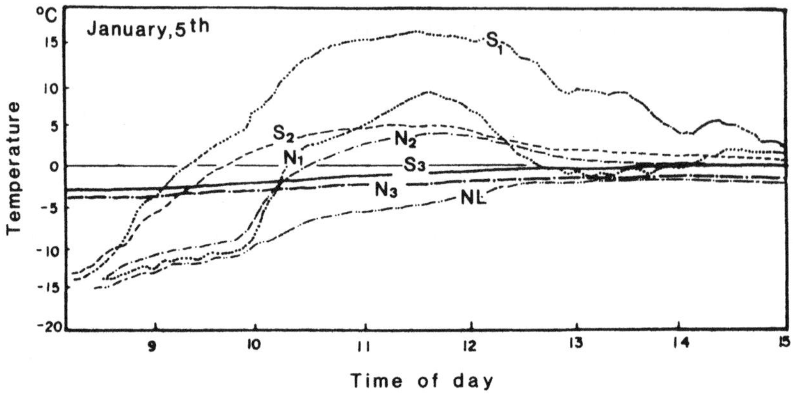

Fig. 1.24. Temperature fluctuations in saplings of *Abies sachalinensis* wintering on SE- and NW-facing slopes in E. Hokkaido. Southern slope: S_1 stem 10 cm above ground; S_2 bud 60 cm above ground; S_3 soil temperature at 10 cm depth. Middle part of the northern slope: N_1 stem 10 cm above ground; N_2 bud 60 cm above ground; N_3 soil at 10 cm depth. *NL* buds of plants at the lower part of north-facing plants. (From Sakai 1970b)

although the bark temperature of the south side (S) rose to about 17 °C. The temperature of the heartwood of a large tree (E), nearly constant throughout the day, ranged from -0.5° to -2 °C during winter, at a mean air temperature in January and February of around -5.5 °C. Daily temperature fluctuations in slender stems at various heights above the snow surface are shown in Fig. 1.23. The rise in bark temperature due to sunshine increased with decrease in height above the snow surface. The bark 5 to 10 cm above the snow surface, especially on the south side, is subjected to the largest temperature fluctuations. A very similar trend was observed in young trees near the ground without snow cover (Horiuchi and Sakai 1978).

In E. Hokkaido severe frost and dry air prevail throughout winter. For 3 months or more the soil freezes to a depth of about 40 cm on the southern slopes and to about 70 cm on the northern slopes of a small hill. During the day, the temperatures of stems and buds of young plants of *Abies sachalinensis* on the southern and northern slopes approach about 20° and 10 °C, respectively (Fig. 1.24). During the night the trees attain minimum temperatures of -20° to -25 °C. These are conditions which favour dehardening and winter desiccation of shoots on slopes facing south.

2. The Freezing Process in Plants

2.1 Freezing of Water and Aqueous Solutions

2.1.1 Ice Nucleation

Ice formation is the passage of water molecules from the random arrangement of a liquid state to an ordered one. Many liquids, including water, do not invariably freeze at the melting point of the solid phase. Such liquids can be supercooled to several degrees below the melting point of the solid phase and will freeze only upon the spontaneous formation of, or addition of, a substance that acts as a catalyst for the liquid-solid phase transition. Catalysts for the water-ice phase transitions are known as ice nuclei. Two general types of ice nuclei exist: homogeneous and heterogeneous.

In *homogeneous nucleation*, the nuclei are formed spontaneously in the liquid without intervention of foreign bodies. Rasmussen and MacKenzie (1972) demonstrated that pure water droplets of about 10 μm in diameter supercooled to −38.1 °C provided no heterogeneous nucleators were present. Water in plant cells may undercool from −41° to −47 °C because cell solutes depress the spontaneous nucleation temperature as shown in the following equation:

$$T_H = 38.1 + 1.8\,T_m\ ,$$

where T_m is the melting point depression for the solution (in K) and T_H is the homogeneous nucleation temperature (in °C). Catalysis of ice formation in water involves a transient ordering of water molecules into a lattice resembling ice. The molecules of pure water tend to aggregate into larger icelike clusters with decreasing temperature. At very low temperatures approaching −40 °C, random grouping of water molecules to icelike clusters attain the size of the critical embryo (1.13 nm radius), the embryo containing about 190 molecules which can efficiently trigger homogeneous ice formation (Fletcher 1970).

In *heterogeneous nucleation* nonaqueous catalysts for ice formation are required for the water-ice phase transition. The probability of such nucleation increases in every case in which the volume of water increases, or in which the temperature is lowered. The number and size of the crystallization units formed depend first on the rate of nucleation and the rate of crystal growth, both of which depend on the cooling rate and on the concentration of the medium; thus increasing concentration results in a steep decrease in growth rate. If the freezing temperature lies within a range in which the rate of nucleation is high and the rate of growth is low, the units will be small but there will be a large number of them. Conversely, if the material freezes at a temperature at which the growth rate is high and the nucleation rate is low, the number of crystallization units will be small and the units will be large (Fig. 2.1).

Potent heterogeneous nucleators have been detected in the form of bacteria bearing an *ice-nucleating gene*. Microbially-mediated ice nucleation was first reported by Schnell and Valli (1972), who found that decaying tree leaves were an important source of ice nucleation particles. Suspensions of *Pseudomonas syringae* isolated from decaying

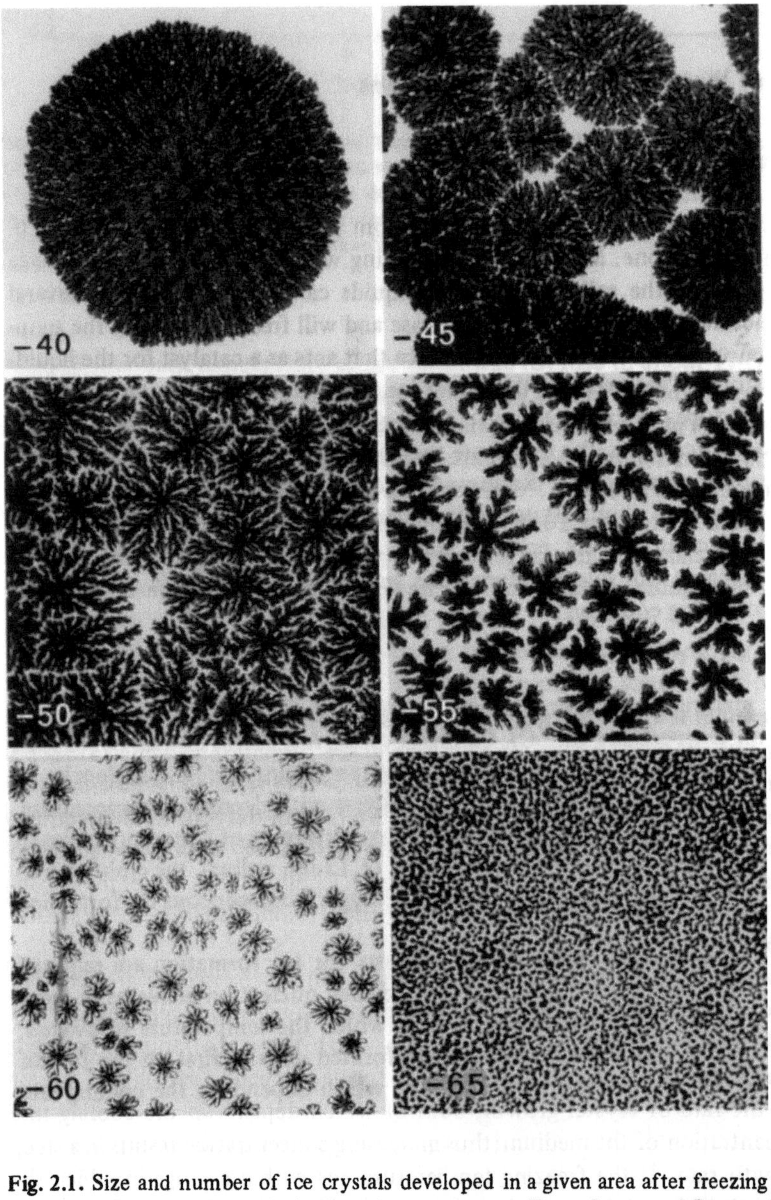

Fig. 2.1. Size and number of ice crystals developed in a given area after freezing of a 50% solution of polyvinylpyrrolidone at the indicated temperatures. (From Luyet and Rapatz 1958)

Table 2.1. Ice-nucleating activity of bacterial cultures. (From Maki et al. 1974)[a]

Culture	Ice-nucleating temperatures (°C)	
	T_1	T_{90}
Pseudomonas syringae C-9	− 2.9	− 3.5
P. syringae	− 3.2	− 3.9
P. aeruginosa	− 7.5	−17.8
Staphylococcus epidermidis	− 6.9	−19.5
Escherichia coli	− 8.3	−17.1
Enterobacter aerogenes	− 9.6	−17.0
Proteus mirabilis	− 8.0	−19.4
P. vulgaris	− 7.8	−17.0
Bacillus subtilis	−10.6	−18.0
B. cereus	− 6.9	−17.0
Uninoculated medium	− 9.2	−17.0

[a] Thirty 0.01 ml drops of test material were placed on a controlled temperature surface and the temperature was slowly lowered from ambient temperature to −25 °C. The temperatures at which 1% (T_1) and 90% (T_{90}) of the drops froze were recorded.

leaves of *Alnus tenuifolia* were found to be highly active in initiating ice nucleation at relatively high temperatures (Maki et al. 1974; Table 2.1). Strains of two species of epiphytic bacteria, *Pseudomonas syringae* and *Erwinia herbicola*, are particularly efficient ice nucleators between −2° and −5 °C. Leaves of many plants collected from several geographically different areas and during different seasons of the year bore a substantial number of ice nucleation active (INA) bacteria. Frost injury to field-grown maize leaves at −5 °C was directly proportional to the logarithm of INA-bacterial populations on these leaves (Lindow et al. 1978; Lindow 1982). Roos and Hattingh (1983) demonstrated that suspensions of *Pseudomonas syringae*, sprayed on sweet cherry leaves, entered the stomatal cavities and into the intracellular space of epidermal cells, remained viable and multiplied. Bacterial masses emerging from the substomatal cavities were enmeshed in strands of unidentified composition, possibly polysaccharide slime. Lindow (1983) supposed that components of the cell membrane of INA-bacteria may be involved in the expression of ice nucleation activity.

Ice nuclei may, of course, form without inoculation from the exterior. Salt and Kaku (1967) found that such ice nucleation took place at sites associated with the cell walls and was not catalyzed by nucleators suspended in the water. Internal nucleators in leaves of *Veronica persica* and *Buxus microphylla* have been indicated to be responsible for the high nucleating rate of these plants (Kaku 1973).

2.1.2 Freezing of Solutions

According to Raoult's Law, the freezing point of a solution is depressed in proportion to the concentration of the particles of solute. An ideal, nondissociated 1 M solution begins to freeze below −1.858 °C, but does not freeze completely at this temperature. Since only the pure solvent in a solution solidifies initially, the remaining solution becomes progressively more concentrated, with a resultant steady drop in freezing point. Thus, in an aqueous solution in the process of freezing, liquid and solid phases co-exist over a broad temperature range. Only when freezing exceeds the eutectic point does the entire solution freeze, with the formation of hydrated crystals. Sudden cooling to below the eutectic point prevents the separation of ice and liquid phases and results in amorphous solidification of the solution. In principle, these basic laws also apply to the freezing of the liquids in plant cells: however, it should be considered that cellular liquids contain a mixture of dissociated, nondissociated, agglutinating and colloidal substances and therefore deviate in their behaviour from that of an ideal solution. Since some solutes, such as sugars and polyols, tend to prevent the crystallization of others, the plant cell cannot be expected to have a specific eutectic point.

2.1.3 Sequence of Events in the Freezing of Liquids

If a liquid continuously loses heat, its temperature drops below the freezing point until it reaches the supercooling point. As soon as ice forms, at the nucleation temperature, heat of crystallization is set free causing the temperature to jump to a maximum value, the *exotherm* peak. In the case of the freezing of pure water the exotherm peak and the freezing point are identical (Fig. 2.2a). The temperature of the sample remains at 0 °C until all of the water has solidified to ice (isothermal plateau at the equilibrium phase change state), and only then does the temperature drop once more until it equals the external temperature. In solutions, too, when freezing sets in, the temperature rises to the freezing point (Fig. 2.2b) which, in this case, is lower than 0 °C, depending upon the depression of the freezing point for the particular solution. The crystallization of ice and the resulting concentration of the rest of the solution brings with it an immediate further depression in freezing point so that the exotherm peak value is only very briefly maintained. The duration of the freezing process, i.e. the time elapsing between the beginning and the end of freezing, depends on the parameters governing the quantity of crystallization heat set free and the speed with which heat is lost. In effect, these are the volume, the water content and the form (surface to volume ratio) of the sample, the temperature difference and the parameter for heat transfer between the sample and the cooler surroundings. Large bodies containing a high proportion of water can, if heat loss takes place slowly, remain in the equilibrium phase change state for hours (see Fig. 4.2).

2.1.4 Vitrification of Water and Solutes

A vitreous state of liquids, i.e. a state of matter in which cohesion and hardness are of the same order as in solid bodies, but in which the molecules are not arranged in a crystalline pattern, is well known as being the normal state of silicate glasses. Criteria for the formation of vitreous ice were presented by McMillan and Los (1965), who carefully deposited water out of the vapour phase onto a copper surface held at the

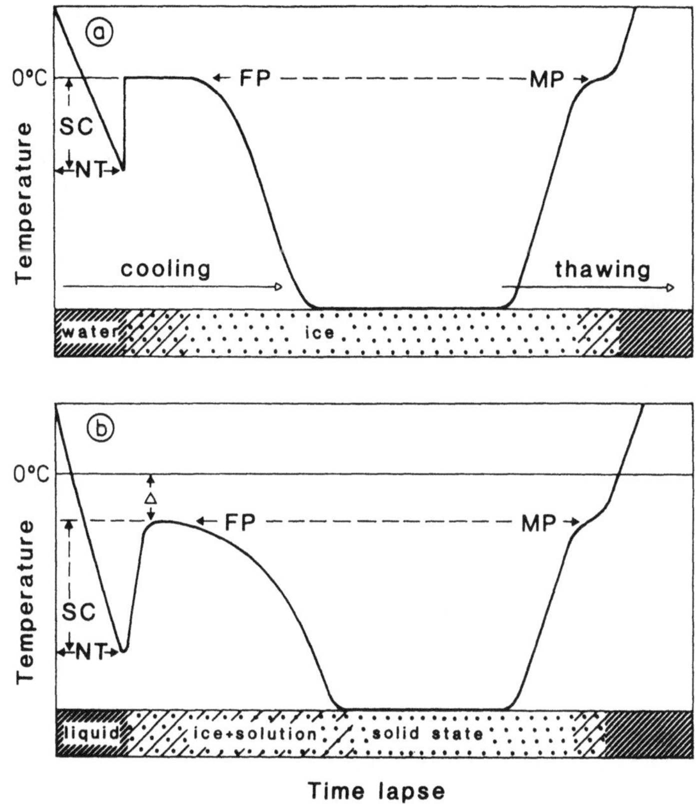

Fig. 2.2a,b. Temperature and phase changes during freezing and thawing of (**a**) pure water and (**b**) an aqueous solution. *NT* nucleation temperature; *SC* supercooling range; *FP* freezing point; *MP* melting point; △ depression of freezing point. (From Larcher 1985a)

temperature of liquid nitrogen. Differential thermal analysis (DTA) in situ of the samples obtained revealed a characteristic glass transition at -134 °C.

During the rewarming process a transition occurs from the vitreous phase to a crystalline state due to a sudden change in thermodynamic properties. Devitrification during the rewarming of rapidly cooled samples (30 μl) of a glycerol solution was demonstrated by Luyet (1967). The differential temperature curve (DT) and the directly recorded temperature curve (T) are shown in Fig. 2.3. The DT curve shows two changes, one marked G (glass transition) at -123 °C, which is characterized by changes in physical properties, such as specific heat, the other marked C (devitrification) at about -112 °C. Devitrification, being a crystallization, is an exothermic process generally marked by a high peak in the curve. Luyet and Rasmussen (1968) revealed the occurrence of three temperatures of transformation: post-devitrification, antemelting and incipient melting. At a still higher temperature, the small crystals lose their stability since their surface-to-volume ratio is too high: they melt and their molecules are transferred to large crystals which grow even larger. This phenomenon was designated migratory recrystallization by Luyet (1967). The temprature at which migratory recrystallization occurs, recognized by the sudden opacity of the specimens,

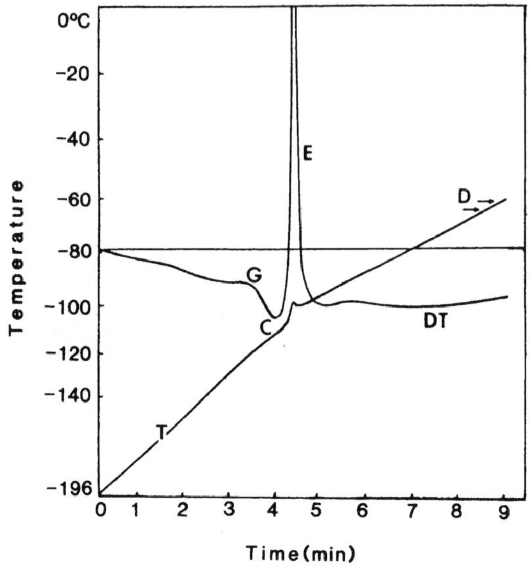

Fig. 2.3. Curves for differential temperature (*DT*) and directly recorded temperature (*T*) obtained during the rewarming of rapidly cooled microsamples of a 45% glycerol solution. *G* Onset of glass transition; *C* heat-releasing crystallization begins; *D* temperature range of grain growth of ice crystals; *E* exotherm. (From Luyet 1967)

is nearly the same as that for incipient melting. The recrystallization process is apparently undetectable by DTA.

2.2 Freezing of Plant Cells

The formation of ice in tissues and the appearance of frozen plant cells are well documented in the literature. Extensive studies employing optical microscopy have been published by Molisch (1897), Asahina (1956, 1978), Modlibowska (1961), Idle (1966), Mittelstädt (1969), Samygin (1974), Sakai (1982a), Ishikawa and Sakai (1982), Chaw and Rubinsky (1985); microcinematographic investigations have been reported by Modlibowska and Rogers (1955), Hudson and Brustkern (1965) and Steponkus et al. (1982). The submicroscopic appearance of frozen plant cells is described in the publications of Moor (1964), Krasavtsev and Tutkevich (1970) and Pearce and Willison (1985a,b).

The appearance of freezing plant cells when viewed under the microscope depends largely upon the speed of cooling and the method of ice inoculation, as well as on the capacity for supercooling and the permeability properties of the cells. Freezing of a small piece of plant tissue under a microscope does not, of course, necessarily represent the actual process of freezing in the intact whole plant under natural climatic conditions. However, with care in interpretation the observations on artificial freezing may certainly contribute to the understanding of freezing of plants under natural conditions.

2.2.1 Extracellular Freezing

Extracellular freezing is defined as ice formation on the surface of the cell or between the protoplast and the cell wall (extraplasmatic freezing). It consists in withdrawal of

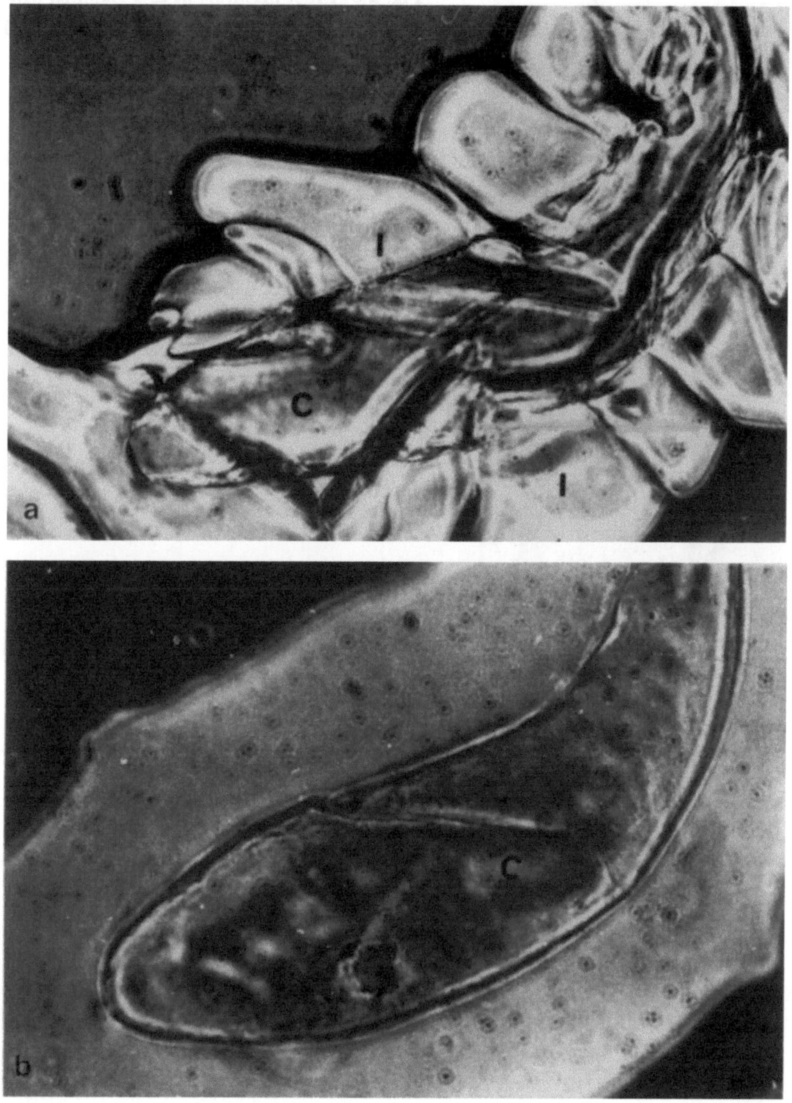

Fig. 2.4a,b. Extracellular freezing at −6 °C of a cultured tobacco cell suspended in silicon oil. **a** Covered with external ice and **b** after thawing. *I* ice crystals; *C* cell. (From Asahina 1978)

water from the cell due to the growth of ice crystals on its external surface (Sachs 1860; Müller-Thurgau 1886; Molisch 1897). This can be easily observed in suitable objects, such as isolated cells and filamentous or unilayer tissues (Fig. 2.4). In staminal hairs of *Tradescantia* under slightly supercooled conditions, ice forms between the protoplast and the cell wall, withdrawing water from the protoplast (Fig. 2.5). This phenomenon may be referred to as 'frost plasmolysis'. Beck et al. (1984) were able to identify frost plasmolysis in leaves of afroalpine giant rosette plants in situ.

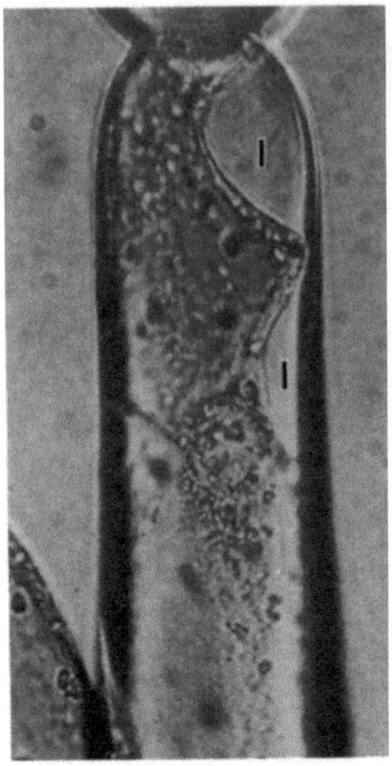

Fig. 2.5. Frost plasmolysis in a staminal hair cell of *Tradescantia*. *I* ice cap between cell wall and protoplast. (From Asahina 1956)

Table 2.2. Vapour pressure (mbar) at equilibrium with ice and supercooled water as a function of temperature. (From Häckel 1985)

°C	0	0.1	0.2	0.3	0.4	0.5	0.6	0.7	0.8	0.9
	Ice									
–30	0.38	0.34	0.31	0.27	0.25	0.22	0.20	0.18	0.16	0.14
–20	1.0	0.94	0.85	0.77	0.70	0.64	0.57	0.52	0.47	0.42
–10	2.6	2.4	2.2	2.0	1.8	1.7	1.5	1.4	1.2	1.1
– 0	6.1	5.6	5.2	4.8	4.4	4.0	3.7	3.4	3.1	2.8
	Water									
–30	0.51	0.46	0.42	0.38	0.35	0.31	0.28	0.26	0.23	0.21
–20	1.3	1.2	1.1	1.0	0.88	0.81	0.74	0.67	0.61	0.51
–10	2.9	2.6	2.4	2.3	2.1	1.9	1.8	1.6	1.5	1.4
– 0	6.1	5.7	5.3	4.9	1.5	4.2	3.9	3.6	3.3	3.1

The vapour pressure of the cell water is higher than that of ice at the same temperature (Table 2.2). Consequently, cell water will diffuse through the plasma membrane to the extracellular ice (Fischer 1911). Due to the loss of water the cell will contract. After rewarming, the cells, if they have not been injured, can soon reabsorb water and regain full turgor, otherwise they would remain collapsed.

The movement of cell water due to the difference between the chemical potential of water in the supercooled state and that of the extraplasmatic ice (see Fig. 2.5, I) results in cell dehydration. If the rate of cooling is sufficiently low, the final temperature is not too low, and the total duration of freezing is short, even less hardy cells can survive extracellular freezing (equilibrium freezing; Olien 1981).

The rate at which the volume of intracellular water changes with temperature was expressed by Mazur (1963) as follows:

$$dV/dT = k\ A\ RT/V_i^o\ ln\, p_e/p_i\ ,$$

where k is the permeability constant of the cell (μm^3 water per μm^2 cell membrane surface per minute per difference in osmotic pressure between inside and outside of the cell); A, the cell membrane area; V_i^0, the molar volume of water; p_i and p_e, the vapour pressures of supercooled water inside and that of water at thermodynamic equilibrium outside the cell; R, the gas constant; T, the temperature.

From this formula, the rate of water loss in extracellularly frozen cells as the temperature decreases is mainly determined by three parameters: (1) permeability constant, (2) surface area of the cells and (3) difference in the vapour pressure between the supercooled water inside and the ice outside. The rate of diffusion of water to ice outside the cells is limited by the permeability of plasma membrane lipids. Therefore, if the temperature drops rapidly enough, the diffusion to the extraplasmatic ice cannot occur with sufficient speed. As a consequence, the increase in the concentration of the cell solutes cannot keep pace with the thermodynamic effect of the temperature lowering which eventually leads to lethal nonequilibrium freezing (Olien 1981).

As a rule, inoculation with ice crystals is a prerequisite for the initiation of extracellular freezing of a cell that is not already in a considerable state of supercooling. Whether the cell freezes extracellularly or intracellularly depends upon whether or not the freezing inside the cell is effectively prevented by the protoplasma membrane (Chamber and Hale 1932). Hardy cells are characteristically resistant to the penetration of ice crystals. This was observed in thin tangential sections of mulberry twig cortex under the microscope during freezing after ice seeding (Sakai 1958). Owing to the remarkably high permeability of hardy cells to water (Asahina 1956; Stout 1979), large ice crystals usually develop on their surface while, at the same cooling rate, less hardy cells easily freeze intracellularly. The size of the ice crystals is one of the main factors preventing their ready passage through cell membranes (Ashworth and Abeles 1984). This must function very effectively in hardy cells. Since water in a slightly supercooled state freezes as discoid or fern-shaped crystals, the crystal front is certainly too large to penetrate the protoplasmic membranes or even the cell wall (Asahina 1956).

2.2.2 Intracellular Freezing

Intracellular nucleation generally does not occur spontaneously unless the cells are supercooled to at least -10 °C (Mazur 1977). At the instant of intracellular freezing, cells are killed as a rule, probably due to the mechanical destruction of biomembranes resulting from the fast growth of ice crystals in the protoplast (Maximov 1914). The manner of intracellular freezing of parenchymal cells can be divided into two distinct patterns, i.e. flash and nonflash type (Asahina 1956). The former is a sudden freezing characterized by an instantaneous darkening of whole cells (Fig. 2.6), whereas the latter is a slow freezing with clearly visible ice growth in the cell. A high cooling rate and a high degree of supercooling favour cell freezing of the flash type. Furthermore, the type

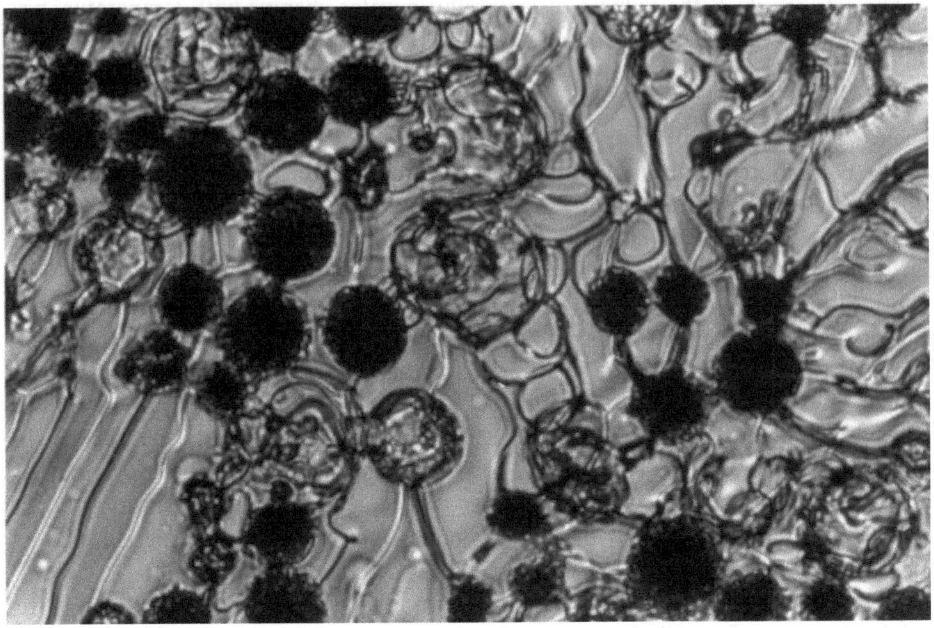

Fig. 2.6. Intracellular freezing of *Marchantia* protoplasts. *Dark cells* contain intracellular ice, *light cells* are extracellularly frozen. Freezing temperature: –15.2 °C; cooling rate: 2.5 K min^{-1}; suspending medium: 0.7 M mannitol. (Photo: Y. Sugawara)

of freezing is affected by the character of the cell concerned. Under normal cooling conditions intracellular freezing kills the cell, as observed in xylem ray parenchyma of deciduous trees (Tumanov and Krasavtsev 1959; Quamme et al. 1973) and florets in flower buds (Graham 1971; Graham and Mullin 1975) after deep supercooling.

Plant cells can survive intracellular freezing provided that very fine ice crystals, which are innocuous to the cells, form intracellularly and then melt before they reach a harmful size (Sakai and Yoshida 1967; Sakai and Otsuka 1967). Innocuous ice formation of this kind occurs only if cooling is extremely rapid, i.e. above 10,000 K min^{-1}, or more (Moor and Mühlethaler 1963; Sakai et al. 1968). Luyet (1937) formulated the hypothesis that even in the normally hydrated state, plant cells that are killed by a slight frost, can nevertheless survive immersion in liquid nitrogen (LN_2) if the rates of cooling and rewarming are ultrarapid. Luyet and Thoennes (1938) achieved these extremely high rates by plunging onion cells directly from room temperature into liquid air (about - 183 °C), followed by direct transfer to warm water (25° to 30 °C). They first considered this to be a vitrification process. This conclusion was based on the absence of the double refraction which is characteristic of crystals when viewed under the polarization microscope. Later investigators revealed the presence of submicroscopic crystals. Although ultrarapid cooling and subsequent rapid rewarming did not permit less hardy cells, such as onion cells, to survive immersion in liquid nitrogen, this method paved the way for preservation of the viability of hardy cells (Sakai 1956, 1966b) and nonhardy cells in the presence of cryoprotectants at the temperature of LN_2.

2.2.3 Freezing of Intact Tissues

Freezing of a small piece of tissue under the microscope differs in some respects from freezing of intact organs or the whole plant. Under natural conditions, the intercellular spaces are filled with air, and extracellular freezing has to start from water films or droplets on the cell surfaces. By employing low-temperature scanning electron microscopy Pearce and Beckett (1985) were able to visualize water droplets in barley leaves in the position occupied in vivo. The droplets, of about 2 μm diameter, were found mainly on the vascular bundle sheath, the guard and subsidary cells and on some mesophyll cells around the substomatal cavity. In leaves of well-watered plants from environments with high air humidity the droplets occurred abundantly, whereas in drought-stressed, wilting leaves they were absent.

The freezing may be gradual enough to prevent the ice from spreading throughout the plant. Thus, ice may be confined to specific regions, forming large masses as much as 1000 times or more the size of a cell, at the expense of water diffusing from relatively distant unfrozen regions (Sachs 1860; Prillieux 1869; Müller-Thurgau 1886). In *Buxus* leaves in late spring Hatakeyama and Kato (1965) observed that the formation of ice masses between veins and spongy tissues split the two apart. Ice masses were also observed between cambium and xylem of the basal stems of young trees of tea and citrus plants. The ice grew at the expense of water migrating from the unfrozen roots, and caused splitting of the living bark (Fig. 2.7a). Terumoto (1960) observed large ice masses in the concentric vascular regions of table beet root frozen at −4° to −5 °C (Fig. 2.7b). If the extracellular freezing did not injure the cells, the ice was free of electrolyte and betacyanine. Further examples are reported by Larcher (1985a). As a rule, extracellular ice which accumulates in the intercellular spaces of plant tissues, give a translucent appearance to the tissues during freezing.

In just the same way as in aqueous solutions, the process of freezing can be followed in plant tissues, either by direct measurement of exotherms or by differential thermal

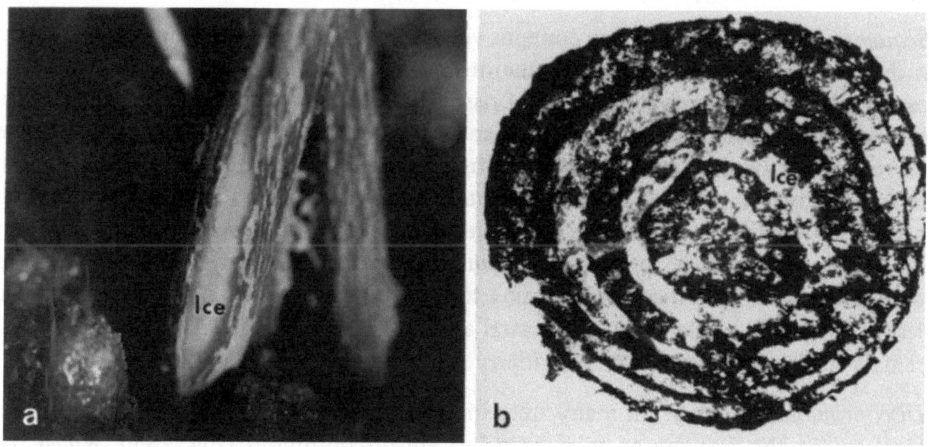

Fig. 2.7a,b. Formation of ice masses in plant tissues. **a** Bark of a young tea plant after freezing in early winter. (From Nakayama and Harada 1973). **b** Concentric accumulation of ice in the vascular bundle rings of a frozen beet root at −5 °C. (From Terumoto 1960)

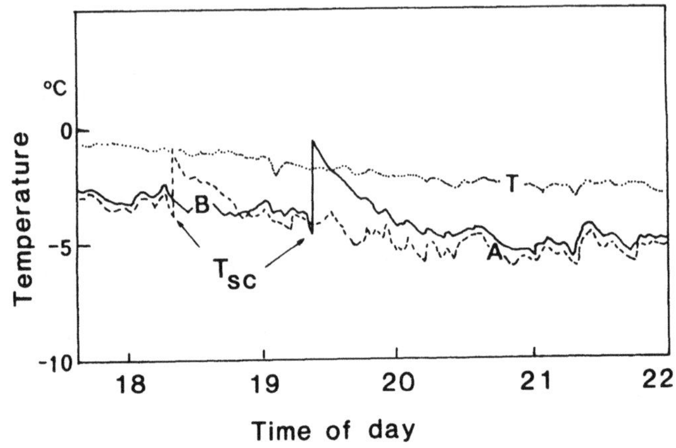

Fig. 2.8. Temperature-time curves at the beginning of freezing of *A* a leaf of *Aucuba japonica* and *B* the stem of a seedling of *Abies sachalinensis* under field conditions. *T* screen temperature; T_{sc} supercooling point. (From Sakai 1966c)

analysis. The initiation of freezing under field conditions of the stem of *Abies sachalinensis* seedlings and of leaves of *Aucuba japonica* are shown in Fig. 2.8.

The pattern of freezing of a particular plant organ depends on its anatomical peculiarities, its state of hardening, its water content and the speed of cooling. Characteristic types can be distinguished:

Continuous Freezing: In homogeneous tissues (e.g. parenchyma) and in plant parts of fairly uniform structure (e.g. leaves with coherent intercellular cavities) freezing is continuous once ice begins to form. The freezing curve therefore exhibits a broad exotherm and is similar to the one-peak freezing curves of aqueous solutions and of cell sap obtained from plant tissues. The breadth of the exotherm plateau is determined by the cumulative heat losses during freezing of the different portions of tissues (Krasavtsev 1972; Brown and Reuter 1974). In very sensitive temperature recordings, therefore, numerous small freezing peaks can be recognized (Rottenburg 1972; Brown et al. 1974).

Sequential Freezing: If an organ contains vessels filled with water, this interstitial water is the first to freeze. The broad exotherm that indicates freezing of the cells is in this case preceded by a steep but very brief rise in temperature (Fig. 2.9a). Freezing curves of this type are characteristic for roots and for shoot axes of herbaceous plants. The curves exhibit two or more peaks if the freezing behaviour of different kinds of tissue or areas of tissue in an organ differ (e.g. septate leaves; Fig. 2.9b). The successive freezing of different regions of a leaf gives frost-sensitive leaves a time advantage. This is seen in the highly septate leaves of *Cinnamomum* and *Laurus*: following a temperature drop to slightly below the threshold freezing temperature it takes hours before the whole leaf is frozen. As a rule, only intercoastal areas are affected (Fig. 2.10), so that the leaf remains partially functional despite injury.

Discontinuous Freezing: In many trees in a winter state the temperature changes accompanying the freezing of their branches are discontinuous (Fig. 2.9c). Whilst the bark parenchyma freezes at around −10 °C, with the development of one large exotherm (high temperature exotherm, HTE), the cells of wood parenchyma, because they are protected from heterogeneous nucleation, supercool to lower temperature before

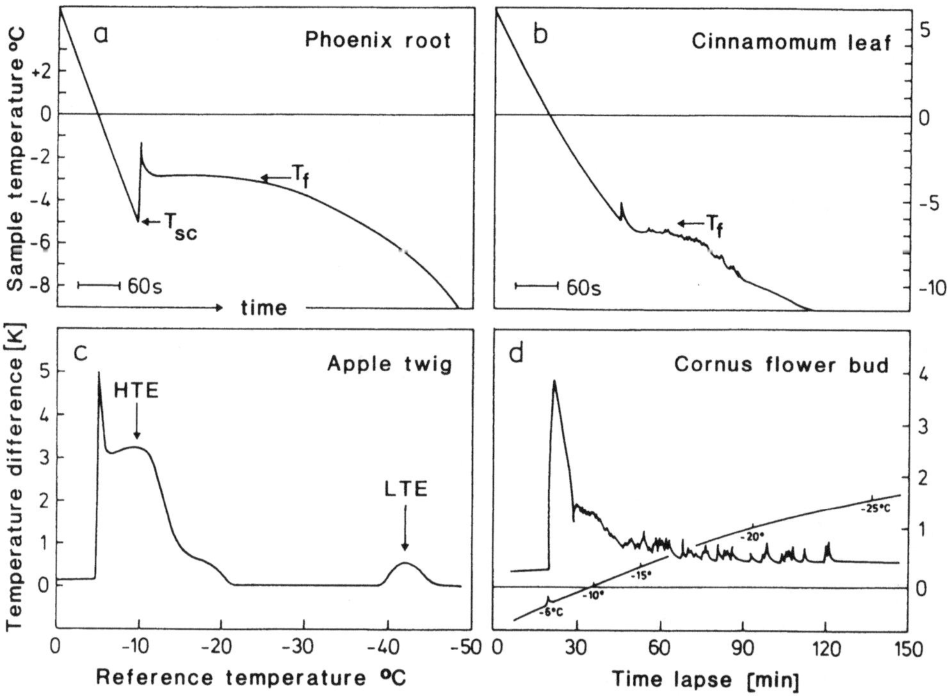

Fig. 2.9a–d. Freezing patterns of various plant organs. **a** Exotherm peak and freezing plateau of a root of *Phoenix canariensis*. **b** Sequential freezing of a septate leaf of *Cinnamomum glandulosum*. **c** Discontinuous freezing of a hardy apple shoot with development of a high temperature exotherm (*HTE*) and one low temperature exotherm (*LTE*). **d** Discontinuous freezing of a flower bud of *Cornus florida* with an HTE during ice formation in the bud scales, serial LTEs indicating freezing of supercooled flower primordia. **a**, **b** represent directly recorded temperature-time curves; **c**, **d** are differential thermograms. (From Larcher 1985a)

freezing individually or in small groups. The freezing of deep-supercooled portions of tissue is indicated by small low temperature exotherms (LTE) which can best be detected by differential thermal analysis. Certain buds and seeds also freeze in a very similar manner to that described above (Fig. 2.9d).

2.2.4 Threshold Freezing Temperature and Supercooling of Plant Tissues

The *threshold freezing temperature* T_f, i.e. the highest temperature at which freezing of living plant tissues occurs, depends on the specific properties of the plant and the tissue; it varies according to the cell sap concentration, the state of maturity and degree of hardening. A survey of T_f is given in Table 4.5. Different parts of one and the same plant (Fig. 2.11) and even different locations on one organ may exhibit considerable differences in threshold freezing temperature. A decrease in water content to 60–70% of saturation usually depresses the freezing temperature by 1–2 K. In bud meristems the T_f is lowered by about 10 K following the withdrawal of 40–70% of its water (Dereuddre 1978, 1979). Accumulation of water-binding substances inside the cell,

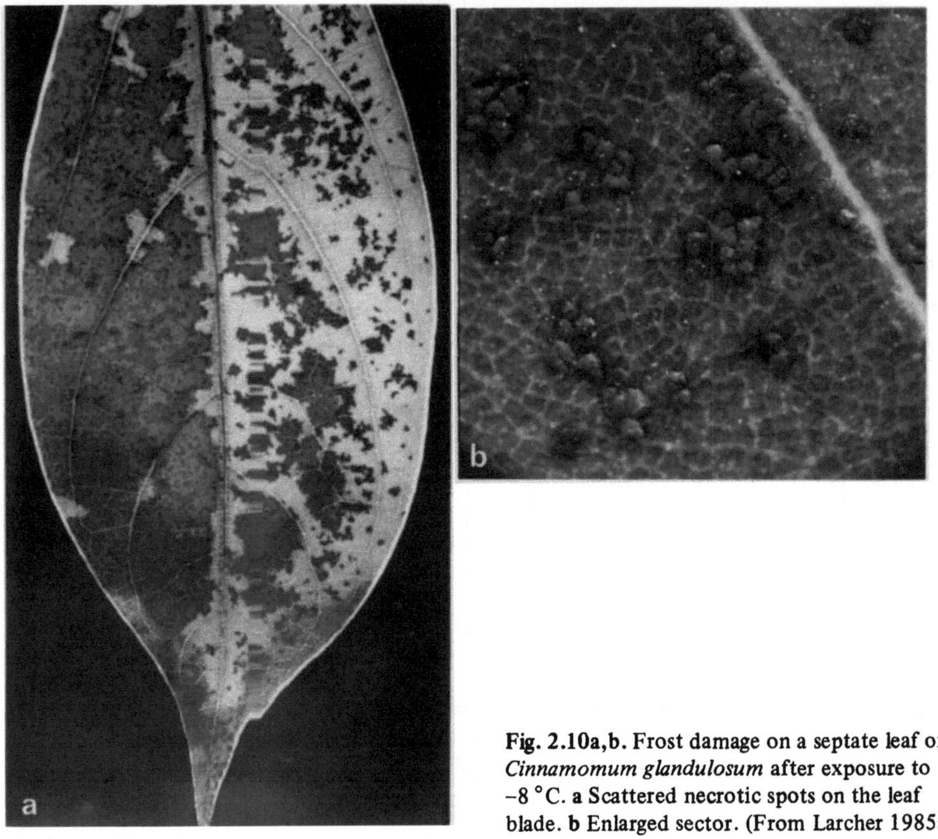

Fig. 2.10a,b. Frost damage on a septate leaf of *Cinnamomum glandulosum* after exposure to –8 °C. **a** Scattered necrotic spots on the leaf blade. **b** Enlarged sector. (From Larcher 1985a)

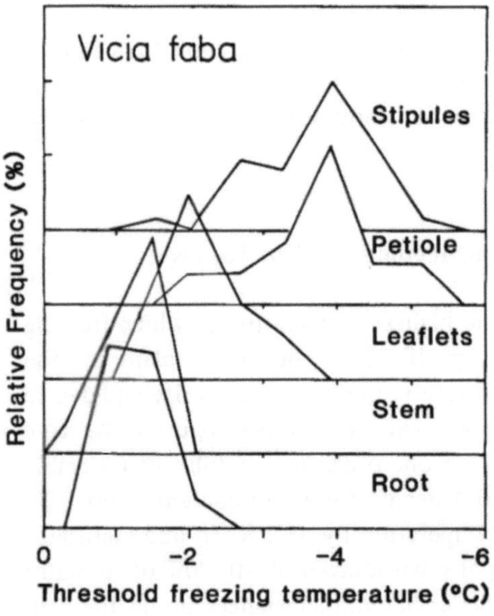

Fig. 2.11. Variation of threshold freezing temperatures within a plant. Frequency distribution of T_f values obtained with young plants of *Vicia faba*. (Based on measurements of J. Fierer, from Larcher 1985a)

particularly water-soluble carbohydrates, depresses the freezing point of the cell sap and thus the T_f of the living cells.

The *threshold supercooling temperature* T_{sc} is the lowest subfreezing temperature attained before ice formation occurs in a plant tissue. Kaku (1964) defined the supercooling point as the temperature at which the supercooled state in a system breaks down spontaneously. The most common type of supercooling was described by Modlibowska (1956) as *transient* supercooling. Most plant tissues can be transiently supercooled to between -4° and -12 °C (see Table 4.5). Transient supercooling and the threshold freezing temperature are directly proportional to each other (Yelenosky and Horanic 1969; Larcher 1985a). Decreasing water content and increasing concentration of osmotically active substances in the cell sap therefore not only effect a depression of T_f, they also lower T_{sc} and prolong supercooling. Certain cells and tissues do not freeze even at low environmental temperatures, or if they do, then only after a considerable time. This type of supercooling, termed *persistent* by Modlibowska (1956), occurs in woody parenchyma and in buds and seeds, as well as in certain leaves with special anatomical characteristics (see Sect. 4.2.1).

The necessary qualifications for effective supercooling are not fully understood, but they include (1) small cell size; (2) little or no intercellular space for nucleation; (3) relatively low water content; (4) absence of internal nucleators; (5) barriers against external nucleators; (6) a dispersion of cells into independently freezing units which allows for supercooling; and (7) the presence of antinucleator substances which oppose the formation of nucleation (Levitt 1980; Hong and Sucoff 1980; Ishikawa 1984).

The temperature at which freezing commences also depends upon whether the plant surface is wet or dry (Lucas 1954; Hendershott 1961, 1962). Ice formation in detached mulberry leaves began at leaf temperatures below -5 °C if their surface was dry (Kitaura 1967a). Some leaves remained supercooled even down to about -7 °C (Fig. 2.12a). However, they readily froze at -2 °C if ice particles were in contact with their surface (Fig. 2.12c). On two frosty nights with conspicuous dew, several leaves even froze at

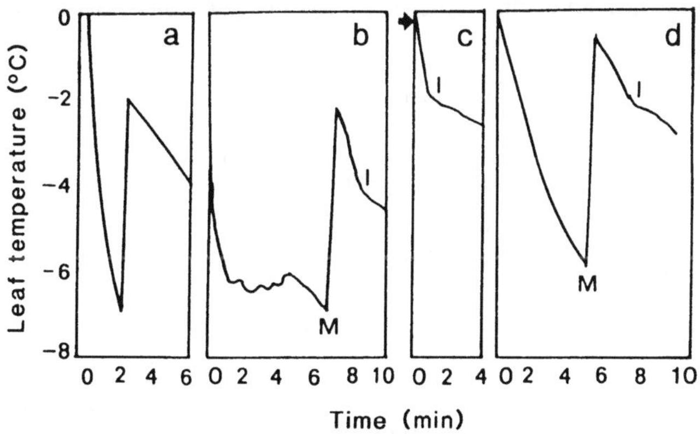

Fig. 2.12a–d. Freezing thermograms for detached leaves of mulberry trees. **a** Leaf with dry surface artificially cooled to -12 °C; **b** leaf cooled under natural conditions, the leaf temperature being 2 K below air temperature; **c** leaf cooled artificially to -8 °C in contact with ice particles (*arrow*); **d** leaf with wet surface cooled to -7 °C. *I* ice formation inside the leaves; *M* freezing of water on the surface of leaves. (From Kitaura 1967a)

leaf temperatures near −0.5 °C and the number of frozen leaves increased as the leaf temperature decreased to −2.5 °C. Ice formation on the surface of wet leaves was always followed by freezing inside the leaf (Fig. 2.12b,d). Leaves of *Eucalyptus urnigera* supercool to −8 °C to −10 °C if the leaf surface is dry, but only to −2° or −4 °C if it is wet (Thomas and Barker 1976). This may explain the advantage of species with glaucous or pubescent water-repellent leaves as compared with plants with easily wettable leaves.

2.2.5 Initiation and Progress of Freezing in the Whole Plant

In trees, whether in leaf or in the defoliated state, ice normally crystallizes first in the water-conducting system. Vessels of large diameter tend not to supercool, and their dilute sap has the highest freezing point of any other solute in the plant (Zimmermann 1964). Once ice forms in the vessels, it will spread throughout the plant body, and extend at the expense of the water vapour in the intercellular air and the surface film of water on the cell walls.

The velocity of freezing through shoots was measured by Kitaura (1967a) in mulberry trees (Fig. 2.13). Freezing proceeded along the vessels from a few nucleation points and reached all parts of the shoots at a relatively high velocity of about 60 cm min^{-1} at −3 °C. A similar rate (74 cm min^{-1}) was observed by Yelenosky (1975) in stems of unhardened citrus trees. The velocity of ice propagation increased with decrease in temperature and was proportional to the degree of supercooling. In the winter-bud stage the progress of ice formation was slower than in both the budding or the foliated stage (Fig. 2.14). The same holds for citrus trees (Yelenosky 1975). From the measurements in the early foliated stage, the relation between the shoot temperature (x) and the freezing velocity (y) could be expressed by the formula:

$$y = -34.7x - 34.9 .$$

Peeling off a strip of bark of 3 cm width halfway up the stem and sealing with vaseline did not alter the velocity of ice propagation. Furthermore, the progress of freezing

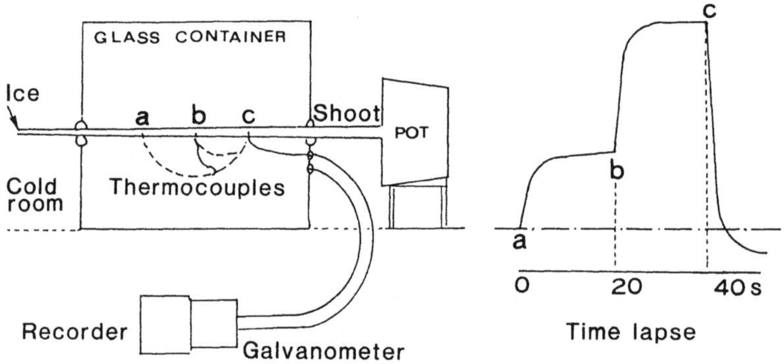

Fig. 2.13. Determination of freezing velocity in plant shoots. *Left:* Experimental device. *Right:* Example of records. After cooling at −1.5° to −3 °C a piece of ice (*arrow*) was placed tightly on the cut surface at the top of the shoot and temperature changes were recorded. *a, b* Measuring junctions; *c* reference junction of thermocouples. (From Kitaura 1967a)

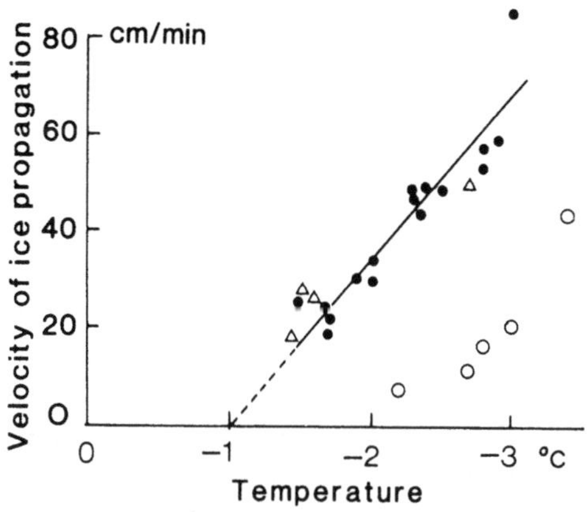

Fig. 2.14. Effect of supercooling (shoot temperature) on the velocity of ice propagation through the shoot of *Morus alba* in the states of (○) winter rest, (△) budding and (•) unfolding of the 1st to 3rd leaf. (From Kitaura 1967a)

through the shoots was as fast as the freezing velocity of pure water in a U-tube (Shinozaki 1954). It is thus evident that freezing in shoots proceeds along the xylem vessels.

Ice formation in a shoot plays an important role in the freezing of the leaves attached to it. Potted mulberry trees were cooled in a cold room (Kitaura 1967b). When the temperature of the leaves was between $-4°$ and -6 °C before the temperature of the shoot axis approached $-3°$ to -4 °C, a leaf on one of the shoots was inoculated with ice. As shown in Fig. 2.15, ice inoculation in a supercooled leaf at -4.6 °C (P1, leaf I)

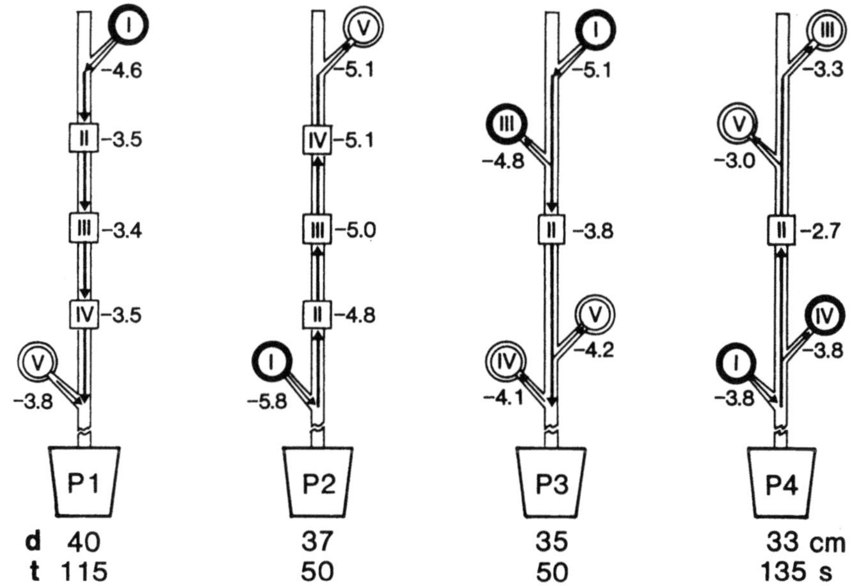

Fig. 2.15. Progress of freezing along the vessels of mulberry shoots. After inoculation of leaf *I* on the top of tree *P1*, ice formation in the shoot extended to the basal stem and leaf *V*. In tree *P4* a basal leaf (*I*) was inoculated. ○ Wet leaf; ◎ dry leaf; (*d*) distance between top leaf and basal leaf; (*t*) time required for ice propagation between terminal leaves. Temperatures were determined by thermocouple measurements in the stem. (From Kitaura 1967b)

caused not only immediate freezing in the leaf itself, but also subsequent freezing in the shoot. Then, the ice formation proceeded downwards to the basal part of the stem and to the lowest leaf (P1, leaf V). Experiments on shoots with more leaves (P3 and P4 in Fig. 2.15) yielded similar results. In mulberry trees with double shoots, freezing in the inoculated shoot reached the lower stem and then extended to the other shoot and its leaves. These results demonstrate that in supercooled trees in the field, freezing proceeds along the vessels from a few nucleation points and reaches all parts of shoots within a relatively short time. This may prevent supercooling and the risk of intracellular ice formation, and is especially effective in hardy plants. Detached leaves and shoots of citrus trees showed a higher degree of supercooling than the intact plants (Cooper et al. 1954; Hendershott 1961, 1962). In addition, the cutting of a mulberry shoot into shorter pieces resulted in lowering of the supercooling point by a few degrees (Kitaura 1967a,b). The T_f, too, can be influenced by excision: the freezing point of intact leaves of *Lobelia keniensis* and *L. telekii* in situ was $-1°$ and -1.4 °C respectively, whereas the freezing point of excised leaf fragments was $-3.6°$ and -2.5 °C. In *Dendrosenecio keniodendron*, on the other hand, there was no such difference between attached and cut leaves (Beck et al. 1982).

3. Freezing Injuries in Plants

3.1 Typology of Freezing Mechanisms

Two basically different categories of frost-killing mechanisms in plant cells can be distinguished:

1. *Immediate Injury of Freezing-Sensitive Cells.* Plant cells and tissues subject to immediate freezing injury, which in the classical descriptive terminology (Molisch 1897) were called "freezing-sensitive", are killed as soon as intensive ice formation becomes evident (Fig. 3.1a). Damage may result from freezing following heterogeneous nucleation, or, in cells capable of persistent supercooling, following homogeneous nucleation. Under natural conditions, in most tissues sudden lethal freezing sets in after a considerable deviation from thermodynamic equilibrium (nonequilibrium freezing; Olien 1978, 1981). Appreciable quantities of ice appear in freezing-sensitive tissues only when, under the stress of a large water potential difference between cell solutes and the adjacent ice, the semipermeability of the cell membranes is lost, permitting the efflux of fluid from the cells. In determining the specific temperature limit for lethal nonequilibrium freezing the stability and integrity of the biomembranes play a decisive role (Dowgert and Steponkus 1983; Lindstrom and Carter 1985). The sudden freezing can thus be regarded as a consequence of the preceding disorganization and as the final violent event in the process of freezing.

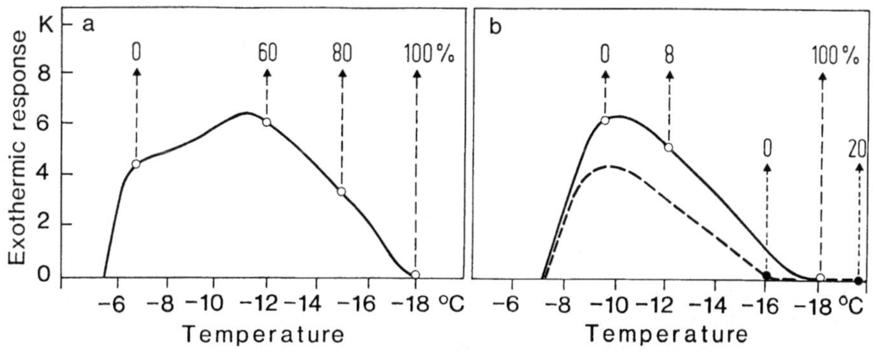

Fig. 3.1a,b. Progress of damage in leaves of *Rhododendron ferrugineum* during freezing: **a** Unhardened, freezing-sensitive leaves; **b** freezing-tolerant leaves after cold hardening in autumn (*solid line*) and mid-winter (*broken line*). *Numbers* indicate the degree of injury as percent of leaf area. (From Larcher and Nagele 1986, and unpubl. data of E. Ralser)

2. *Freezing Injury Following Dehydration of Freezing-Tolerant Cells.* In tissues with tolerance of equilibrium freezing (Olien 1978, 1981) freezing injury appears at temperatures much lower than the temperature at which ice formation begins (Fig. 3.1b). The extent of freezing tolerance can be expressed by the difference between the temperature at which freezing takes place (appearance of an exotherm) and the temperature at which initial injury appears (incipient killing temperature LT_i). The decisive criterion indicating the ability to tolerate freezing is the retention of cell viability despite the presence of large quantities of extraplasmatic ice. In the case of slow cooling, continuous adjustment of the thermodynamic equilibrium enables freezing-tolerant cells to survive cooling down to a specific subfreezing temperature. Cellular death ensues, if at all, at temperatures corresponding to a specific threshold for dehydration tolerance. The degree of dehydration can be expressed either as the fraction of liquid cell water or as the depression of the cell water potential.

According to Gusta et al. (1975) the plot of liquid water per unit dry matter (L_T) versus temperature (T) gives a hyperbolic correlation (Fig. 3.2). The function becomes linear if the amount of liquid water is plotted against 1/T:

$$L_T = L_0 \Delta Tm/T + k ,$$

where L_0 is the liquid water content before freezing, ΔT_m is the freezing point depression and k is a constant representing the amount of unfreezable water. At temperatures below about $-10\,^\circ C$ very small changes in liquid water content become responsible for

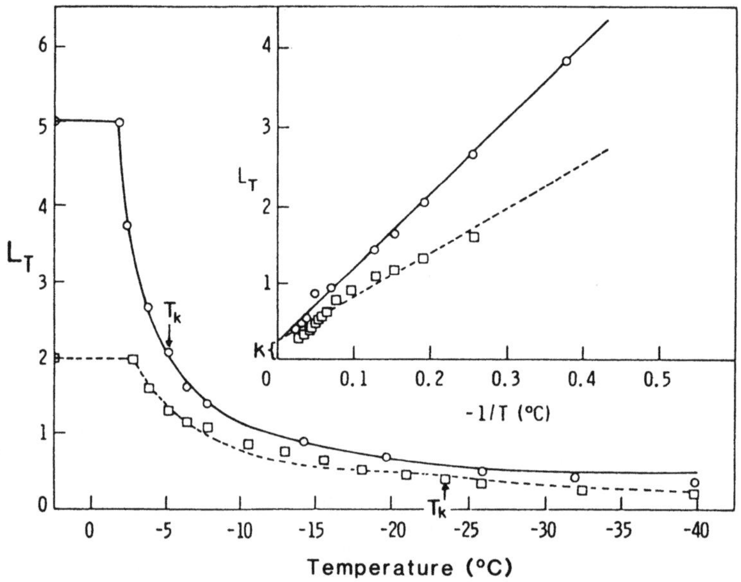

Fig. 3.2. Freezing curves for (□) cold acclimated and (○) unhardened crowns of *Secale cereale* obtained by using nuclear magnetic resonance spectroscopy. The liquid water content, L_T, is expressed in gram liquid water per gram dry sample. L_T is plotted versus temperature and the reciprocal of temperature. The *lines* drawn are the best fitting hyperbolae in the L_T versus T plots and the best fitting straight line in the L_T versus 1/T plots for the results between −2.5° and −40 °C. In the inset *k* indicates the amount of liquid water which does not freeze; T_k killing temperatures. (From Gusta et al. 1975)

the limitation of the resistance to equilibrium freezing. From the above formula an equation can be derived in which the dehydration effect is expressed in terms of cell water potential (Rajashekar and Burke 1982). The water potential (ψ) that is in equilibrium with extraplasmatic ice at the same temperature can be obtained by the equation

$$\psi_T = (RT/V_w) \ln (e_{ice}/e_w) ,$$

where R is the gas constant, V_w the molar volume of water, e_{ice} the vapour pressure over pure ice, e_w the vapour pressure over liquid water at the temperature T (Jones 1983). Below 0 °C the value of $\psi_{(T)}$ decreases by approximately 1.2 MPa K^{-1}.

3.2 Causes of Death by Freezing

From the very beginning, research into the processes involved in freezing injury in plants has been directed primarily towards the elucidation of causes of cell death. Of the manifold hypotheses and theories that have been propounded over the years, none has proved entirely satisfactory. The main reason for this lies in the diversity of aspects involved in freezing injury. Ice formation and the ensuing injuries take different courses depending upon species, state of hardiness and the conditions of freezing, so that no single, generally effective mechanism, can be said to be responsible for cell death and survival (Steponkus 1981). It is, however, clear that the plasma membrane plays a central role in cellular behaviour during a freeze-thaw cycle, and disturbance of the semipermeable characteristics or lysis of the plasma membrane and/or the tonoplast is a primary cause of freezing injury (Ziegler and Kandler 1980).

There is considerable evidence that irreversible alterations of the biomembranes of extracellularly frozen cells occur in the frozen state at lethal temperature (Greenham 1966; Krasavtsev 1967; Stout et al. 1980). According to NMR studies on freezing plant tissues, Rajashekar et al. (1979) showed that plasma membranes lost their semipermeability to Mn^{2+} immediately after freezing at a critical temperature. Palta and Li (1978) reported that the semipermeable properties of onion cells remained intact following incipient freezing injury, whereas the active transport properties were damaged. They maintained that the freezing injury is caused by enhanced K^+ efflux accompanied by a small but significant loss of Ca^{2+} following incipient freezing injury. Based on the results, Palta and his colleagues proposed a hypothesis that the active transportation system is the primary site of freezing injury (Palta et al. 1982). Recent studies are focussed on molecular mechanisms of biomembrane alterations, especially of the plasma membrane, to freezing injury (Steponkus and Wiest 1979; Singh and Miller 1980, 1982; Palta et al. 1982; Sikorska and Kacperska 1982; Gordon-Kamm and Steponkus 1984; Yoshida 1984b; Yoshida and Uemura 1984; Pearce and Willison 1985a,b; Review: Steponkus 1984; Schmitt et al. 1985; Fujikawa and Miura 1986).

Phospholipid degradation catalyzed by phospholipase D was found in poplar bark tissues frozen below the critical temperature (Yoshida 1974; Yoshida and Sakai 1974). Phospholipase D is assumed to be intercalated or assembled in membrane structures as a functional protein. In cortical tissues of poplar twigs in autumn frozen below

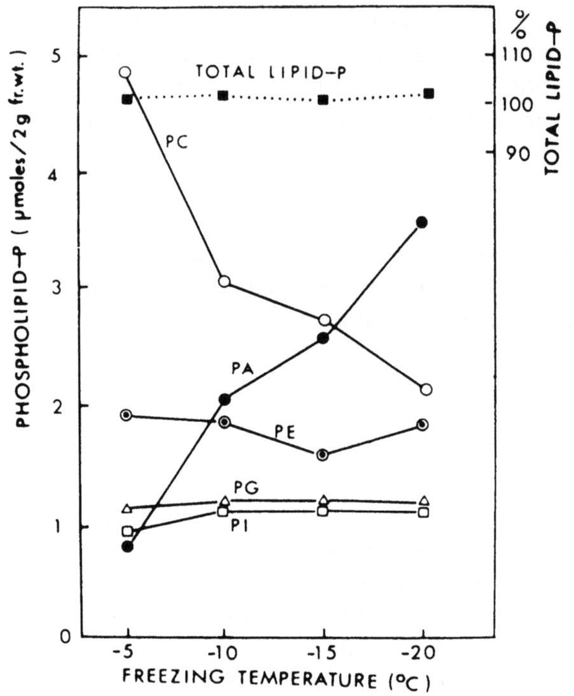

Fig. 3.3. Changes in phospholipids and total lipid phosphorus in cortical tissues of *Populus* x *euramericana* cv. *'gelrica'* during freezing. The samples, collected on Oct. 4, resisted freezing to −5 °C. The amount of total lipid phosphorus is expressed as the percentage of unfrozen control. *PC* phosphatidylcholine; *PE* phosphatidylethanolamine; *PI* phosphatidylinositol; *PG* phosphatidylglycerol; *PA* phosphatidic acid including a trace amount of diphosphatidylglycerol. (From Yoshida and Sakai 1974)

−10 °C, a decrease in phosphatidyl choline was accompanied by a concomitant increase in phosphatidic acid, while little or no change was observed in other phospholipids (Fig. 3.3). The time course of the increase in freezing at −10 °C was related to the decrease in phosphatidyl choline. Thus, the decrease in phosphatidyl choline during freezing seems to be intimately associated with freezing injury of tissues (Yoshida and Sakai 1974). The following observation strengthens this view. Very hardy cortical tissues of poplar twigs in winter survived slow freezing to −70 °C or immersion in LN_2. In these deeply frozen tissues, little or no change in phospholipid components was detected. Even after incubation at 27 °C for 2 h following thawing, almost all of the phospholipid components still remained unchanged (Fig. 3.4). On the other hand, the tissues frozen by direct immersion in LN_2 from room temperature (−196 RF in Fig. 3.4) were killed, and browning of the tissues was observed immediately after thawing. In these rapidly frozen tissues, however, a drastic change in phospholipid components, including phosphatidyl inositol and phosphatidyl ethanolamine, and a slight decrease in the total amount of lipid phosphorus occurred. On thawing, the changes are strongly accelerated. That phospholipid degradation is associated with freezing injury has been confirmed in some herbaceous plants (Rodinov et al. 1973; Wilson and Rinne 1976; Horvath et al. 1979). In yeast cells, which were injured by sudden freezing in LN_2 or rapid dehydration by freeze drying, Souzu (1973) found a degradation of phospholipids following thawing or rehydration. He postulated that sudden removal or addition of water molecules, which affects hydrophobic bondings between lipid and protein components of biomembranes, may be a dominant factor resulting in rupture of the membrane lipoprotein structure and in subsequent degradation of the phospholipids, thus leading to biomembrane dysfunction.

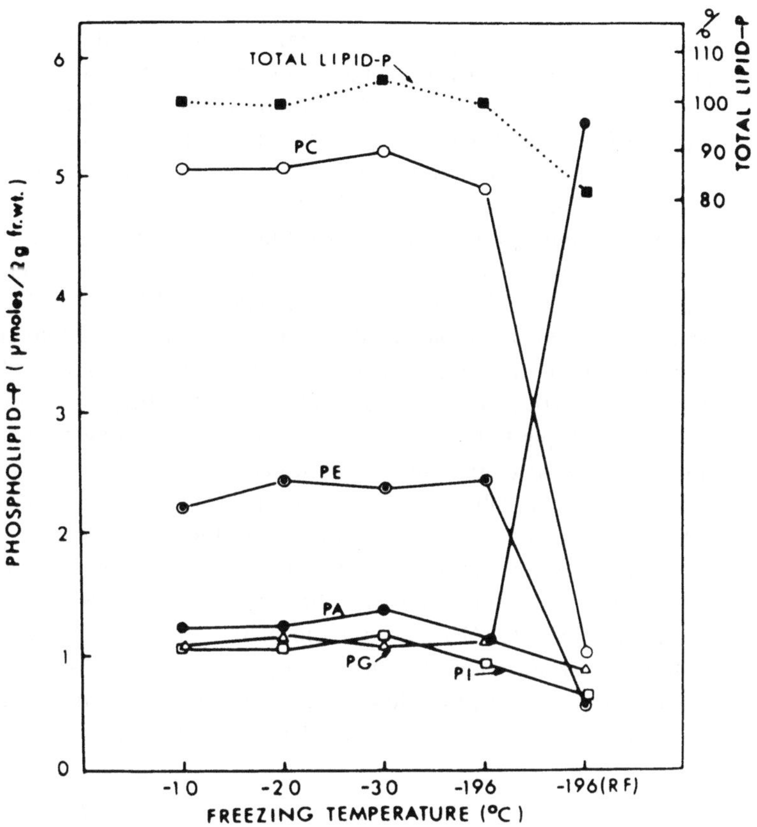

Fig. 3.4. Changes in phospholipids and total lipid phosphorus in hardy cortical tissues of *Populus* x *euramericana* after freeze-thawing. The samples, from early November, resisted slow freezing to −30 °C as well as to the temperature of liquid nitrogen (−196 °C). Tissues directly immersed in LN_2 from room temperature (−196 RF) were killed. Lipid extractions were performed after incubation at 27 °C for 2 h following thawing. The amount of total lipid phosphorus is expressed as the percentage of the unfrozen control. *Abbreviations* as in Fig. 3.3. (From Yoshida and Sakai 1974)

Temperature-dependent phase transition of membrane lipids from a lipid-crystalline state to a gel state and further to a nonlamellar, hexagonal II phase is now a well-known phenomenon over a wide range of temperatures (Lyons and Raison 1970; review by Quinn 1985). This phase transition leads to an alteration in the molecular architecture of membrane lipids which ultimately evokes injury through subsequent events, including dysfunction of cell membranes and metabolic imbalance (Lyons et al. 1979). In most cases, the effects of low temperatures on plant membrane-bound enzymes have been interpreted in terms of changes in their activities as a result of the phase transition of membrane lipids, probably involving conformational changes of the enzyme proteins (Raison et al. 1971a,b). Yamaki and Uritani (1973, 1974), however, reported that in chilling-stored sweet potato, phospholipids are released from the mitochondrial membrane due to decrease in the capacity of the membrane protein to bind to the phospholipids. From the results of their experiments they conclude that changes in the hydrophobic protein moiety play an important role in the initiation of chilling injury. Asahi et al. (1982) also proposed that the discontinuity in the Arrhenius plot for cytochrome-c-oxidase is ascribable to a conformational change of the enzyme protein rather than a change in the fluidity of phospholipid surrounding the proteins.

Only limited information is available as to the phase transition of membrane lipids and its possible connection with *frost* injury of plant cells. Fey et al. (1978) have reported that temperature-dependent membrane phase separation occurs in moderately cold-hardened wheat leaves under a subzero temperature below which the leaves sustained injury. Using electron spin resonance (ESR), Singh (1979, 1980) and Singh and Miller (1982) demonstrated that isolated plant cells undergo an irreversible reorganization of lipid bilayers to a less ordered amorphous state after freezing at a lethal temperature. However, as pointed out by Briggs et al. (1982), when intact or isolated protoplasts are used for ESR studies, the problem remains as to whether or not the fatty acid spin labels were selective for the plasma membrane. They showed that most of the labels were partitioned throughout the cells and that therefore the ESR signals reported an average from all membranes in the cells.

To gain more exact information on the role of membrane proteins and lipids with respect to the thermal properties of plasma membranes under subzero temperature, Yoshida (1982, 1984b) used purified plasma membrane vesicles from mulberry bark and orchard grass tissues. The thermotropic transition temperatures of mulberry plasma membrane vesicles shifted downward as cold hardiness increased (Yoshida 1984b). In the experiment, the fluorescence polarization of 1,6-diphenyl-1,3,5-hexatriene (DPH)

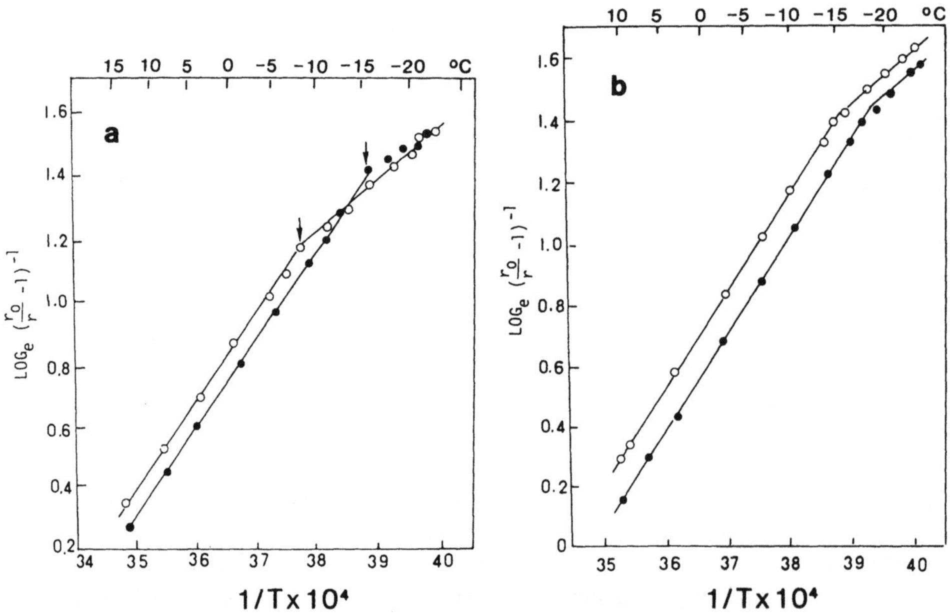

Fig. 3.5a,b. Arrhenius plots of fluorescence anisotropy parameters of DPH enbedded into **(a)** plasma membranes and **(b)** liposomes of total lipid extracts from plasma membranes of seedlings of *Dactylis glomerata*. (o) Samples from early October (hardy to −7.5 °C); (•) samples from early December (hardy to −15 °C). *Arrows* indicate thermotropic transitions of membrane lipid bilayers. *Ordinate:* anisotropy parameter. (From Yoshida 1984a)

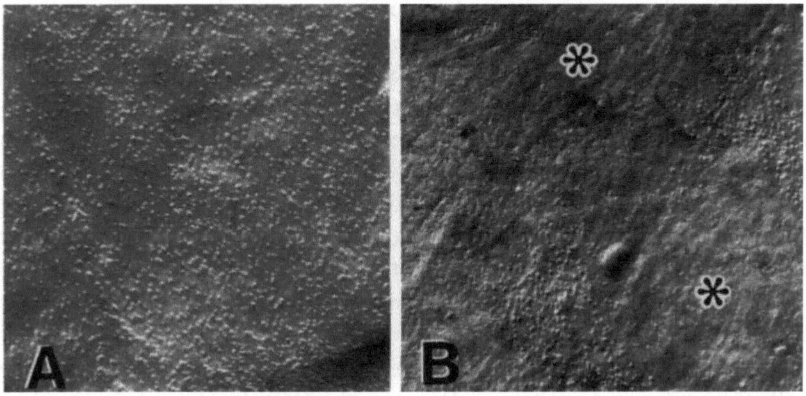

Fig. 3.6A,B. Protoplasmic face of plasma membranes of frozen cortical cells of *Morus bombycis*. **A** Frozen to −20 °C; all cells remained alive. **B** Frozen to −40 °C; all cells were killed. *Asterisks* indicates area free of intramembrane particles; × 30.000 (T. Niki, unpubl.)

was measured in 40% ethylene glycol and, therefore, the probe reported the physical state of the membrane structure as mainly affected by the low temperature per se, and not by a freeze-dehydration effect. The thermotropic transition of the membrane samples isolated from orchard grass seedlings in early October and early December showed a distinct inflection on the Arrhenius plots around −8° and −16 °C respectively (Fig. 3.5a). These inflection temperatures appear to coincide with the killing temperatures. Comparable results were obtained with rye seedlings (Yoshida and Uemura 1984) and *Helianthus tuberosus* (Ishikawa and Yoshida 1985). However, the thermotropic transitions of the liposomes prepared from total lipids of plasma membrane vesicles occurred at much lower temperatures, i.e. around −16° and −18 °C respectively, as compared with those of intact membrane vesicles (Fig. 3.5b). Following treatment with pronase, the thermotropic transition also shifted downward (Yoshida 1984b). Thus, the thermotropic properties of the plasma membrane appear to be dependent on the membrane proteins.

Yoshida (1984b) postulated that alterations of protein conformation and concomitant changes in lipid-protein interactions in plasma membranes at critical temperatures result in redistribution of membrane proteins due to lateral displacement, in apparent loss of certain proteins due to vertical displacement and in activity changes in some enzymes. Pearce and Willison (1985b) found areas free of intramembrane particles (IMP) frequent in the plasma membranes of wheat tissues frozen extracellularly to a damaging temperature, but absent from plasma membranes of supercooled tissues and of tissues frozen to a nondamaging temperature. The same IMP-free areas associated with freezing injury were observed in mulberry bark cells (Fig. 3.6) and hardened cultured cells from brome grass (T. Niki, unpublished). This change was observed to be irreversible. The quantitative relationship between the degree of freezing injury and occurrence of areas of plasma membrane free of IMP strongly indicates the primacy of changes in the plasma membrane in explaining tissue injury by extracellular freezing (Pearce and Willison 1985b). Each intramembranous particle comprises several simple or conjugated proteins (Verkleij and Ververgaert 1978). Thus, the formation of patches free of IMP probably involves a redistribution of membrane proteins. Pearce and Willison (1985b) proposed four explanations for IMP-free patches: lipid patches might exclude IMP if they were (1) in the gel phase, or (2) in the hexagonal II phase, (3) the aggregation of the particles may be a consequence of direct cross-linking between them, or (4) a failure in the mechanism which maintains the usually even scattering of particles in the membrane may occur; a hexagonal II phase was not observed. In their results, the IMPs did not appear clumped. Even when 20-30% of the membrane was free of IMPs they were fairly evenly scattered as the IMPs had been led or pushed aside. This observation leads to the postulation that IMP-free patches may be fluid rather than gel-like. From these results Pearce and Willison (1985b) proposed the hypothesis that a lesion or change in the cytoskeletal-IMP-tethering system is involved in freezing injury. This change might lead to more general irreversible membrane reorganization, because the areas free of IMP become centers for processes of structural alterations.

Karow and Webb (1965) hypothesized that death by freezing occurs primarily as a result of extraction of bound water from vital cellular structures. A similar vital water hypothesis has been proposed by Weiser (1970). Mazur et al. (1981) found that the fraction of solution remaining unfrozen has a major effect on cell survival over the excessive salt concentration or cell shrinkage. Chen et al. (1984) suggested that the observed increase in salt concentration (Fennema 1973) and reduced cell volume (Meryman 1968) may not be the direct cause of freezing injury.

Dehydration-induced phase transitions in biological membranes, in particular the lamellar-to-hexagonal II phase transition, have been assumed to be a possible cause of injury during extreme cellular dehydration (Crowe and Crowe 1982; Crowe et al. 1983). Gordon-Kamm and Steponkus (1984) reported lamellar-to-hexagonal II phase transition in the plasma membrane of isolated protoplasts from nonacclimated rye leaves after freeze-induced dehydration, but not in the supercooled protoplasts. However, Pearce and Willison (1985a) did not observe the hexagonal phase II after freeze fixation at lethal temperature, even by employing a freeze-fixation technique (Deamer et al. 1970) which is used for retaining the hexagonal II phase in model systems.

For protoplasts isolated from nonacclimated leaves (LT_{50}: -5 °C) and cooled to -5 °C, the predominant manifestation of injury is lysis of protoplast during osmotic expansion after thawing of the suspending medium. This form of injury occurs because of irreversible endocytotic vesiculation of the plasma membrane during freeze-induced osmotic contraction (Steponkus et al. 1982, 1983); it does not appear in acclimated protoplasts even at -25° to -30 °C. Steponkus et al. (1983) also observed that osmotically induced injuries in protoplasts at 0 °C or room temperature resulted in changes in membrane morphology that are characteristically different in nonacclimated and acclimated protoplasts. In nonacclimated protoplasts, the surface area contraction of the plasma membrane is achieved by endocytotic vesiculation, whereas the contraction of acclimated protoplasts is achieved by exocytotic extrusions of membrane material. Many of the extrusions appear as a tethered sphere. Unlike nonacclimated cells, in acclimated protoplasts the changes are readily reversible by exposure to hypotonic solutions (Steponkus et al. 1983).

A recent innovation in the study of the mechanisms of freezing injury and tolerance is the use of isolated protoplasts (Siminovitch et al. 1978; Wiest and Steponkus 1978; Steponkus and Wiest 1979; Singh and Miller 1980; Gordon-Kamm and Steponkus 1984) which, however, suffers one important disadvantage. The protoplasts require the presence of an external osmoticum to prevent cell lysis. Upon initiation of freezing of protoplasts in suspension, channels of high concentrations of solute form within the ice. A majority of the protoplasts accumulate in these channels, thus subjecting the cell to a plasmolytic stress and the plasmalemma to a very high external solute stress unlike cells in normal intact tissues. Furthermore, in the presence of a cell wall, the plasma membrane collapses closely attached to this cell wall during freezing. Thus, the decrease in the surface area of plasma membrane may be comparatively smaller than that of isolated protoplasts. In such cases, whether or not the protoplasts are subjected to freezing stress equivalent to that of the tissues in situ may become questionable.

It can be concluded that the plasma membranes experience complex types of stress during freezing, and it might therefore be difficult to distinguish the specific factors responsible for membrane damage. These factors may involve a physical effect of the low temperature per se, freeze-induced reduction in the surface area and solute concentration effects, freeze dehydration of the protoplasm, molecular packing of membrane constitutents induced by cell shrinkage and/or a combination of these, as well as changes in pH and ionic strength. The current position is discussed by Heber et al. (1979), Levitt (1980), Steponkus (1981, 1984) and Hincha et al. (1984).

3.3 Phenomenology of Frost Damage

Freezing damage in the field takes on many forms and is not always immediately recognizable for what it is. Injuries can only be ascribed unequivocally to freezing if their distribution indicated a cooling gradient (e.g. injuries at the surface of a stand; damage to the base of the stem; cf. Chap. 1) or if damage can be connected with a particular frost event (e.g. appearance after a late frost). Towards the end of winter, however, the causes of a damage may be more complex (Parker 1963a). The phenomenology of frost damage has been described exhaustively and illustrated in handbooks and textbooks of plant pathology, agriculture, horticulture and forest sciences. Most observations made in the field of applied science are of considerable value to the ecologist.

3.3.1 Symptoms of Injury

A distinction can be made between destructive events resulting from the freezing stress in tissues, which implies cell death, and from mechanical effects of the accumulation of ice which causes rupture of tissues. The following symptoms may appear immediately after freezing and thawing:

Discolouration: Due to decay of biomembranes, cell contents that were previously confined to their own compartments now come into contact with each other and with oxygen. This leads to coloured reaction products which adhere to the cell proteins and cell walls. Discolouration in any part of a plant is the most frequent and clearest indication of freezing injury.

Bleaching: Injured tissues of green parts of plants that do not form necrogenic pigments turn pale green, yellow or even white. Furthermore, following a severe winter, chlorophyll formation may remain defective, so that chlorotic leaves are produced.

Tissue Shrinking and Dieback: Decay of frozen tissue leads to the formation of holes in leaves, to constrictions in roots, stems, leaf and flower stalks of herbaceous plants, and to withering and dieback of shoot tips and roots. The necrotic parts of buds dry out.

Rupturing Due to the Mechanical Effects of Frost: The pressure exerted on the immediate surroundings as a result of the expansion of ice masses forming in tissue spaces leads, by mechanical means, to tearing of the tissues (frost blisters, frost boils).

Malformations: These result from injury to meristematic and incompletely differentiated tissue, which subsequently give rise to organs exhibiting distortions, stunting or fragmentation. Frost damage to leaf and flower primordia in winter buds may result in malformations, flower sterility or the buds may even fail to open altogether.

Heterochronism: Another abnormality that may result from freezing of buds is a delay in the onset of the various stages of development, e.g. retarded sprouting and flowering.

The various parts of a plant and the different life forms exhibit characteristic kinds of damage: these have been described, with illustrations, by Larcher (1985a). The following section gives a detailed description of some forms of injury typical for trees and discusses their ecological significance.

3.3.2 Freezing Damage Peculiar to Trees

Woody plants exhibit a variety of frost injuries specific to their particular morphology. Such peculiarities are a direct consequence of the highly differentiated structure of their shoot system, of the cambial growth with its distinct annual rhythm of activity and their longevity.

Damage to basal stems of young trees occurs in frost basins, flat and open lands, plateaus and sunny slopes. Field observations in central Japan revealed that basal stem damage of junvenile *Cryptomeria* (1.2 cm basal diameter, 30–50 cm high) occurred during a period between mid-November and early December, when the air temperature near the ground surface dropped below the hardiness level of the basal stem (Fig. 3.7). Early winter is the critical period for young *Cryptomeria*. The hardiness of the basal stem increased abruptly from mid-December to late January, reaching a level of −20° to −25 °C which was maintained until mid-February. In late March, when severe frost below −10 °C rarely occurs in the plantation, the basal stem was still hardy to −10 °C. The hardiness of the basal stem was much lower than that of the higher parts except

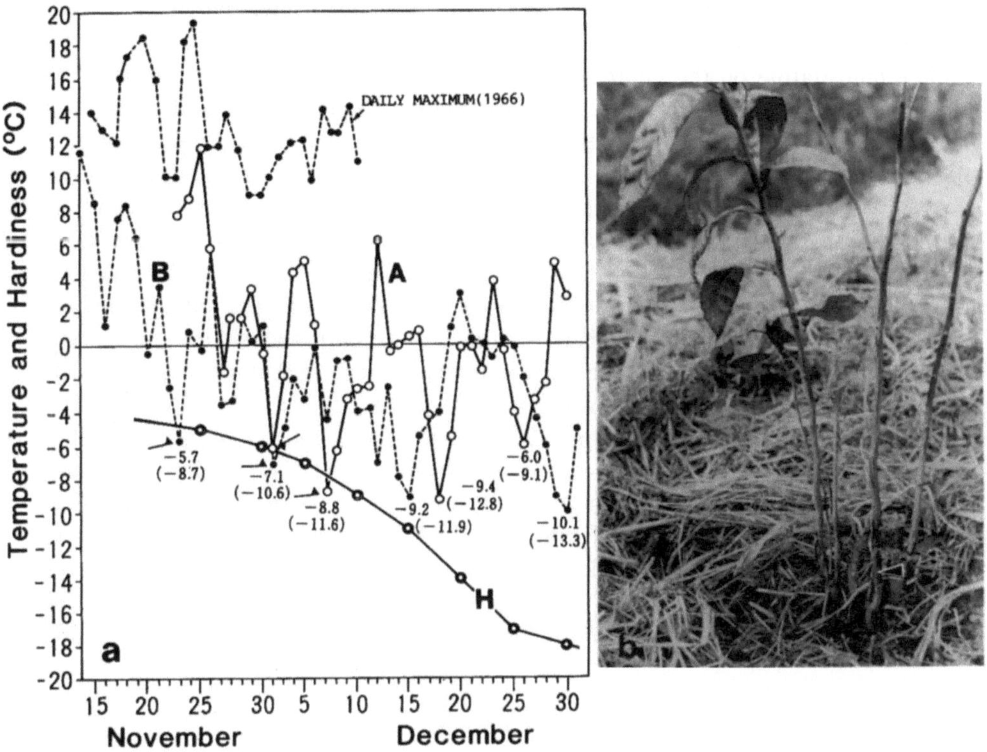

Fig. 3.7a,b. Frost damage to the basal stem of young plants. **a** Development of injuries in *Cryptomeria japonica*: daily minimum air temperatures in 1965 (*A*) and 1966 (*B*); temperatures in *parenthesis* were measured at 30 cm above ground. *H* hardiness of the basal stem; *arrows* indicate that frost damage was observed. (From Horiuchi and Sakai 1978). **b** Localization of frost damage on young tea plants. (Photo: A. Sakai)

Table 3.1. Frost hardiness (lowest survival temperature in °C) of the cambial zone on the south side at different heights on a stem of young *Cryptomeria japonica*. (From Horiuchi and Sakai 1978)

Stem part	Nov. 20th	Dec. 11th
Upper (45 to 50 cm)	–10	–15
Middle (25 to 30 cm)	– 8	–13
Basal (3 to 8 cm)	– 5	–10

for immature tip parts (Table 3.1). Microscopic observations of tissue sections taken from the basal stem revealed that the cambium and the adjacent cells are the least hardy tissues of the stem.

"Sunscald" is a freezing injury to trunks, which is confined to the southern and southwestern sides, ranging upwards to about 2 m in cold areas. The lower limit of the damage depends on the snow depth. In seriously injured stems, the cambium and ad-

jacent tissues of the decayed bark separate from the wood. Winter sunscald is caused by alternating freezing and thawing (Kramer and Kozlowski 1979). On a cold winter day the temperature on the sunlit side of a trunk rises to such an extent (20 to 30 K above the shaded side; cf. Fig. 1.22) that thawing occurs (Harvey 1923; Sakai 1966c; Martsolf et al. 1975). When the sun suddenly disappears behind a cloud or another object, the temperature of the thawed tissue drops rapidly. Freezing may be so sudden as to result in intracellular freezing and cell death (Levitt 1980). However, this explanation has never been confirmed experimentally. Martsolf et al. (1975) determined daily temperature fluctuations of peach trunks in winter. The south side of the bark cooled at 0.03 K min^{-1} between -2° and -18 °C in winter. This cooling rate is too small to cause intracellular freezing. In the bark of trunks of *Populus* and *Kalopanax* freezing was initiated in the evening at -1° to -2 °C without supercooling below -5 °C (Fig. 3.8), probably because of the spread of ice from the shady side or central part of the trunk which remained frozen even in the daytime. Spontaneous intracellular freezing generally does not occur unless the cells are supercooled to -10 °C or below, especially in hardy tissues (Mazur 1977). From the results obtained, it may be deduced that intracellular freezing is not the cause of sunscald. The following experiment also supports this consideration. A piece of trunk (20 cm long, 20 cm diameter) of *Acer mono* was frozen in a cold room at -20 °C for 2 days. Then the south side of the trunk was warmed to 5° or 10 °C with an infrared lamp, after which the heating source was turned off.

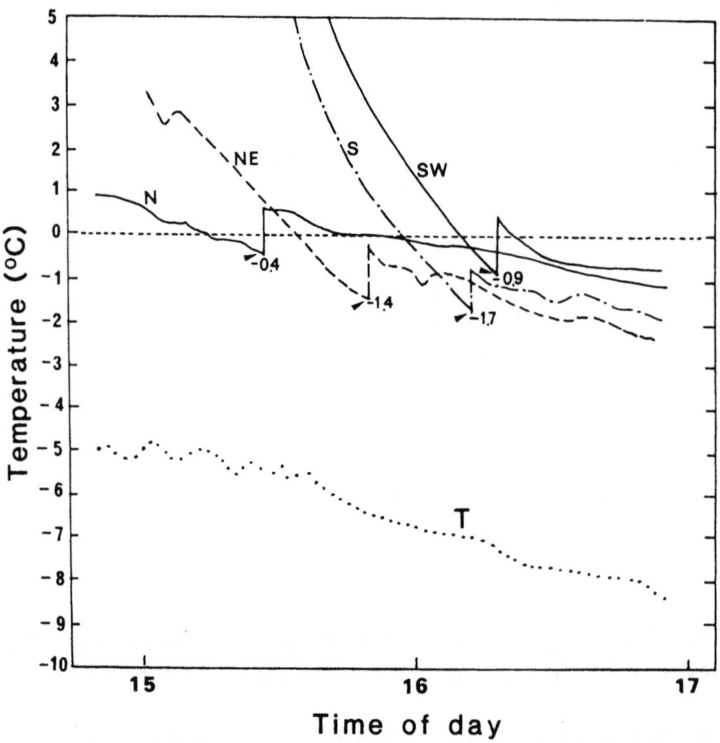

Fig. 3.8. Initiation of freezing (*arrows*) in the bark of a poplar trunk of 15 cm diameter. *S* south; *SE* southeast; *SW* southwest; *NE* northeast and *N* north side of the trunk; *T* air temperature. (From Sakai 1966c)

The cooling rate between $-1°$ and $-10\ °C$ was $0.3\ K\ min^{-1}$. Since the warming-up was too short for dehardening, no injury was observed in the bark tissues. The same result was obtained in trunks of *Aesculus turbinata, Kalopanax septemlobus* and *Ulmus laciniata* (Sakai and Horiuchi 1972). Thus, sunscald may occur when bark tissues and cambium are unhardened, or dehardened due to sunshine, in early or late winter and afterwards cooled to unusually low night temperatures.

Frost splitting of trunks develops as radial splits from the centre of a tree trunk to the bark during extremely severe frost (Fig. 3.9). In a survey made in Hokkaido by Ishida (1963), about 500 trees or 8% of all tress examined were found to have one or more frost cracks. *Abies sachalinensis* is most frequently frost-cracked with a percentage

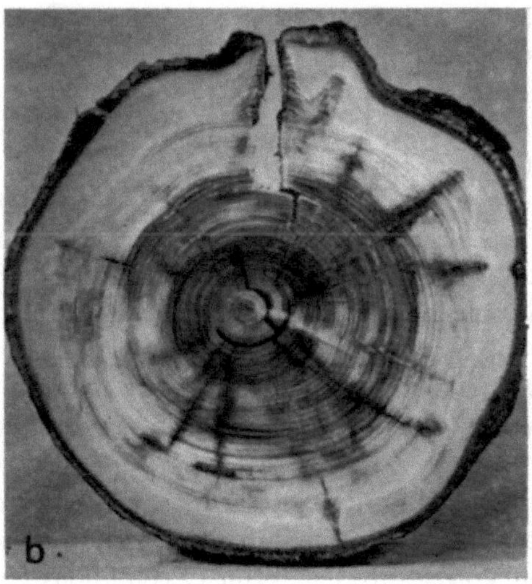

Fig. 3.9a–c. Frost splitting in trees of *Abies sachalinensis*. **a** Frost crack extending 8 m above the ground. **b** Cross-section of a log with internal cracks extending radially or tangentially, partly healed. Wetwood appears dark. **c** Swan boards from a frost-cracked log. Wetwood is outlined with black chalk. (From Ishida 1963)

Table 3.2. Relation between the amount of wetwood and the frequency of occurrence of frost cracks appearing on sections of trunks at 1 m height above the ground. (From Ishida 1963)

	Wetwood area (%)					
	0	1–10	11–20	21–30	31	Total
Abies sachalinensis						
Number of trees	29	78	27	11	9	154
Number of frost-cracked trees	0	11	9	8	8	36
% cracked trees	0	14	33	73	89	23
Picea jezoensis						
Number of trees	51	4	0	0	0	55
Number of frost-cracked trees	0	0	0	0	0	0
% cracked trees	0	0	0	0	0	0

of 23% (Table 3.2). Of deciduous trees, *Acer, Betula* and *Quercus* species are most frequently damaged by frost cracks, followed by *Fraxinus, Ostrya* and *Ulmus* species. The frost cracks increased with increase in the diameter of the trunks. Frost cracks on angiosperm trees appeared from ground level to nearly the top of the trunks, up to 20 or 25 m, but a large number of frost cracks were located within 3 m of the ground. In conifers no general tendency in direction was observed. Moreover, many individual trunks had two or more cracks on different sides of the trunk. In contrast, Chiba (1965) found that frost splitting of poplar cultivars and *Alnus hirsuta* was always on the south or southwest side of trunks. The same result was recently reported in hybrid poplar trees in Canada (Popovich 1984). These facts indicate that the problem is connected with the internal peculiarities of a tree trunk. The frost cracked trunks of *Abies sachalinensis* always contain wetwood (Fig. 3.9b,c) and frost cracks developed only at the places where wetwood was found (see Table 3.2). Frost splitting seldom occurs in the trunks of *Picea jezoensis* and *P. glehnii*, in which little or no wetwood is found. The water content of wetwood varies considerably from 90 to 200%. There was a close relationship between the width of the frost crack and the amount of ice-packed internal cracks of standing trees (Fig. 3.10). Small logs with frost cracks were cut out from standing timbers in winter and thawed at a room temperature of 10° to 15 °C. After thawing the frost cracks closed, but when the logs were put into a cold room at −15 °C, they opened again. An increase in width of the frost cracks was almost only seen in connection with freezing of the inner parts (T_3 in Fig. 3.11), but not if freezing affected the outer parts (T_2) of the logs. The increase in width terminated with the end of freezing of the inner parts of the log. These results clearly show that the principal internal cause of frost cracks is the existence of wetwood and its freezing. Wetwood of *Abies sachalinensis* tends, in general, to be situated in peripheral parts of the trunk; its distribution in the inner part of the trunk is very complicated. Root wetwood is located at the basal portion of the trunks. Branch-borne wetwood, on the contrary, tends to be scattered in each branch, especially dead branches, but it often extends towards the trunk axis.

Why frost cracks occur in a particular tree trunk or a particular part of the trunk and even in a certain species can be understood if it is accepted that wetwood, wounds and internal shakes (Shigo 1972; Butin and Volger 1982) are primary conditions for

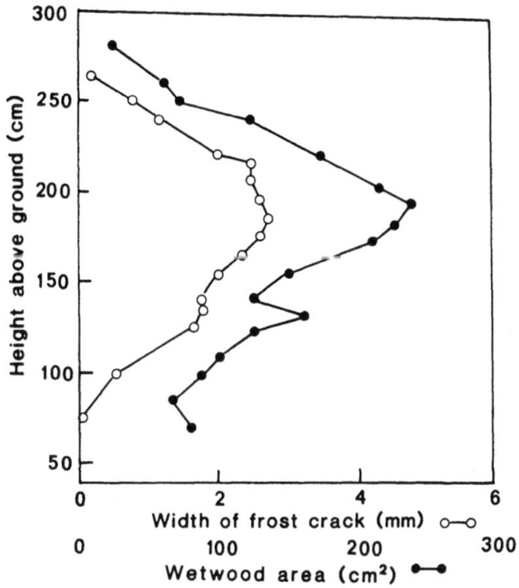

Fig. 3.10. Relationship between the amount of wetwood and width of frost crack in a trunk of *Abies sachalinensis* at various heights above the ground. (From Ishida 1963)

their development. An interesting observation in this connection is that wetwood in *Abies alba* (but not in other conifer species) is a characteristic phenomenon related to other symptoms of the European forest disease (Schütt 1981). This kind of wetwood is infected by bacteria (Bauch et al. 1979), which could act as ice nucleators. The external condition for the occurrence of frost splitting is low temperature. The width of the frost crack is most closely correlated with the temperature in the outer part of the trunk (r = 0.959), followed by that of the air temperature (r = 0.868). It thus becomes clear that there must be an internal strain due to temperature changes in the trunk of a standing tree even if there is no frost crack. As shown in Fig. 3.12, at a certain low temperature (T_0), squeezing does not appear in a wood disc free from wet-

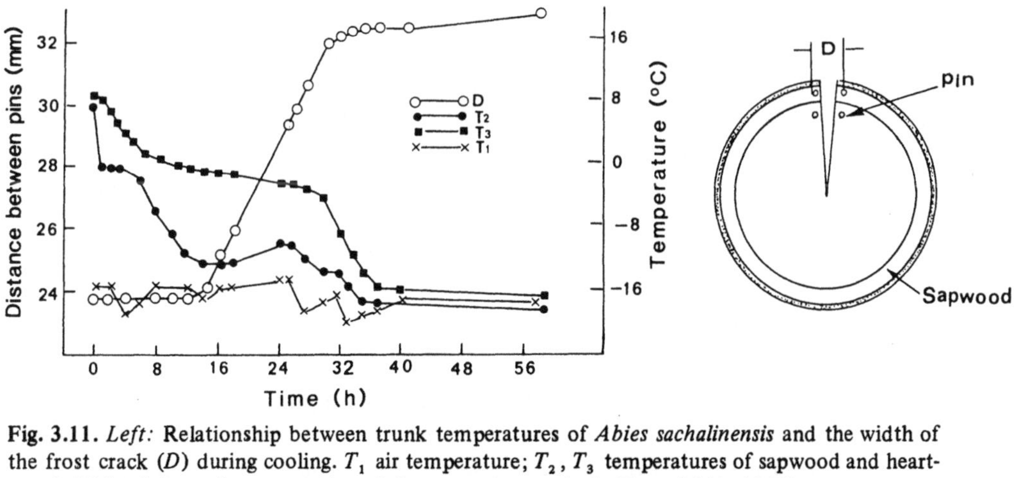

Fig. 3.11. *Left:* Relationship between trunk temperatures of *Abies sachalinensis* and the width of the frost crack (*D*) during cooling. T_1 air temperature; T_2, T_3 temperatures of sapwood and heartwood. *Right:* Schematic presentation of the measuring method. (From Ishida 1963)

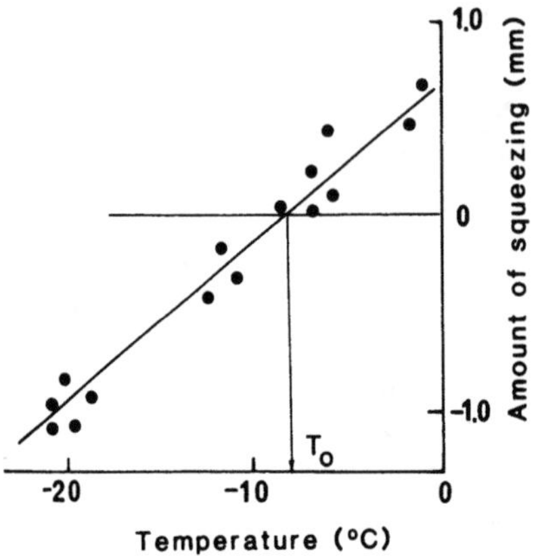

Fig. 3.12. Relationship between subfreezing temperatures and the amount of squeezing in an uninjured trunk of *Abies sachalinensis*. (From Ishida 1963)

wood and any other defects, while the amount of strain has positive values above the temperature T_0, and has negative values below it. Therefore, above the temperature of nonsqueezing, compression occurs in the tangential direction at the outer part of the disc, and tension below the nonsqueezing temperature (about − 10 °C). In general, the nonsqueezing temperature in the presence of wetwood was higher than that in normal wood.

Conifers and deciduous hardwoods native to interior Alaska, Canada and E. Siberia are tolerant to frost splitting. The frequency of frost-splitted trees seems to increase near northern or altitudinal distribution limits, for example, of *Cryptomeria* and *Abies mariesii* in northern Japan, and *Tsuga plicata* in Alaska. Thus, frost splitting appears to be one of the important factors setting northern or altitudinal boundaries.

3.4 Results of Freezing Injury and Chances of Recovery

Depending on the extent of the frost injury and the importance of the damaged tissue to the plant as a whole, partial or even complete repair is possible. On the other hand, partial injury weakens the plant, diminishes at least temporarily the yield of dry matter production and of reproduction, and may thus reduce the competitive vigour of a plant. In assessing the possibility of a plant's survival after injury its powers of recovery have to be known. Numerous observations after freezing have been published in agricultural, horticultural and sylvicultural journals, but only few of them were the results of experiments primarily desgined to answer the above question. For literature sources, we refer to reviews by Alden and Hermann (1971), Kozlowski (1971) and Larcher (1973, 1981a, 1985a).

The degree of final damage depends on the localization and the extent of the injuries, and on the possibility of restitution of lost plant parts.

Damage of Organs Important for Reproduction. In phanerogams, where the reproductive organs are the most susceptible plant parts (see Chap. 6), the first step in partial frost damage, as a rule, consists in the temporary loss of the ability of the plant to propagate. In severe winters the flower primordia in the buds may be killed, although

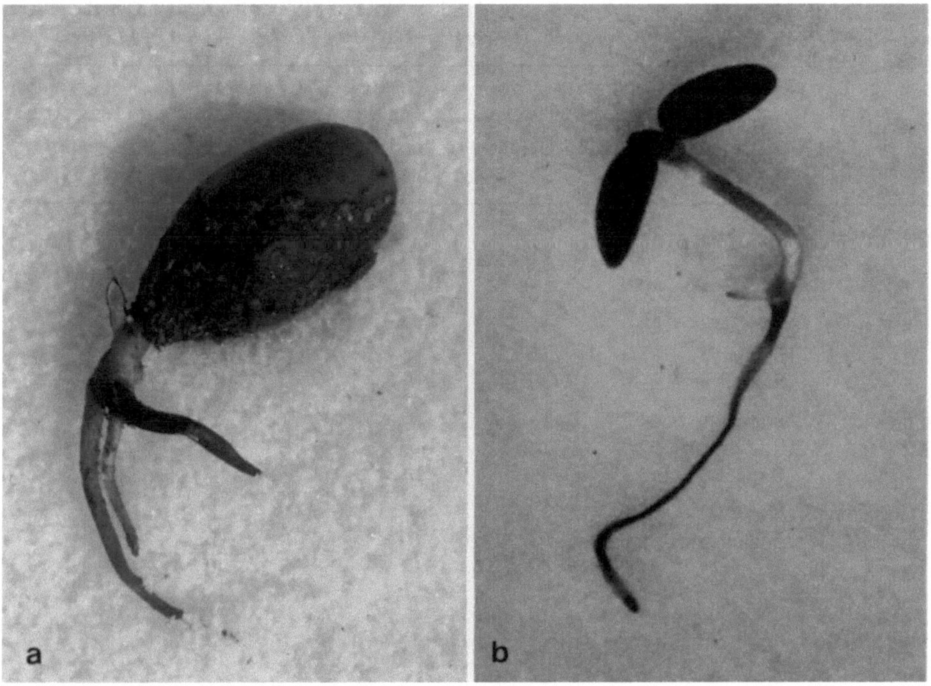

Fig. 3.13a,b. Recovery of germinating seeds and seedlings after partial frost injury. **a** *Quercus ilex* 2 weeks after freezing at −2 °C; the primary root was killed and appears black. **b** *Dryas octopetala* 1 week after freezing at −6 °C; the primary root was killed; it is soon replaced by a lateral root. All shoot parts survived the frost uninjured. (Photos: W. Larcher and M.Th. Eccher)

vegetative shoots will develop in the following growing season. After spring frost, flowers whose pistils have been injured are either shed or form fruits with a reduced number of seeds. Germinating seeds are particularly susceptible to frost; however, their capacity to recover is conspicuous too: after injury to the primary root and the extending shoot tip these organs are quickly replaced by secondary roots and the outgrowth of cotyledonar buds (Fig. 3.13).

Reduced Productivity and Regrowth Due to Defoliation. The effect of a loss of leaf area on the production capacity of plants depends on the extent and localization of freezing injuries. Leaves with a particularly sensitive vascular system (especially of the petiole) are shed shortly after frost even if the entire assimilation parenchyma remained intact. In frost injuries restricted to limited areas of the mesophyll the leaves remain attached to the plant and are able to supply the plant with photosynthates in proportion to the undamaged area. Reports on frost damage after severe winters and from defoliation experiments describe growth reduction in woody plants after loss of foliage. Conifers seem to be more sensitive to foliage losses than broad-leaved trees. When the entire foliage is lost, in many cases no annual ring is formed in the following season; at best, the growth ring will attain 50% of the normal width. Complete defoliation may eventually result in tree death (Monange 1961; O'Neil 1962; Kozlowski 1971; Ikeda 1982; Table 3.3). In contrast to the situation in woody plants, the loss of assimilatory

Table 3.3. Injury and recovery of 2- to 9-year-old citrus trees of different varieties after the severe winter 1976/77 in W. Japan with minimum temperatures to −10.4 °C. (From Ikeda 1982)

Cold hardiness	Variety	April			August		
		Foliage killed[a]	Defoliation[b]	Wood killed[c]	Recovery rating[d]	Dead trees (%)	Yield of next year[e]
Most hardy	Trifoliate	0	0	0	0	0	H
	Yuzu	0	1.0	0	0	0	H
	Troyer	0	5.0	0	0	0	L
Hardy	Miyagawa	3.0	3.0	1.0	0.2	0	H
	Kiyomi	3.3	3.3	2.0	0	0	H
	Clementine	4.0	3.8	2.0	1.0	0	H
Somewhat hardy	Sanbôkan	4.5	5.0	2.5	0.9	0	H
	Kishûmikan	2.3	2.2	1.0	1.0	0	H
	Minneola	3.7	3.0	2.0	1.0	0	H
Intermediate	Hassaku	5.0	5.0	4.0	2.6	0	L
	Amanatsu	5.0	5.0	5.0	3.7	0	L
	Lee	5.0	4.0	3.0	2.8	0	L
Somewhat less hardy	Valencia	5.0	5.0	5.0	4.1	0	L
	Suzuki navel	5.0	5.0	5.0	3.6	0	L
	Natsumikan	5.0	3.6	2.4	3.6	20	L
Less hardy	Iyo	5.0	5.0	5.0	6.1	33	N
	Dancy	5.0	3.2	4.3	6.3	50	N
	Ponkan	4.5	5.0	4.5	5.9	50	N
Least hardy	Banpeiyu	5.0	5.0	5.0	6.0	63	N
	Lisbon	5.0	5.0	5.0	7.3	82	N
	Cleopatra	5.0	2.0	4.0	6.8	75	N

[a] 0 = no damage – to → 5 = 100% leaf killed.
[b] 0 = no leaf fall – to → 5 = 90 to 100% leaf drop.
[c] 0 = no damage – to → 5 = secondary scaffold branch killed.
[d] 0 = no damage – to → 8 = tree dead.
[e] H = heavy crop L = light crop; N = no crop.

leaf area in herbaceous plants, especially grasses, can readily be compensated by formation of new leaves and therefore rarely leads to persistent disturbances.

Deficiencies in Shoot Formation from Damaged Buds. Loss of foliage is seldom a very serious event for a plant if the buds are not damaged; if they have been injured may the consequences be fatal. Normal refoliation is only to be expected if the buds are more resistant than the leaves, which is, in fact, the case in the majority of woody plants. An exception is provided by certain conifers, e.g. some *Cupressaceae, Abietoideae* and *Pinus* species, in which the leaf primordia in the terminal buds are no more frost resistant than the needles. Such species can therefore only partially replace their needles after heavy losses. Growth from partially injured buds is still possible, although irregular and retarded, as long as some of the bud meristem remains intact. Regularly recurrent injury to terminal buds causes trees to become stunted due to loss of apical dominance of the main shoot (see Fig. 7.7). Trees with a tendency to a shrublike form, and shrubs, have better prospects of survival in habitats exposed to frost.

Consequences of Severe Injuries to Supporting Organs. Total damage occurs when an entire organ system is lost. Extensive freezing at the frost-sensitive lower stem and rootstock of woody plants, the crown meristems in grasses, as well as total damage of the root system also causes the death of the plant.

After a partial injury to the shoot system, trees may die in subsequent years if the damage is irreparable and eventually spreads. Recovery is to be expected if the cambium remains intact or largely undamaged. Observations after severe winters and freezing experiments with fruit trees have revealed that the prognosis for recovery is doubtful, if approximately 50% of the cambium has been killed (Kemmer and Schulz 1955; Pisek and Eggarter 1959; Lapins 1965). Recovery experiments with 3-year-old seedlings of *Abies alba* and *Larix decidua* have shown that the probability of restitution and thus the long-term survival of frost-damaged plants varies depending largely on seasonal differences in the localization of frost injuries (unpublished results of G. Bendetta as cited in Larcher 1981a). After frost in spring and summer, when the cambium is the most

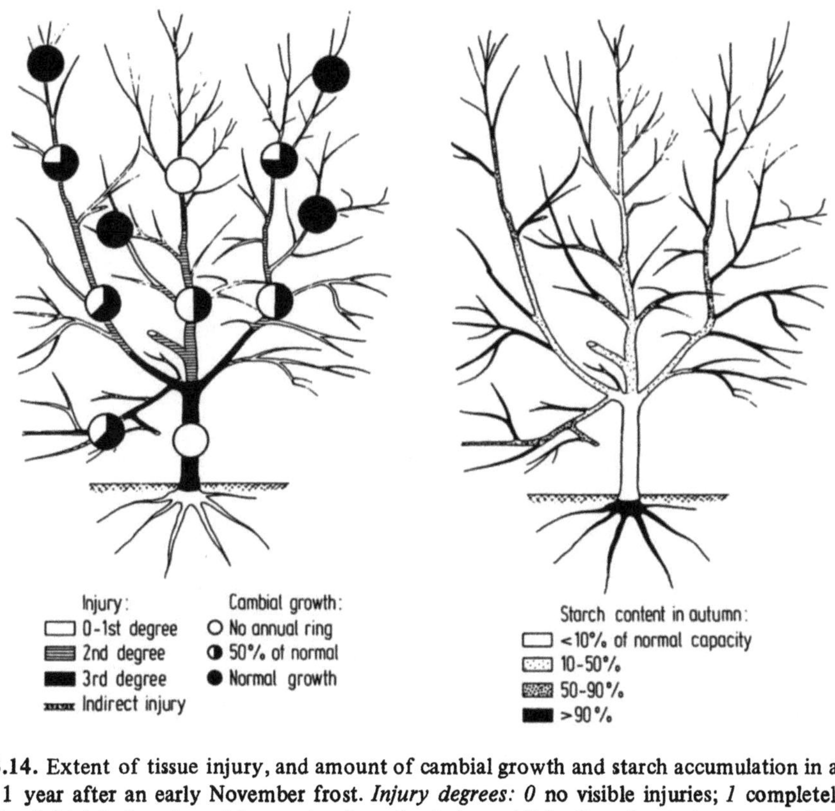

Fig. 3.14. Extent of tissue injury, and amount of cambial growth and starch accumulation in apple trees 1 year after an early November frost. *Injury degrees: 0* no visible injuries; *1* completely reparable slight injuries of cambium and xylem, frost rings visible; *2* moderate to severe xylem and bark injuries, restitution possible by intact cambium; *3* irreparable injuries, no or little cambial activity. *Indirect injuries:* secondary dieback of twigs. (From Larcher 1981a)

sensitive shoot tissue, the plants were killed if this tissue was damaged to an extent of 50% or more. During winter, when the cambium was extremely frost-tolerant, the xylem and the needles of fir were injured first (at $-30°$ to -36 °C); plants with 50% xylem damage were able to survive depending on the amount of intact leaves, but no recovery was observed if less than one quarter of the leaf area remained intact.

After severe frost damage, productivity and growth of woody plants are reduced for years. A secondary effect of wood necrosis and narrow tree ring formation is the lack of storage capacity for reserve materials (Fig. 3.14). In autumn much less starch can be deposited in frost-damaged trees. As a consequence, in the following spring carbohydrate mobilization will be insufficient and may become limiting for the new growth; the leaves remain smaller, less leaf area develops and the productivity of the tree is further diminished.

4. Mechanisms of Frost Survival

4.1 Components of Frost Survival

Survival capacity, i.e. the ability of a plant to withstand adverse weather conditions (Larcher 1973), is a highly complex phenomenon. Although frost survival capacity depends primarily on the specific frost resistance of a plant, various mechanisms are involved which help the plant to mitigate, prevent or escape excessive low temperature stress. Furthermore, recovery from injuries must be taken into consideration: in order to survive unusually severe frost the plant need not necessarily remain completely undamaged, but repair must be still possible (see Sect. 3.4). Therefore, the chance for frost survival of a plant in a given environment consists in mitigating or excluding the freezing risk, in frost resistance and in recovery after frost damage.

4.1.1 Frost Mitigation

The brief hazard presented by nocturnal radiational cooling is met by a variety of protective mechanisms affecting the thermal balance of plants:

Active Frost Protection. Nyctinastic movements of leaves and petals which minimize the heat loss of protected organs (buds, pistils) are an active mechanism delaying the radiational heat loss during night. Particularly efficient is the effect of forming a night bud with tightly folded leaves in giant rosette plants of tropical high mountains (see Sect. 7.5.2). Sleep movements of leaves of Fabaceae result in a reduction of heat loss to the extent of about 1 K (Schwintzer 1971; Enright 1982); even a temperature difference of 1-2 K can be decisive in determining the extent of injury in the case of highly sensitive plants.

Insulation. Those parts of plants that are either underground or covered by a layer of litter are the most effectively screened from frost (see Fig. 1.11). The innermost and more deeply situated parts of cushion plants and tussocks are also less exposed to radiational cooling (Hedberg and Hedberg 1979). Similarly, a dense covering of attached dead leaves protects the most frost-sensitive parts of the stems of tropical high-mountain megaphytes, particularly the cambial zone and the conducting tissue, from nocturnal freezing (Goldstein et al. 1985b; see Fig. 7.33). In tree stems with thick bark, under conditions of alternating freezing and thawing, the diurnal temperature fluctuations are mitigated and freezing is delayed in the inner parts (see Fig. 1.22).

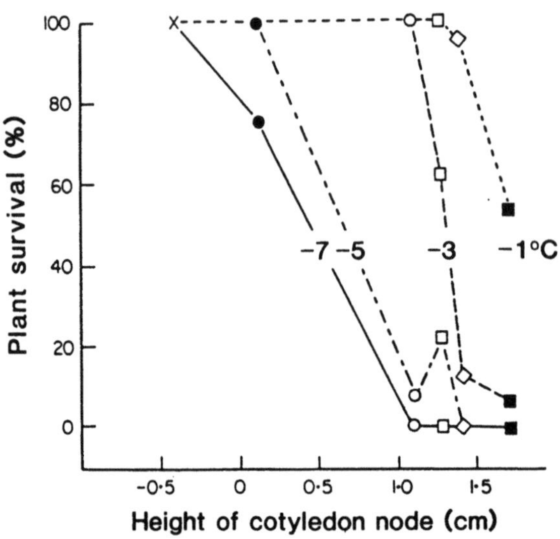

Fig. 4.1. Leaf survival of *Centrosema virginianum* accessions related to height of the cotyledon node (*different symbols*) after frost at the indicated temperatures. (From Clements and Ludlow 1977)

Storage of Heat by the Plant. Plants with a large volume to surface ratio, such as cacti and other stem succulents, store heat taken up during the day with the result that during the night the lowering of the internal temperature lags 2–3 h behind that of the surrounding air (Mooney et al. 1977; Nobel 1980). The interior of plant cushions remains several K warmer during night than their surface (Ruthsatz 1978). Thick tree trunks also cool down much more slowly than slender stems due to thermal buffering.

Heat Uptake from the Surroundings. Certain growth forms, such as rosette, prostrate and cushion plants, are favoured by heat conduction from the soil. The parts of the shoots near the ground and especially the axillary buds of rosette plants remain, on a clear night, 2–6 K warmer than the freely radiating upright parts (see Fig. 1.8). Even very slight morphological differences can have important consequences, as the following example shows (Clements and Ludlow 1977; Fig. 4.1): Ecotypes of the subtropical Fabacea *Centrosema virginianum* from higher latitudes, with short hypocotyls (less than 0.5 cm), suffer less damage from nocturnal frost than those from lower latitudes (with hypocotyls longer than 1 cm). However, the leaves of both ecotypes freeze between −1° and −2 °C if they are exposed to frost under identical laboratory conditions, and in neither can they be hardened. This shows that the advantage of the ecotype with the shorter hypocotyls is due solely to frost mitigation by heat conduction from the ground. The recognition of this fact illustrates the general principle that a proper interpretation of observations made in the field can only be made on the basis of a thorough causal analysis.

Heat Production During Freezing. The heat of crystallization (331.6 J g^{-1}) released during the freezing of the water infiltrating and adhering to leaf bases, leaf sheaths and attached litter, keeps the temperature at 0 °C until all of the water is frozen. In cushion plants the freezing of the water-saturated litter may take up to several hours, so that the temperature in the region of the living parts of the plant does not drop to a dangerous level during the night (Fig. 4.2). For further examples, see also Sect. 7.5.2.

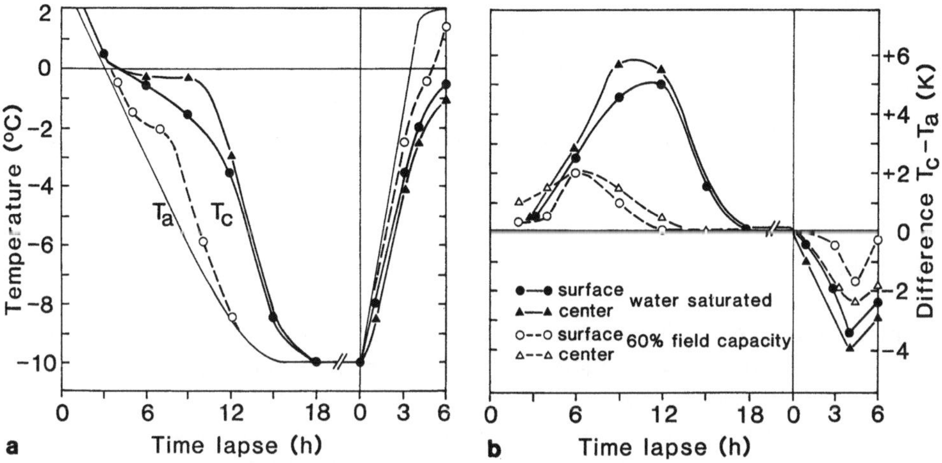

Fig. 4.2. a Thermal buffering of a cushion of *Silene acaulis* of 7 cm diameter and 4 cm height. b Temperature difference between cushion (T_c) and surrounding air (T_a) during controlled cooling and rewarming. The changes in the air temperature were chosen to simulate environmental conditions during a night in spring or autumn at the alpine habitat of the plant: cooling rate 0.6 K h^{-1}, rewarming rate 2.5 K h^{-1}; air flow rate 0.2 m s^{-1}. Thermocouples were inserted into the leaf layer (3 mm below surface) and the center of the cushion. In the water-saturated state the core temperature remained above −2 °C for 5 to 7 h after the air temperature had decreased below this level; thermic equilibrium with the air was only approached after 15 h freezing. In the same cushion, after drying to 60% field capacity, the buffering effect was much reduced, but still recognizable for at least 10 h (W. Larcher, unpubl.)

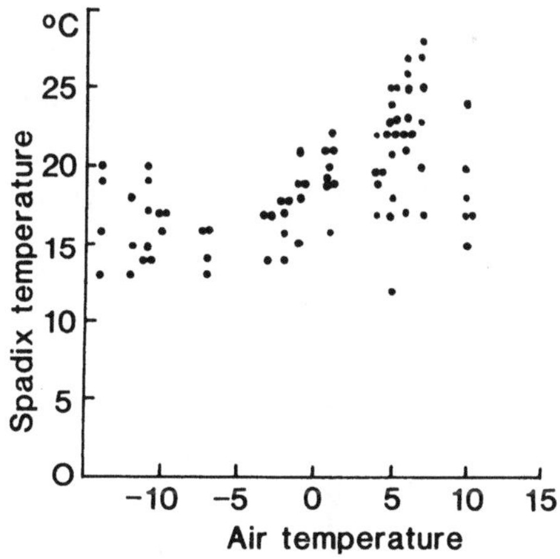

Fig. 4.3. Inflorescence temperatures of *Symplocarpus foetidus* at various environmental temperatures. (From Knutson 1974)

Metabolic Thermogenesis. A further source of heat that may be utilized to delay cooling and thus to diminish the severity of frost is the heat released by respiration. Knutson (1974) found that the temperatures of the fleshly-petaled and tightly packed flowers on the spadices of *Symplocarpus foetidus*, a small North American Aracea, which appear in early spring, remain above +10 °C, surprisingly even if the air temperature drops to −15 °C (Fig. 4.3). This is achieved by doubling the respiration rate from 12 °C to subfreezing temperature levels. Cyanide-resistant respiratory pathways (Meeuse 1975; Lambers 1982) and high catalytic activities of phosphoenol pyruvate carboxylase (ap Rees et al. 1981) are responsible for heat production by the thermogenic tissues of Araceae species. To what extent the heat developed in respiration is exploited by other plants (Vojnikov et al. 1984), and its significance for frost survival, deserve further investigation.

4.1.2 Frost Exclusion

Spatial Frost Exclusion. Thermal protection of plants or plant parts against severe winter frost can be achieved by adaptation of the life form. The system of "life forms" introduced by Raunkiaer (1910) is based on the adaptations to the complex constraints placed upon plants by the direct and indirect action of frost in winter, i.e. low temperatures and winter desiccation. There is a clear correlation between frost resistance and protection from frost (Fig. 4.4). Plants and parts of plants that are shielded from severe cold by snow or litter, require little resistance (Fig. 4.5). Spatial exclusion of severe frost is of very obvious ecological significance in mountains where snow is unevenly distributed in winter (see Sect. 7.5.1). The perennating buds of chamaephytes and hemicryptophytes in the understorey of forests of the temperate zone are considerably less hardy than the buds on exposed overwintering shoots (Till 1956; Yoshie and Sakai 1981a; Table 4.1a); rhizomes are less resistant the deeper they are situated in the ground (Till 1956; Kappen 1964; Table 4.1b). Swamp and water plants, despite acquiring very little frost resistance (Table 4.2), suffer hardly any damage even in the most severe winters because the temperature below the ice cover does not drop below 0 °C.

Temporal Frost Exclusion. Plant parts and ontogenetic stages which are especially susceptible to frost may escape danger by exact timing of their life cycle and develop-

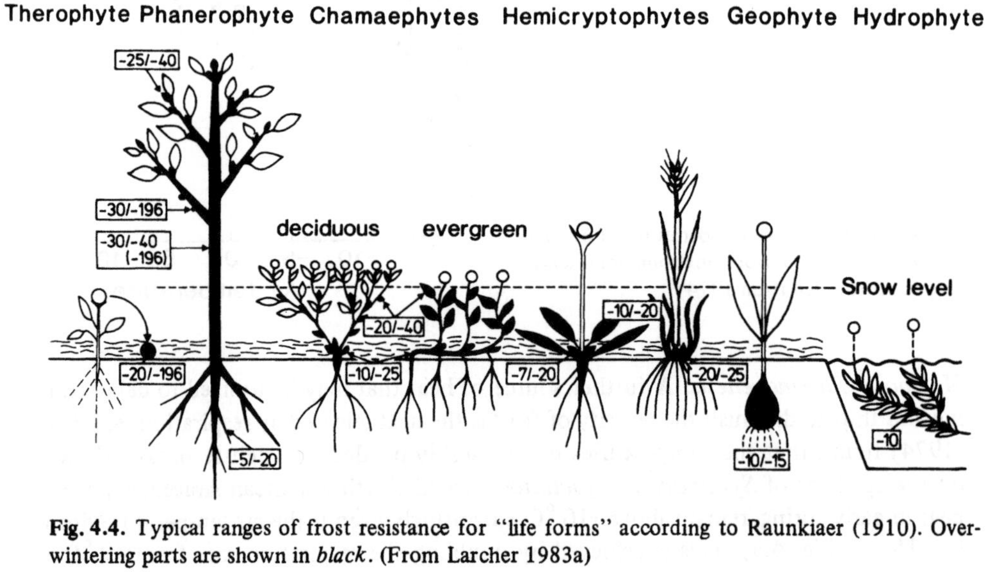

Fig. 4.4. Typical ranges of frost resistance for "life forms" according to Raunkiaer (1910). Overwintering parts are shown in *black*. (From Larcher 1983a)

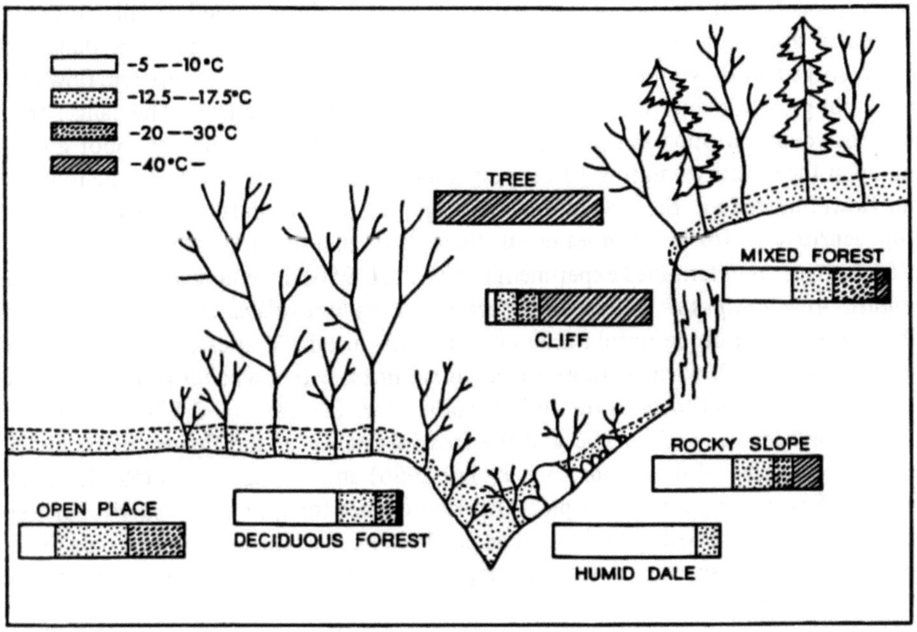

Fig. 4.5. Relationship between habitat distribution and potential frost resistance of fern rhizomes in Hokkaido. All of the epiphytic ferns and most of the species growing on cliffs survive −40 °C, whereas the rhizomes of 40–60% of the species occurring on forest floors and more than 80% of ferns in a humid dale are already frost damaged at −5° to −10 °C. (From Sato 1982)

Table 4.1. Frost resistance (°C) in winter of chamaephytes and hemicryptophytes. (From Till 1956; Kappen 1964)

a) Phanerograms of the forest floor		b) Fern rhizomes	
Evergreen leaves:		below litter	−12
3 to 5 cm above litter	−11.5 to −14.5	1–3 cm below ground	− 9 to −10
5 to 10 cm above litter	−11.5 to −18.0	3–5 cm below ground	− 7 to − 9
10 to 20 cm above litter	−13.0 to −20.0	12 cm below ground	− 2.5
Buds:			
below litter	− 7.0 to −11.5		
close to litter	−12.5 to −18.0		
3 to 20 cm above litter	−15.5 to −19.5		

Table 4.2. Frost resistance (°C) of swamp and water plants in winter. Samples were cold acclimated at 0 °C for 20 days before testing (A. Sakai and O. Shibata, unpubl.)

Species	Leaves and buds	Stems	Roots
Cicuta virosa	−10	− 7	− 7
Menyanthes trifoliata	−10 to −12	−10 to −12	−10
Nymphaea tetragona	− 5 to − 7	− 5 to − 7	− 5 to −7
Potamogeton distinctus	− 3		− 3
Sparganium stoloniferum	− 5		− 5
Utricularia japonica	− 8 to −10		
Vallisneria asiatica	− 5		− 3

mental processes. By the early termination of growth, the autumn leaf fall and the induction of dormancy in the woody plants of winter-cold regions, both the danger of frost damage and of winter desiccation are reduced. Late bud break, foliation, flowering and shoot growth reduce the hazard of damage from late spring frost. The danger from late frost depends upon the probability of its occurring at the time of shoot growth and upon the sensitivity to frost of the young shoots and flowers. Although late foliation and flowering reduce the risk of frost damage, plants with delayed development are more sensitive to frost, if it does occur, than plants that develop early. This is obvious in forest trees: as determined experimentally by Till (1956) the young leaves of *Carpinus betulus, Acer campestre, Acer pseudoplatanus, Corylus avellana* and *Alnus glutinosa*, which in central Europe unfold around mid-April, sustain injuries at −2.5° to −5.5 °C (mean value −3.8°), whereas those of *Fagus sylvatica, Quercus robur* and *Fraxinus excelsior*, which unfold towards the end of April, freeze at −2.0° to −2.5 °C. A clear connection between flowering time and the frost resistance of the flowers exists also in understorey herbs of deciduous forests (Till 1956) and in bog ericads (Reader 1979; Fig. 4.6). Seasonality timing is of particular significance for spring frost survival of trees at their latitudinal distribution limit (Langlet 1937) and at the alpine timberline (Tranquillini 1979a). In species of wide geographical distribution, genotype differentiation

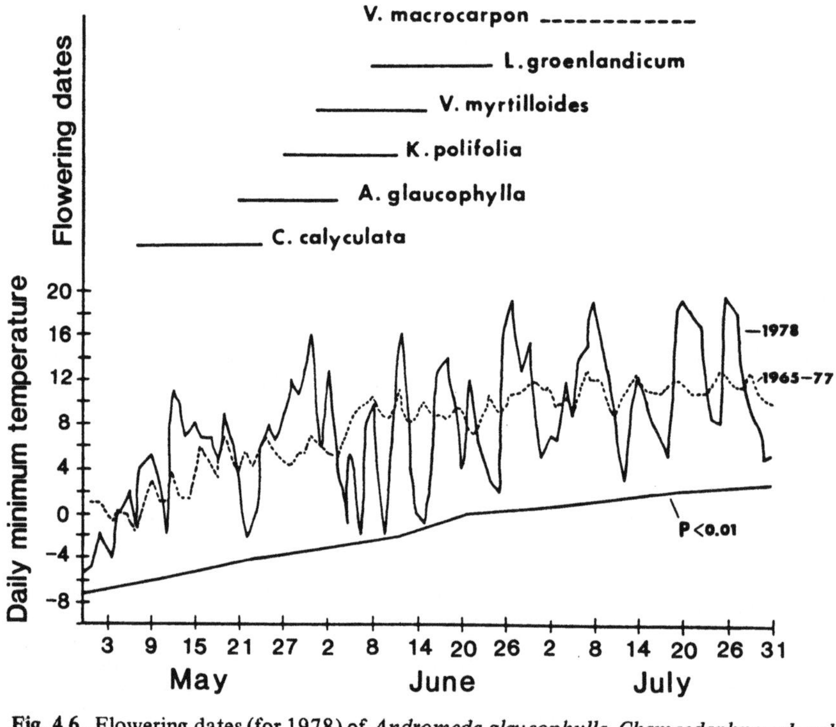

Fig. 4.6. Flowering dates (for 1978) of *Andromeda glaucophylla, Chamaedaphne calyculata, Kalmia polifolia, Ledum groenlandicum, Vaccinium macrocarpon* and *V. myrtilloides*, and daily minimum air temperatures recorded at a marsh bog in southern Ontario. *Dotted line:* low temperature average; $P < 0.01$: lowest probable temperature. The periodic occurrence of temperatures below −7 °C would select against May flowering of most of the bog ericads, except *C. calyculata*. (From Reader 1979)

and selection leads to the evolution of "climatic races" (Münch 1923). Such rhythmo-ecotypes respond to the temperature thresholds and photoperiods that act as signals for the approach and the end of winter at the particular locality to which they are adapted.

4.1.3 Components of Frost Resistance

Frost resistance is the ability of a plant to survive subfreezing temperatures (Larcher 1973). Resistance to freezing stress results from the ability of the protoplasm to tolerate the various strains exerted by ice formation in the tissues and from the effectiveness of avoidance mechanisms in delaying or preventing the attainment of thermodynamic equilibrium between cell and surrounding ice. In the concept of Levitt (1958, 1980) a distinction is made between avoidance of any freezing of plant cells (due to supercooling, lowering of the freezing point and absence of freezable water in desiccated cells) and avoidance of intracellular freezing (due to extracellular or translocated ice formation). Freezing tolerance, according to Levitt, under natural conditions can thus only be tolerance of extracellular freezing (which corresponds to avoidance of intercellular freezing).

4.1.3.1 Categories of Freezing Resistance

From a comparative viewpoint, different categories of freezing can be distinguished on the basis of the relative importance of the various components of frost resistance and on hardening capacity (Larcher 1981b, 1982, 1985a; Tables 4.3 and 4.4):

Permanently freezing-sensitive plants are protected against frost damage only by freezing point depression and supercooling. In general, tropical plants, in particular all chilling-sensitive plants, cannot be frost-acclimated at all. There are, on the other hand, freezing-sensitive plants (or tissues) which enhance their freezing avoidance by lowering T_f and by increasing the range and duration of supercooling. The acclimation effect can be up to 3-5 K in leaves (Table 4.5). It is greatest in bud meristems and xylem parenchyma which develop deep supercooling capacity during winter (Table 4.6).

Freezing-tolerant plants are able to survive considerable freezing in their organs. In seasonally freezing-tolerant plants the resistance to equilibrium freezing is a *temporary* state linked with reduced growth activity in perennial herbaceous plants of temperate regions (Gusta and Fowler 1979; Sikorska and Kacperska-Palacz 1979) and with dormancy in woody plants (Tumanov 1967, 1979; Weiser 1970). In such plants freezing tolerance is lost during periods of intensive growth. Certain high mountain plants (and probably also arctic plants: Yoshie and Sakai 1981b) sustain extracellular ice formation even during the growing season (*permanently* freezing-tolerant species: see Sect. 7.5).

The individual organs and tissues of any one plant may react differently to frost, so that a distinction has to be made between plants that are either totally or only partially tolerant to freezing.

Partially freezing-tolerant plants, even in their frost-hardened state, contain freezing-sensitive tissues and can therefore only acquire a limited degree of frost resistance.

Table 4.3. Typology of freezing and frost resistance of plants. (From Larcher 1981b)

Category	Injury at °C	Type of freezing	Resistance mechanism	Hardening
Nonhardening, freezing-sensitive plants or tissues	– 1 to – 3	Nonequilibrium freezing by heterogeneous nucleation after transient supercooling	Freezing point depression of cell sap	None
Freezing-sensitive plants or tissues with hardening capacity	– 3 to –10		Lowering of T_f by accumulation of solutes, persistent supercooling	Induced by low temperatures
Deep-supercooling, freezing-sensitive tissues	–10 to –50	Intracellular freezing by homogeneous nucleation	Barriers to heterogeneous nucleation	Mainly seasonal
Limited freezing tolerance	Down to about –50	Dehydration and deformation of cells due to equilibrium freezing	Molecular and ultrastructural changes in biomembranes and proteins, accumulation of cryoprotective solutes, prophylactic redistribution of water within and out of the organ	Seasonal and responsive to environmental temperatures, dormancy-linked in woody plants
Unlimited freezing tolerance	None	None		Seasonal and responsive to environmental temperatures, dormancy-linked in woody plants
Desiccation-induced freezing tolerance	None	None	Extremely low water content	By water loss

Table 4.4. Components of frost resistance and hardening capacity of various plants. (From Larcher 1982)

Species and plant part	Temperature below which tissue freezing occurs			Frost resistance			Freezing tolerance	References
	T_f^{unh}	T_f^h	Acclimation effect	LT_i^{unh}	LT_i^h	Acclimation effect	$LT_i^h-T_f^h$	
	°C	°C	K	°C	°C	K	K	
Solanum tuberosum, leaves	– 3	– 3	0	– 3	– 3	0	0	Chen et al. (1976), Chen and Li (1980a)
Trachycarpus fortunei, leaves	–10	–11	1	–10	–11	1	0	Larcher and Winter (1981)
Olea europaea, leaves	– 5	–10	5	– 5	–10	5	0	Larcher (1963a)
Quercus ilex, leaves	– 5	– 8	3	– 5	–13	8	5	Larcher and Mair (1969)
Abies alba, needles	– 4	– 7	3	– 4	–30	25	23	Pisek et al. (1967)
Malus sylvestris, bark	– 5	– 6	1	– 5	–30...–60	25–55	25–55	Larcher and Eggarter (1960), Mittelstädt (1969), Tyurina et al. (1978)
Malus sylvestris, xylem	– 8	–38	30	– 8	–30...–35	Up to 30	0	Larcher and Eggarter (1960), Quamme (1976), Krasavtsev and Khvalin (1978)
Betula papyrifera, xylem		–10			–196	Unlimited	Unlimited	George et al. (1974b)

T_f = Threshold freezing temperature; LT_i = temperature causing initial injuries; unh = unhardened state; h = hardened state of the plant.

Table 4.5. Freezing temperatures of various plant organs. (From data of numerous authors; for references, see Larcher 1985a)

Plant part	Threshold freezing temperature (°C)		Transient supercooling to °C
	Unhardened	Cold acclimation	
Shoots of herbaceous plants	−1 to −3	−3 to − 5	−4 to −12
Succulent leaves	−1 to −3	−3 to − 8	−6 to −10
Leaves of deciduous woody plants	−3 to −5		
Evergreen leaves	−3 to −5	−4 to −10	−6 to −12
Conifer needles	−2 to −5	−4 to − 8	−8 to −15
Cortex of woody twigs	−3 to −5	−6 to − 8	−6 to −14
Xylem	−3 to −5	−8 to −10	−7 to −10
Flowers	−1 to −3		About− 3
Fruits	−1 to −3		−1 to − 4
Roots	−1 to −2		−3 to − 6

Table 4.6. Deep supercooling of plant organs

Plant part	T_{sc} (°C)	References
Winter buds		
Shoot primordia (conifers)	−20 to −30	Sakai (1978b, 1979a), Dereuddre (1978), Pierquet et al. (1977)
Flower primordia	−10 to −25	Graham and Mullin 1976), Ishikawa and Sakai (1981), Kaku et al. (1980), Quamme (1978), Rajashekar and Burke (1978), Sakai (1979a)
Xylem		
Warm-temperate trees	−16 to −25	Kaku and Iwaya (1978, 1979), Sakai and Hakoda (1979)
Cold-temperate trees	−40 to −48	George et al. (1974b), Krasavtsev and Khvalin (1978), Pierquet et al. (1977)

Partial freezing tolerance is found in woody evergreen plants of regions with mild winters, where, although the stems show the beginnings of freezing tolerance, the leaves remain sensitive to freezing even when the plant is in the cold-acclimated state. This type of differentiation is seen in the more resistant species of Aurantieae (Yelenosky 1975), and especially in many of the Mediterranean sclerophyllous species (Larcher 1970). In winter, most of the woody angiosperms of the temperate zone develop freezing tolerance in their bark tissue, cambium and vegetative buds, whereas the xylem parenchyma undergoes deep supercooling and thus remains sensitive to freezing.

Total freezing tolerance, extending to all parts of the plant is developed by perennial herbaceous plants that grow on habitats severely exposed to frost, and certain woody plants of the boreal regions. The highest degree of perfection with respect to frost hardiness is attained by those species with total resistance, which is also unlimited. To this group belong certain conifer species, and species of *Betula, Populus* and *Salix* that grow far north and well into the inner continental regions, as well as certain arctic and alpine dwarf shrubs that overwinter in localities with little snow cover. Such plants survive the lowest recorded temperatures without suffering damage.

Desiccation-induced freezing tolerance has been observed in desiccation-tolerant thallophytes and cormophytes in the air-dry anabiotic state. Unlimited resistance has been demonstrated in bacteria (Lozina-Lozinskii 1974), aerial algae (Edlich 1936), lichens (Kappen and Lange 1970), certain mosses (Lipman 1936), ferns (Kappen 1966) and in the desiccation-tolerant phanerogams *Ramonda myconi* (Kappen 1966) and *Myrothamnus flabellifolia* (Vieweg and Ziegler 1969). Also dry seeds, spores and pollen and desiccated tubers can be resistant to ultralow freezing temperature (see Sect. 4.5.1).

4.1.3.2 Ecological Significance of the Components of Frost Resistance

Mechanisms of freezing avoidance are favourable for plants living in habitats where slight frosts occur during periods of metabolic and developmental activity. Freezing point depression, which can be increased after a few days of cold weather, is a prompt and in most cases sufficiently efficient protection against episodic frosts (Fig. 4.7). Lowering of T_f and T_{sc} ensures maintenance of a high leaf water potential (Goldstein and Meinzer 1983; Beck 1986) and of CO_2-uptake at subzero temperatures (Larcher 1961, 1980c, 1981c; Pisek et al. 1967; Larcher and Wagner 1976) and permit ready dry matter production in the morning after moderate night frost (Moser et al. 1977).

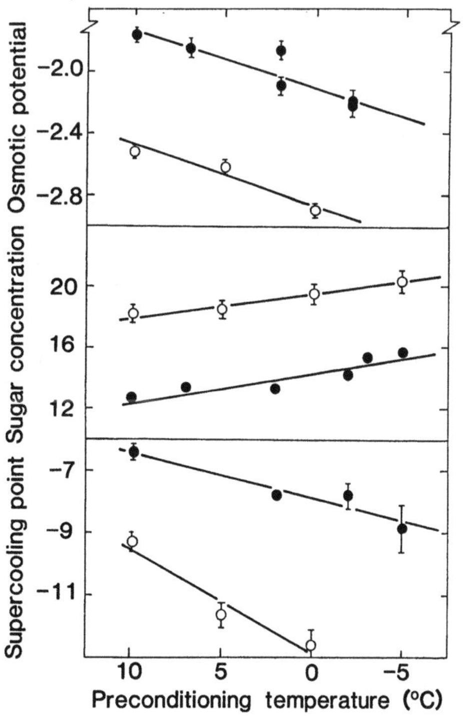

Fig. 4.7. Supercooling points (°C), sugar concentration (% dry weight) and osmotic potential (MPa) of leaves of *Polylepis sericea* from 3900 m a.s.l. in the Venezuelan Andes after preconditioning at various temperatures at the beginning (•) and the end (○) of the dry season. (From Rada et al. 1985a)

Where frost is a regularly recurring, severe and long-lasting threat to plant life, as in high latitudes and in central parts of Eurasia and North America with a continental climate, dormancy-linked freezing tolerance is the only effective survival mechanism. However, since deep and stable dormancy consists in the depression not only of growth but also of metabolic activity, the photosynthesis of evergreen plants in cold-temperate regions is strongly reduced during winter (reviews: Bauer et al. 1975; Larcher and Bauer 1981; Öquist 1983).

4.2 Supercooling as a Survival Mechanism

Supercooling, as a rule, represents a transient, unstable state which may provide protection against *brief* radiation frosts to an extent of 3–8 K below the freezing point of the tissues (Hatakeyama 1960, 1961; Kaku 1975; Marcellos and Single 1979). Plants which exhibit persistent supercooling are able to remain in the supercooled state for long time; floral bud meristems and xylem parenchyma of many temperate trees supercool to low temperatures in some cases down to the homogeneous nucleation point (see Table 4.6). However, it is worth noting that specific, reproducible limits for persistent supercooling exist at any temperature level, even well above the homogeneous nucleation point (see Figs. 4.12 and 7.39).

4.2.1 Persistent Supercooling of Leaves

A moderate improvement of frost resistance by persistent lowering of the nucleation temperature down to −10° to −12 °C, and exceptionally lower, was observed in evergreen leaves with narrow intercellular spaces and/or small mesophyll cells (Fig. 4.8). Examples of plants with persistently supercooling leaves (at least in the acclimated state) are certain Mediterranean sclerophyllous woody plants (Larcher 1963a, 1970; see Table 4.4: *Olea europaea*), certain plants of tropical mountains (e.g. species of *Espeletia* and *Polylepis*; Larcher 1975; Larcher and Wagner 1976; Goldstein et al. 1985a,b; Rada et al. 1985a), palms from subtropical and warm temperate regions (Larcher 1980b; Larcher and Winter 1981; see Table 4.4: *Trachycarpus fortunei*) and bamboos (Ishikawa 1984).

In *Sasa senanensis*, a temperate bamboo, persistent supercooling was evidenced by DTA and cryomicroscopic observation. *Sasa* leaf blades have ladderlike veins. As the leaf segment mounted in water was cooled, the water surrounding the segment froze first at −2 °C; then, between −20° and −24 °C, minute areas of leaf tissue, compartmentalized by longitudinal and cross veins, suddenly froze one after the other at random. Supercooled tissue nucleated in the same temperature range (Fig. 4.9). Leaves of *Sasa*

Fig. 4.8a–d. Histology of persistently subcooling leaves showing densely packed mesophyll: **a** *Olea europaea*, transverse section; **b** *Polylepis sericea*, transverse section (Photos: W. Larcher); **c** *Trachycarpus fortunei*, transverse section (Photo: J. Wagner); **d** *Sasa senanensis*, longitudinal section (Photo: M. Ishikawa)

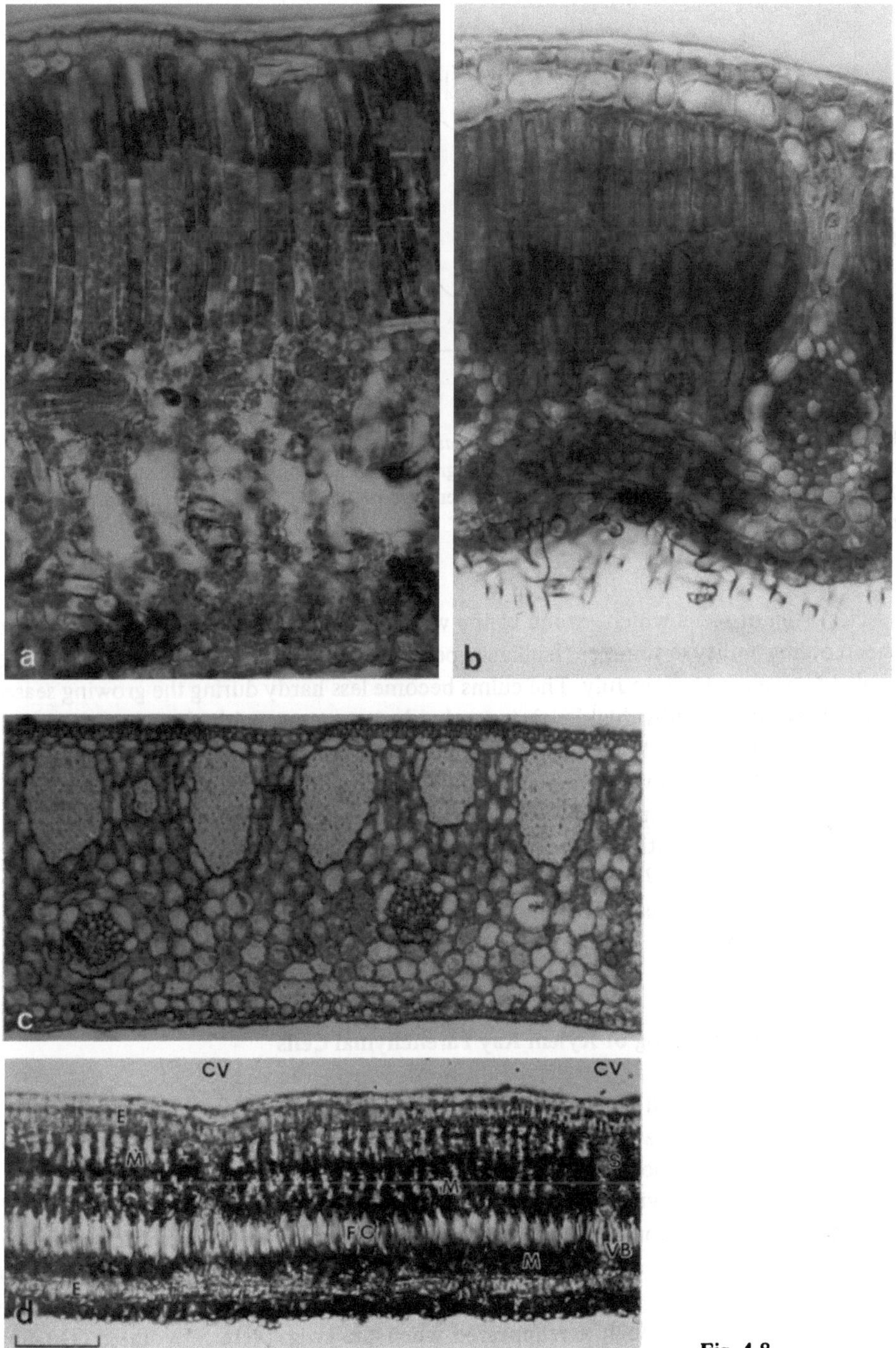

Fig. 4.8

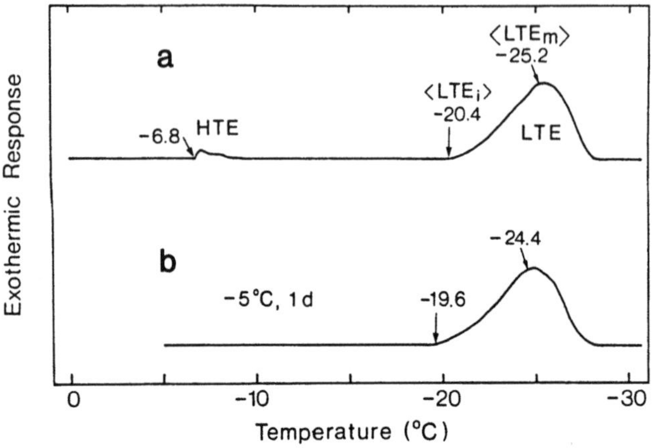

Fig. 4.9a,b. DTA profiles for freezing leaves of *Sasa senanensis*. **a** Leaf segment, collected in December, freezing at a cooling rate of 5 K h^{-1}; **b** segment excised from a leaf blade stored at −5 °C for 1 day before experiment. LTE_i initiation temperature of LTE; LTE_m peak temperature of mesophyll freezing. (From Ishikawa 1984)

species, even of some which extend to the warm temperate zone in Japan, retain a high supercooling ability in summer (Ishikawa, pers. commun.): leaves of *S. senanensis* supercooled to about −16 °C in July. The culms become less hardy during the growing season. Thus, *Sasa*, at the individual level, is nonhardy in summer with the exception of the mature leaf blades. The low temperature exotherm of leaves shifts to −20 °C in December, with a decrease in water content. In winter the buds and the culms survive −15 °C and −20° to −25 °C, respectively, by deep supercooling (Ishikawa 1984).

Very similar observations were made with *Trachycarpus fortunei*, a palm of the warm temperate zone (Sect. 7.1.2). Presumably the peculiar structure of palm and bamboo leaves might be responsible for the maintenance of such a high supercooling ability throughout the year.

4.2.2 Deep Supercooling of Xylem Ray Parenchymal Cells

In winter, bark cells and vegetative buds of apple twigs are hardy to −70 °C and even to immersion in LN_2 following prefreezing at −30° or −40 °C; the xylem ray parenchyma, however, is killed as soon as an LTE of about −40 °C appears (Quamme et al. 1972; Fig. 4.10). During thawing there was no endotherm near the temperature at which the LTE occurred, but upon subsequent refreezing, even if the xylem was damaged, the exotherm reappeared at a somewhat higher temperature, if the twigs were rewarmed to at least −2 °C (melting point) before refreezing. However, the LTE was not present in freeze-dried twigs, although it reappeared when the twigs were rehydrated (Quamme et al. 1973). These observations suggested that water which was trapped in xylem ray parenchymal cells remained unfrozen to as low as −40 °C after freezing of the bulk water in twigs. From his study on mulberry twig pieces, Aoki drew the same conclusion as early as 1957, when he observed a second exotherm appearing on the DTA profiles.

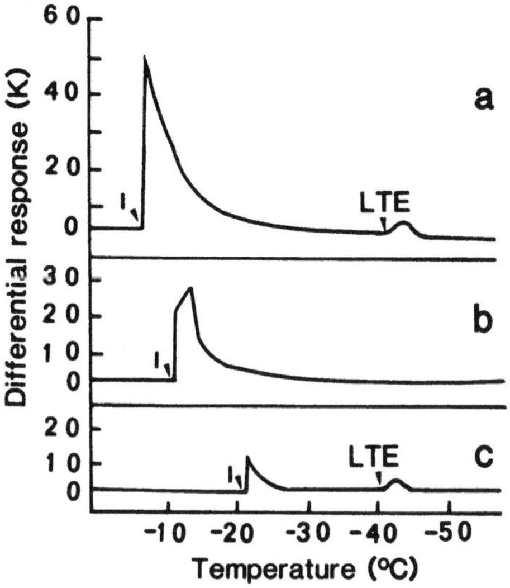

Fig. 4.10a–c. DTA curves of freezing apple twigs in winter. **a** Twig piece; **b** bark only; **c** xylem only. *I* initiation of freezing. (From Quamme et al. 1972)

Using several methods, such as differential scanning calorimetry, pulsed nuclear magnetic resonance spectroscopy and low temperature microscopy, George and Burke (1977a) confirmed that low temperature exotherms in xylem result from the freezing of cellular water in a manner predicted for supercooled dilute aqueous solutions. The supercooled fraction in xylem is found to be extremely stable even at −35 °C, which is slightly above the homogeneous nucleation temperature for water.

The freezing process in the xylem ray parenchyma starts at a single point and spreads rapidly from cell to cell. Under microscopic observation the ray parenchymal cells suddenly appear to darken as a result of intracellular freezing (Hong and Sucoff 1980). The water in xylem parenchyma resists to dehydration when exposed to 80% relative humidity at 20 °C. However, D_2O exchange experiments revealed that only a weak kinetic barrier to water transport exists in the xylem parenchyma (George and Burke 1977a).

It is easier to explain the advantages in the supercooling of the xylem ray cells than the mechanism itself. The xylem ray parenchyma is the least hardy stem tissue in cool-temperate deciduous woody angiosperms (Sakai and Weiser 1973; Sakai 1978a, see Sect. 7.31). Further, supercooling may be the only resistance mechanisms available for xylem cells, since there is essentially no intracellular space for extracellular freezing. Thus, if freezing occurs it must be intracellular, and this type of freezing cannot be tolerated by plant cells. Deep supercooling of these cells is favoured by their small size and their low water content. Xylem parenchyma can acclimate to low temperatures between fall and winter (see Fig. 5.2), the increase in cell solutes probably resulting from conversion of starch to sugar (Sakai 1966a).

In most of the studies on deep supercooling of xylem either thawed samples or samples frozen only slightly below 0 °C have been used. Also, in many of these studies, high cooling rates, 60–120 K h^{-1}, were applied. In order to approach natural conditions more closely, Gusta et al. (1983) used nonthawed stem sections naturally frozen at approximately −20 °C in the field. Care was taken to ensure that the samples were not exposed to temperatures warmer than −15 °C before transfer to the precooled

Table 4.7. Winter hardiness of xylem tissues of very hardy taxa. (From Gusta et al. 1983)

Species	Low temperature exotherm[a]		Lethal temperature (°C)
	Nonthawed	Thawed[b]	
Acer negundo	–[c]	–	Below –60
Prunus padus commutata	–	–	Below –60
Fraxinus pennsylvanica	–	–	–50
Quercus macrocarpa	–	–	–47
Prunus americana	–	–	–45
Vitis riparia	–55	–46	–50
Quercus coccinea	–53	–47	–45
Ulmus americana	–53	–47	–50
Pyrus ussuriensis	–46	–43	–45
Malus robusta cv. *erecta*	–46	–43	–47

[a] Initiation temperature of the low temperature exotherm.
[b] Samples warmed to 21 °C for 15 min before the DTA. When samples are thawed, the cells rehydrate and, with a fast cooling rate, there is insufficient time to allow water to escape from these cells.
[c] No detectable exotherm down to –60 °C.

freezing chamber. The major findings were: (1) during periods of prolonged freezing, LTEs can occur in woody tree species at much lower temperatures (–46° to –55 °C) than have been previously reported; (2) LTEs observed in some very hardy species (e.g. in *Fraxinus pennsylvanica, Quercus macrocarpa, Elaeagnus angustifolia*) in the fall or early winter disappear under continuous low subzero temperatures approaching –38 °C or shift to lower temperatures (e.g. in *Ulmus americana, Pyrus ussuriensis, Malus robusta* var. *erecta*). Thus, during prolonged exposure to low temperatures, water must have escaped from the supercooling cells. Those samples which did not display an LTE in mid-winter were killed in the range of –45° to –50 °C (Table 4.7). Possibly all of the freezable water associated with the LTE was translocated to the sites of ice formation, which would account for the absence of the exotherm. In this case, death of the cells was probably due to dehydration effects. Because xylem cells of these species resist water loss due to their rigid cell wall, they may not have evolved tolerance of the severe dehydrating effects experienced during freezing at low temperatures unlike the freezing tolerant xylem parenchyma of *Salix, Populus* and *Betula*. The boundary between deep supercooling and extracellular freezing has become obscured by the findings of freeze dehydration of supercooled parenchyma in very hardy species. However, the rate of water flow outside supercooled parenchyma during slow cooling seems to be very small. Death of the cells may be due to an irreversible phase transition in the membranes as suggested by Rajashekar et al. (1979) and Yoshida (1984a).

Supercooling of aqueous solutions in plants seems to have its limits at –45° to –50 °C. The distribution of woody plants protected by this resistance mechanism therefore was expected to be confined to the area where minimum winter temperatures never drop much below about –45 °C. George et al. (1974b) found a good correlation between the northern boundary of 49 deciduous tree species in N. America and the minimum temperature isotherm which corresponds to the LTE temperature of the

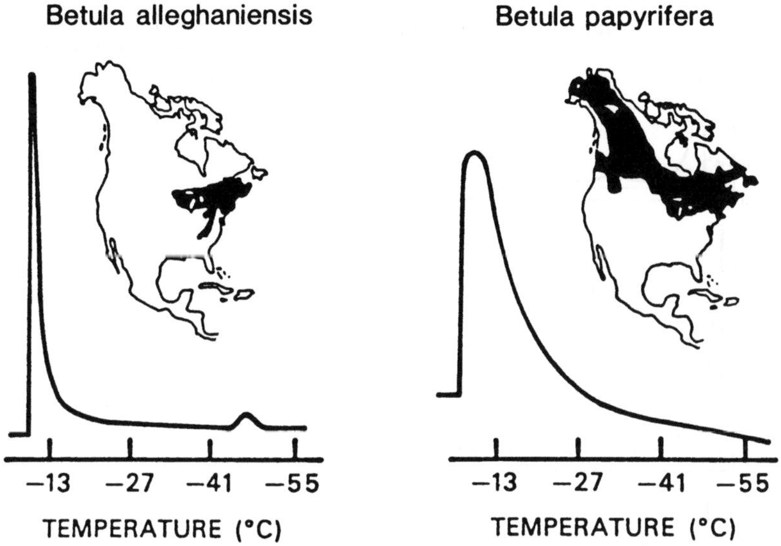

Fig. 4.11. Natural geographic ranges and DTA curves of freezing xylem of *Betula alleghaniensis* (deep supercooling) and *B. papyrifera* (freezing tolerant). (From George et al. 1974b)

winter-acclimated xylem (Fig. 4.11). The -40 °C minimum isotherm as presented in the Hardiness Zone Map of N. America (for reference, see Fig. 1.1) appeared to represent an average borderline for tree species with deep supercooling xylem. There are, however, exceptions since certain species are found in areas where temperatures lower than -45 °C are not uncommon (Gusta et al. 1983). For example, *Ulmus america* and *Fraxinus pennsylvanica* are both native to areas of northern Canadian provinces where temperatures lower than -45 °C have been reported. With very slow freezing conditions, in *Fraxinus pennsylvanica* the low temperature exotherm disappears. This suggests that cell walls in the wood of *F. pennsylvanica* are more elastic than those of trees which maintain an LTE even after long-lasting low subfreezing temperatures.

In very hardy deciduous trees native to E. Asia which extend to inland Hokkaido, Sakhalin, NE China and the Maritime Provinces of Far East USSR, the xylem ray parenchyma survived between -40° and -45 °C, but not at -50 °C, when cooled slowly at 5 K decrements daily (Sakai, unpublished). Extremely hardy species of boreal deciduous trees (*Salix, Populus* and *Betula*) had no LTE and survived freezing at -70 °C or below (Fig. 4.12). These xylem parenchyma presumably were able to freeze extracellularly. The twigs of *Salix sachalinensis* in mid-fall and early spring sustained freezing injury between -20° and -25 °C, and here, too, no LTE was detected. It is known that wood elements of *Salix, Populus* and *Betula* have thin and elastic cell walls, which easily permit freeze dehydration of their cells during freezing. In tropical and subtropical willows, *Salix tetrasperma* in SE Asia and *S. safsaf* throughout Africa (cf. Fig. 7.40), the cortex survives freezing between -20° and -70 °C depending upon provenance, but the xylem does not survive temperatures of -15° to -30 °C. No LTE was, however, detected in association with injury (Sakai 1978c). This suggests that the xylem parenchyma of all *Salix* species possesses the same freezing mechanism, i.e. extracellular freezing, even in less hardy species.

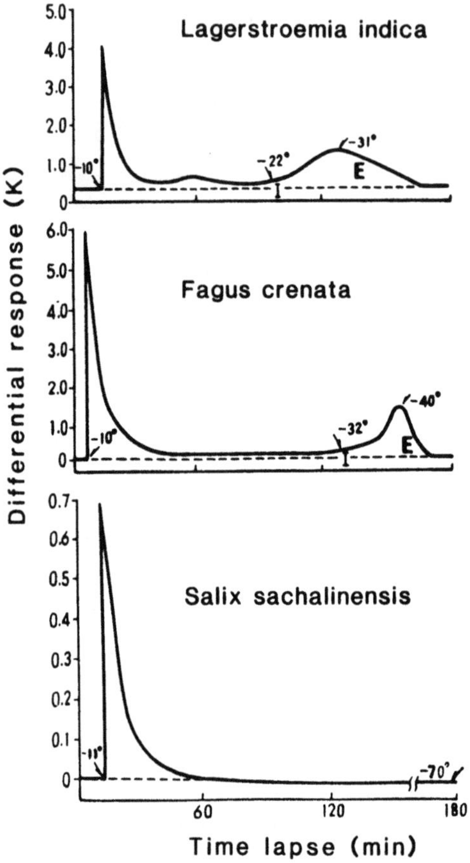

Fig. 4.12. DTA curves of freezing in shoots of woody plants native to Japan. (From Sakai 1978a)

Whether deep supercooling or extracellular freezing occurs in the xylem of deciduous trees appears to depend on the histological and/or physiological characteristics of the parenchyma concerned. In a comparative study on *Betula* species of different geographical distribution in Europe and Asia, Furst (1983) was able to relate the degree of winter hardiness with wood structure: the hardier species *B. pendula* and *B. platyphylla* have smaller vessel diameters with thicker cell walls than the less hardy species *B. korshinskyi* and *B. turkestanica*.

4.3 Extraorgan and Extratissue Freezing

By water translocation from supercooled tissues or organs to nucleation centres in adjacent tissues ("extratissue freezing") or outside the organs ("extraorgan freezing", as defined by Sakai 1979b), a high degree of freezing tolerance can be attained in buds and seeds: In freezing winter buds, water migrates from the shoot primordia to the bud scales and into basal parenchyma (Sakai 1978b, 1979a; Dereuddre 1978, 1979). Redistribution of water in certain seeds from embryonal meristems to distal parts during slow freezing has been described by Keefe and Moore (1981) and Ishikawa and Sakai (1982). In woody stems, it was supposed that water might be redistributed between cambium cells and the adjacent cortical parenchyma which contains large intercellular spaces (Larcher 1982).

Freezing of translocated water ("translocated ice formation"; Larcher 1985a) makes possible a slow and therefore less harmful dehydration of the tissues. A quantitative analysis of this process, including determination of the amount of translocated water, has been carried out in primordial shoots of winter cereals by Krasavtsev and Khvalin (1982).

4.3.1 Deep Supercooling and Extraorgan Freezing of Hydrated Seeds

Freezing avoidance by supercooling of hydrated lettuce seeds and the level of resistance were determined by Stushnoff and Junttila (1978). The first exotherm appeared as a single peak at about -10 °C indicating that it represents freezing of bulk water in the outer layers of the seed coat. Each seed showed a low temperature exotherm which caused death (Fig. 4.13). In hydrated seeds, the mean exotherm temperatures were inversely correlated with the moisture content within the limits of 20 to 40%. However, the freezing pattern changed considerably after disruption of the seed coat. As soon as the radicle emerged, the LTEs disappeared and the first exotherm represented the killing point of the seed. Similarly, embryos excised from the seeds after imbibition for 2 or 12 h produced only one exotherm and did not survive below this nucleation temperature. If the surface of the endosperm was pierced with a needle, the second exotherm was abolished and death was coincident with the first exotherm temperature.

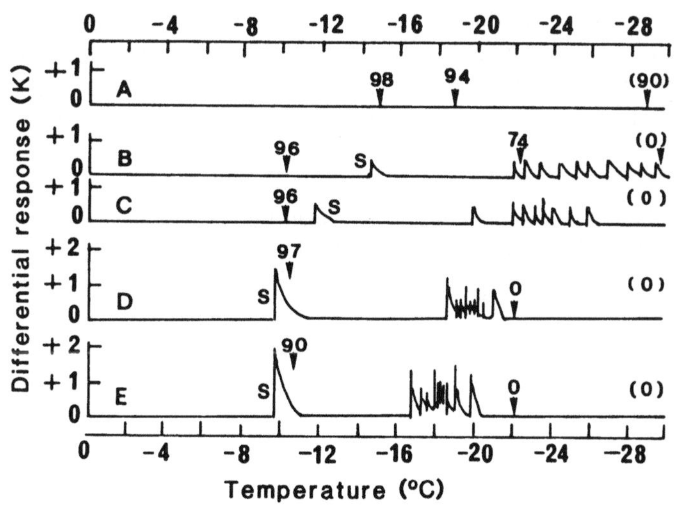

Fig. 4.13. DTA profiles for lettuce seeds with different moisture contents. Ten seeds were used in each sample. *Numbers* and *arrows* show the percentage of survival (germination) in samples removed at temperatures indicated. *Numbers in parentheses* show survival of samples frozen to −35 °C. Cooling rate: 20 K h^{-1}. *S* first exotherm. Water content (*A*) 16%; (*B*) 21.9%; (*C*) 26%; (*D*) 33%; (*E*) 38.2%. (From Stushnoff and Junttila 1978)

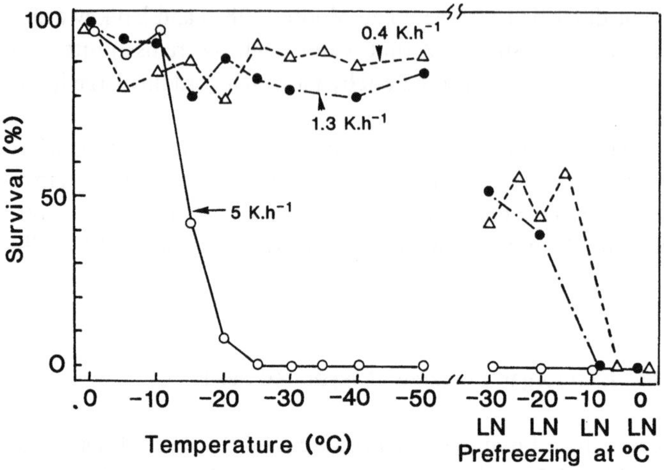

Fig. 4.14. Effect of the cooling rate on survival of hydrated lettuce seeds with a water content of 48.7% (f.w.). *Right:* Seeds precooled to −10°, −20° and −30 °C at various cooling rates prior to immersion in LN_2. (From Ishikawa and Sakai 1982)

Thus, the integrity of the endosperm envelope seems to facilitate deep supercooling which imparts freezing avoidance to −12° to −18 °C in intact imbibed seeds cooled at 12 K h^{-1}.

The supercooling ability of lettuce seeds was observed by Ishikawa and Sakai (1982) to be dependent on cooling rate, as shown in Fig. 4.14. For water-imbibed lettuce seeds (48.7% f.w.) cooled at a rate of 5 K h^{-1}, the 50% killing temperature was −14° to −15 °C, which coincided with the mean LTE of −14° to −15 °C. When the cooling rate was slower than 1.3 K h^{-1}, an LTE was scarcely detected on the DTA profiles and the seeds survived at least −50 °C. Moreover, nearly 50% of the seeds withstood the temperature of LN_2 after being precooled to −20 °C at 1.3 K h^{-1}. In the case of seeds of *Vitis coignetia*, which are larger than lettuce, there was no marked change in the

Table 4.8. Water content (on dry weight basis) of fully imbibed seeds of *Lactuca sativa* after cooling to various temperatures down to −25 °C. (From Keefe and Moore 1981)

Freezing temperature (°C)	Water content (%)
− 5	90
−10	78
−15	64
−20	59
−25	59

Seeds were cooled at a rate of 1 K h^{-1} after ice inoculation. External water which was segregated as ice outside the endosperm during freezing was removed by centrifugation following freezing.

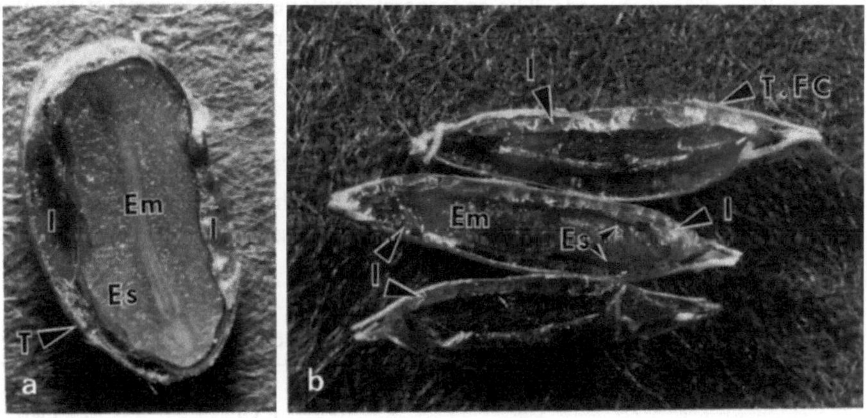

Fig. 4.15a,b. Extraorgan freezing of seeds of **a** *Celastrus orbiculatus* and **b** *Lactuca sativa* cooled to −10 °C. *Em* embryo; *Es* endosperm; *T* testa; *I* ice. (From Ishikawa and Sakai 1982)

exotherm temperature at cooling rates between 0.4 and 5 K h^{-1}. But the storage of the seeds at −5° to −10 °C for a few days shifted the exotherms to a lower temperature. Almost the same results were obtained by Keefe and Moore (1981) who observed in imbibed lettuce seeds that slow freezing resulted in a gradual dehydration of the embryo, which was responsible for the absence of LTEs in slowly (1 K h^{-1}) cooled seeds to −25 °C (Table 4.8).

Microscopic observation revealed that water in the embryo and endosperm froze out during slow cooling or storage at −5° to −10 °C. In lettuce seed, ice was formed between the integument and the endosperm; in *Celastrus*, ice accumulated mainly between the endosperm and testa (Fig. 4.15). This kind of freezing was termed extraorgan freezing (Ishikawa and Sakai 1982). In the case of grape and *Oenothera* seeds, more ice was formed outside the testa than in the space between the testa and the endosperm. Ice in hydrated seeds seemed to be formed in a particular space depending on species. Consequently, the hydrated embryo and endosperm increase their frost resistance to the extent to which they tolerate the dehydrated state. Stushnoff and Junttila (1978) indicated the necessity of the integrity of the endosperm envelope in

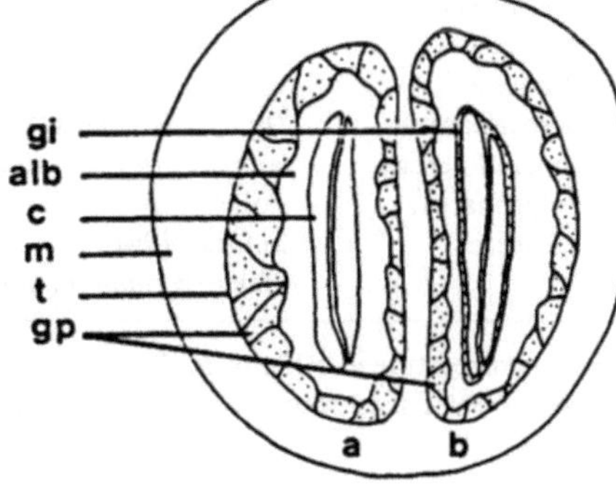

Fig. 4.16. Distribution of ice in a fruit of *Ligustrum vulgare*, frozen at −15 °C, with the embryo (*a*) surviving and (*b*) killed; *alb* albumen, *c* cotyledons, *m* mesocarp, *gp* external ice, *gi* internal ice, *t* seed coat. (From Gazeau and Dereuddre 1980)

supercooling of lettuce seeds. The endosperm envelope seems to protect the embryo by acting as a barrier which blocks ice propagation from the outside, but is permeable to water or vapour in both directions. The frost survival of seeds depends on the extent of water translocation from the embryo: in fruits of ***Ligustrum vulgare*** the embryo was killed if ice was formed between cotyledons and endosperm due to insufficient redistribution of water (Gazeau and Dereuddre 1980; Fig. 4.16).

Extraorgan freezing seems to act as resistance mechanism in many hydrated seeds under natural conditions. Results of DTA studies on seeds of woody and herbaceous plants are summarized in Table 4.9.

Table 4.9. Results of DTA studies on seeds. (From Ishikawa and Sakai 1982)

Family	Species	Range of LTE[a] (°C)		Water content (% f.w.)
Woody plants				
Caprifoliaceae	*Viburnum dilatatum**[b]	−23 to	−27	32.9
	*V. wrightii**	−20	−22	34.8
Styracaceae	*Styrax japonica*	−22	−28	37.7
Ericaceae	*Vaccinium oldhamii*	−14	−22	60.1
	V. bracteatum	− 9	−20	35.1
Araliaceae	*Kalopanax pictus**	−24	−30	−
Vitaceae	*Vitis coignetiae**	−15	−19	28.9
Celastraceae	*Celastrus orbiculatus**	−22	−24	34.9
	*Euonymus alatus**	−19	−23	45.0
Aquifoliaceae	*Ilex crenata*	−19	−21	45.2
	I. pedunculosa	−13	−16	32.2
Coriariaceae	*Coriaria japonica**	−14	−24	−
Rutaceae	*Phellodendron amurense**	−21	−25	−
Rosaceae	*Rhaphiolepsis indica*	− 8	−13	44.0
	*Rosa rugosa**	−11	−25	−
	*R. multiflora**	−16	−27	−
Saxifragaceae	*Hydrangea paniculata*	−18	−23	53.5
Theaceae	*Eurya japonica*	−14	−22	32.0
	E. emarginata	−15	−21	41.4
Berberidaceae	*Berberis sp.**	−17	−24	45.0
Magnoliaceae	*Magnolia obovata**	−14	−15	30.1
Herbaceous plants				
Asteraceae	*Adenocaulon himalaicum*	−15	−17	49.1
	Carpesium abrotanoides	−16	−21	51.2
Gentianaceae	*Tripterospermum japonicum**	−17	−22	54.0
Onagraceae	*Oenothera sp.*	−20	−26	30.9
Buxaceae	*Pachysandra terminalis**	−15	−18	−
Rosaceae	*Geum japonicum*	−12	−22	51.5
Berberidaceae	*Diphylleia cymosa**	−18	−21	−
Amaranthaceae	*Achyranthes japonica*	−14	−18	48.7
Polygonaceae	*Polygonum filiforme*	−15	−18	36.0
Liliaceae	*Smilax china*	−18	−22	38.3
	*Maianthemum dilatatum**	−15	−18	49.7

[a] Low temperature exotherm; cooling rate was about 5 K h^{-1}.
[b] Seeds of asterisked species were within their fruits or arils when used for DTA.

4.3.2 Deep Supercooling and Extraorgan Freezing of Floral Primordia of Woody Angiosperms

Graham (1971) and Graham and Mullin (1976) first observed that low temperature exotherms of flower primordia of *Rhododendron mollis* and *R. mollis* x *R. roseum* coincide with the lethal temperature. It was concluded that freezing avoidance by deep supercooling may be a mode of freezing resistance in azalea flower primordia. Graham's finding was proved by George et al. (1974a) using DTA (Fig. 4.17) and nuclear magnetic resonance spectroscopy, and by Kaku et al. (1985) using NMR relaxation time measurement. Microscopic observation indicated that if ice crystals were introduced into a primordium at slightly below 0 °C, intracellular freezing occurred and no resistance to ice growth existed in the primordium tissue itself (George and Burke 1977b). Similar results were obtained on excised azalea floral primordia (Ishikawa and Sakai 1981). Observation of the freezing process in petal and receptacle tissues showed that these cells, which do not have intercellular spaces, froze intracellularly, resulting in death of the primordia at −2 °C even when massive ice had formed around them. These results suggest that a barrier to thermal equilibrium is required to prevent direct ice growth in the floral primordium. One characteristic of deep supercooling in flower buds, especially of buds with smaller floral primordia, is its dependence on cooling rate. An example is *Cornus officinalis* in early spring. The flower buds have 20 to 25 flower primordia surrounded by four involucral scales (Fig. 4.18). Each individual flower primordium supercooled to −12° to −20 °C when cooled at 5 K h^{-1} as indicated by LTE spikes in Fig. 4.19b. When the bud was cooled slowly to −9 °C at 5 K increments daily, LTEs shifted to a lower temperature and the exotherm became smaller (Fig. 4.19c). During the slow cooling, ice accumulated in the scales (Fig. 4.20), the water content of the flower primordia decreased and that of the scales increased, while other parts of the flower buds showed little change in the water content (Fig. 4.21). Thus, the water in the floral primordia

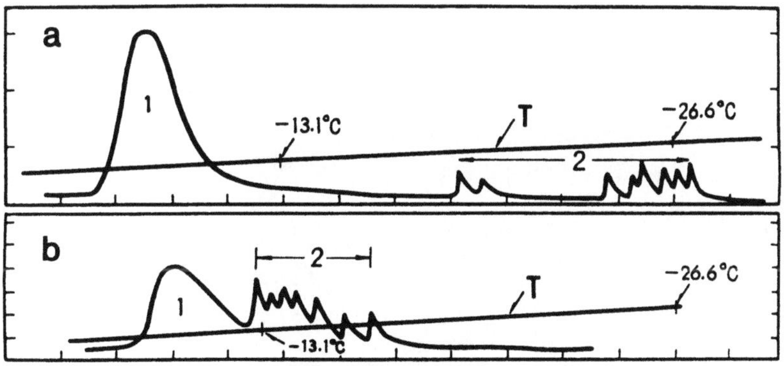

Fig. 4.17a,b. DTA profiles for freezing flower buds of *Rhododendron kosterianum* cooled at 8.5 K h^{-1}. **a** Typical winter behaviour on March 1st; **b** spring type on April 16th with converging exotherms indicating loss of hardiness. Exotherm *1* corresponds to freezing of water in the bud scales and the axis upon which the primordia are attached; *T* reference temperature. (From George et al. 1974a)

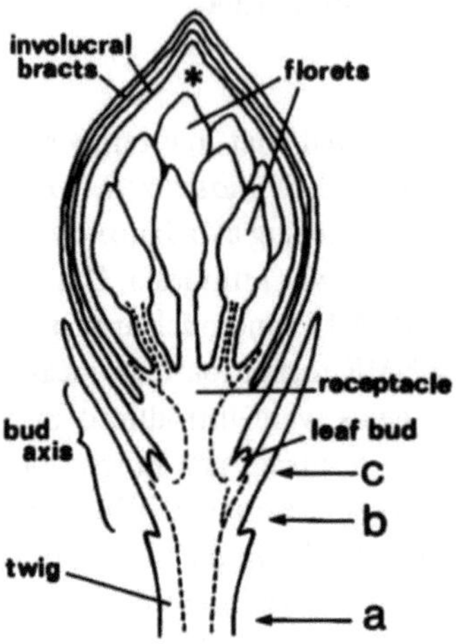

Fig. 4.18

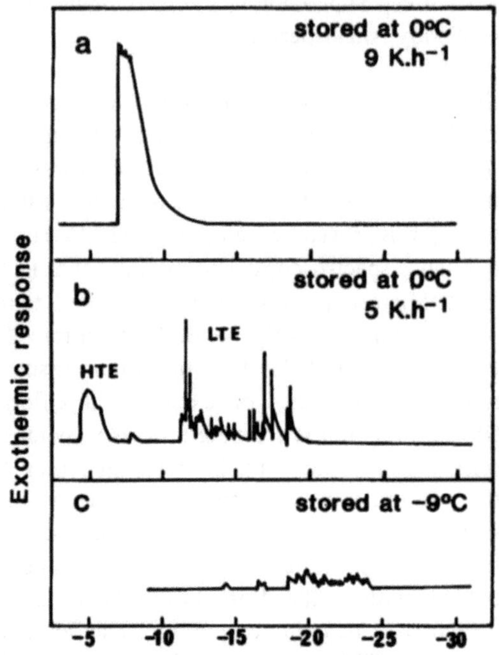

Fig. 4.19

Fig. 4.18. Longitudinal section of a flower bud of *Cornus officinalis*. *Broken lines* indicate vascular tissues; *a, b, c* show where the bud was cut from the twig for DTA. (From Ishikawa and Sakai 1982)

Fig. 4.19a–c. DTA profiles for freezing of floral primordia of *Cornus officinalis* in early April, cooled at different rates. **a** Scales and florets froze simultaneously, as indicated by the large exotherm, when cooled at 9 K h^{-1}. **b** At 5 K h^{-1} scales froze first as shown by an *HTE*, small spikes (*LTE*) between −8° and −18 °C indicate lethal freezing of primordia. **c** Buds precooled to −9 °C within 2 days (scales had already frozen), then cooled further at 5 K h^{-1} without thawing. (From Ishikawa and Sakai 1982, 1985)

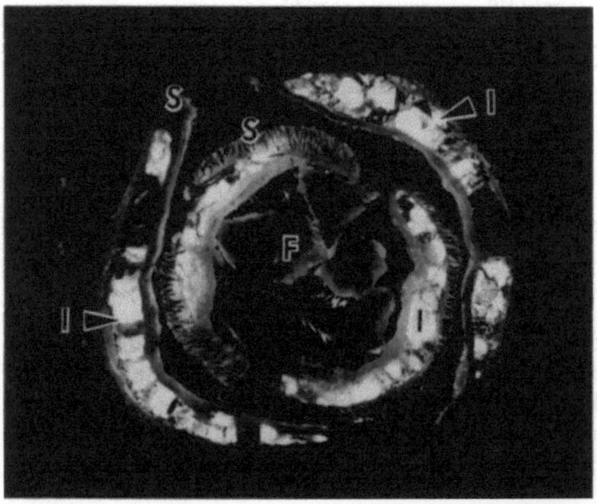

Fig. 4.20. Cross-sections of flower buds of *Cornus officinalis* after extra-organ freezing. The buds were cooled to −9 °C within 2 days and photographed with polarized light. *I* Ice crystals formed in the involucral scales; *S* scales; *F* floral primordium. (From Ishikawa and Sakai 1985)

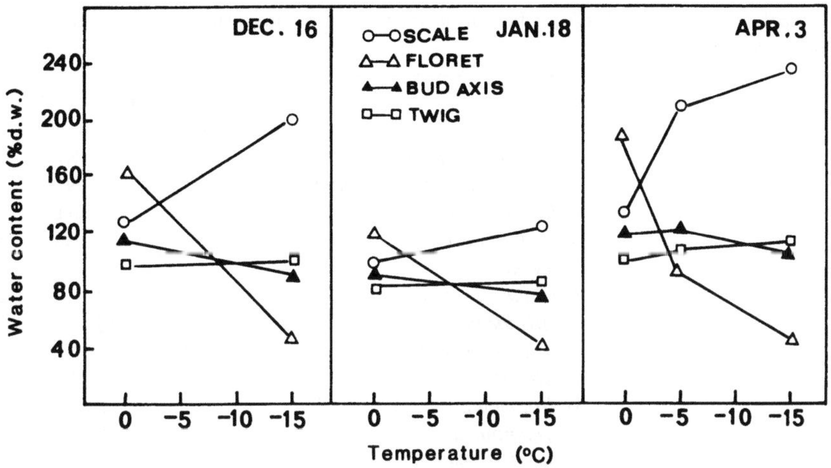

Fig. 4.21. Changes in the water content of flower buds of *Cornus officinalis* in early spring when cooled to −15 °C in 5 K decrements daily. To determine the water content of the frozen buds, each bud was dissected into its components at −5° to −20 °C. Each part was then wrapped in aluminium foil and put into a small plastic bag. (From Ishikawa and Sakai 1982, 1985)

migrates to the scales where it freezes, while the dehydrated primordia increase their supercooling ability during the slow cooling.

A similar redistribution of water from floral primordia to scales was observed in flower buds of several *Rhododendron* species cooled slowly to −15 °C (Ishikawa and Sakai 1981). The supercooling ability of the floral primordia was enhanced as their freezing point depression decreased and was inversely proportional to their water content (Table 4.10). We now have to ask by what route does the water migrate from the floral primordia to the scales. By an experiment with *Cornus officinalis*, in which silicon oil was injected with a syringe into the space between the primordia and the scales, it could be demonstrated that water in the floral primordia migrates by way of vascular systems to the scales rather than through the surface of primordia (Ishikawa and Sakai 1985). Involvement of vascular systems was indicated in another experiment. At the cooling rate of 2 to 3 K h^{-1}, floral primordia in a winter bud of *Cornus officinalis*, cut at line *a* in Fig. 4.18 normally supercooled (LTE: −16° to −23 °C). On the other hand, in the buds cut at line *c*, they did not supercool and froze together with other tissues

Table 4.10. Threshold freezing temperature (T_f) of florets and exotherm temperature (T_e) of whole buds of *Rhododendron japonicum* in winter, depending on cold treatment. The buds were cooled at 5 K daily decrements from 0° to −15 °C and then at a rate of 4–6 K h^{-1}. (From Ishikawa and Sakai 1981)

Precooling to °C	T_f (°C)	T_e (°C)
0	−1.2 to − 1.8	−18.5 ± 0.4
−10	−6.0 to − 7.8	−27.0 ± 0.3
−15	−9.8 to −11.5	−28.2 ± 0.4

Table 4.11. Results of DTA studies on flower buds of woody plants in winter. (From Ishikawa and Sakai 1982)

Family	Species	Range of LTE[a] (°C)		
Caprifoliaceae	*Viburnum furcatum*		ND[b]	
	Sambucus racemosa var. *miquelii*		ND	
Oleaceae	*Syringa vulgaris*		ND	
	Forsythia suspensa	−25	to	−30
	F. koreana	−24		−28
Ericaceae	*Vaccinium smallii*	−12		−20
	V. vitis-idaea	−15		−29
	Gaultheria miqueliana		ND	
	G. adenothrix		ND	
	Andromeda polifolia	−18		−22
	Pieris japonica	−10		−23
	Parapyrola asiatica	−18		−19
	Chamaedaphne calyculata	−21		−26
	Cassiope lycopodioides	−18		−25
	Arcterica nana	−18		−21
	Loiseleuria procumbens	−15		−22
	Phyllodoce sp.	−12		−21
	Tsusiophyllum tanakae	−20		−24
	Rhododendron japonicum	−20		−36
	R. dilatatum	−16		−25
	R. ripense	−11		−13
	R. obtusum	−19		−29
	R. tschonoskii	−20		−24
	R. brachycarpum	−22		−28
	R. dauricum	−27		−31
	R. keiskei	−11		−17
	Menziesia multiflora	−21		−24
	M. multiflora var. *purpurea*	−14		−24
	M. pentandra	−18		−28
	Ledum palustre subsp. *diversipilosum*	−22		−25
Cornaceae	*Cornus stolonifera*		ND	
	C. officinalis	−17		−32
	C. kousa	−19		−30
	C. florida	−21		−30
Hippocastanaceae	*Aesculus turbinata*		ND	
Cercidiphyllaceae	*Cercidiphyllum japonicum*	−19		−24
Ulmaceae	*Ulmus davidiana* var. *japonica*	−25		−33
	U. pumila	−21		−27
Betulaceae	*Betula platyphylla* var. *japonica*		ND	
Salicaceae	*Populus sp.*		ND	

[a] Low temperature exotherm; cooling rate was 2 to 9 K h^{-1}.
[b] LTE was not detected even when amplified 100 times.

(LTE: -6° to -7 °C). When cut at line *b*, the LTEs shifted to slightly higher temperature.

Bud scales and the proper cuticle probably protect the primordia from external ice crystals. Barriers against internal ice propagation at the attachment region of the primordia are also needed. The structure of vascular tissues might play a role in the prevention of internal ice inoculation. Another possibility is that a decreasing water content in the primordial peduncle or bud axis during slow freezing (Quamme 1978) might create a dry region which could serve as a barrier against internal ice propagation. However, a more detailed analysis regarding bud morphology is necessary to elucidate the barriers against ice inoculation.

These features of freezing avoidance in floral primordia suggest that this type of supercooling differs considerably from that of xylem ray parenchymal cells, in that it involves water redistribution during slow cooling and subsequent dehydration as in extracellular freezing. Very hardy floral primordia yielding LTEs at rapid cooling rates might tolerate -50 °C or below upon exposure to continuous low subzero temperatures (Table 4.11). It is conceivable that species of *Prunus* (Burke and Stushnoff 1979; Rajashekar and Burke 1978) and *Cornus*, in which LTEs were not detected and which tolerate -50° to -70 °C, might have tiny florets which undergo extraorgan freezing rather than extracellular freezing. If this were true, the species surveyed in each genus *(Prunus, Cornus)* would share the same mechanism, i.e., extraorgan freezing. The phylogenetic background might be responsible for the involvement of morphological features of the flower buds (the size and/or the developmental stage of primordia), which is very similar in one genus or even subfamily (Ishikawa and Sakai 1981). A genus which has large floral primordia might yield a LTE, while those with tiny primordia might have no LTE. Within the same genus, in general, primordia of hardier species are much smaller than less hardy ones.

4.3.3 Extraorgan Freezing of Conifer Buds

In terminal winter buds of *Abies homolepis* and *A. balsamea* LTEs were observed between -30° and -35 °C in excised buds, while excised primordia of such buds supercooled only to -12° to -15 °C (Sakai 1978b, 1979a; Table 4.12). The degree of super-

Table 4.12. Comparison of exotherm temperatures of excised shoot primordia and of whole terminal winter buds of firs. (From Sakai 1979a)

Species	Exotherm temperature (°C)		Freezing tolerance of excised primordial shoots[b]
	Primordial shoots of excised terminal buds[a]	Excised primordial shoots[a]	
Abies homolepis	–30 to –33	–12 to –14	None
Abies balsamea	–30 to –35	–12 to –15	None

[a] DTA of excised terminal buds wrapped with aluminium foil and each excised primordial shoot enveloped separately with aluminium foil were carried out at 0.6 K h^{-1}.

[b] Excised primordial shoots immersed in a small water drop at 0 °C were inoculated with ice at -5 °C. These primordial shoots did not survive freezing even at -5 °C.

Table 4.13. Frost resistance (°C) of shoot primordia in terminal winter buds of firs cooled at different rates and thawed at 0 °C. (From Sakai 1979a)

Species	5 °C/2 h	5 °C/day	5°C/day to −20°C −20°C 30 days, 10°C/day to −60°C	Native habitat
Abies firma	−23	−30	–	Temperate (Japan)
A. spectabilis	−20	−25	–	Temperate (3,900 m, Nepal)
A. homolepis	−30	−35	–	Cool temperate to sub-boreal (Japan)
A. sachalinensis	−30	−40 (−45)[a]	−50[a]	Sub-boreal (Hokkaido)
A. balsamea	−35	−50	−60, −30 LN[b]	Sub-boreal (Canada)

[a] Lateral buds along the twig.
[b] Immersion in liquid nitrogen from −30 °C.

cooling of the excised primordia increased with decreasing water content, and no exotherm was detected in the primordia with a water content below about 20% (f.w.). DTA profiles of primordia of whole buds showed that LTEs shifted markedly to a lower temperature when cooling took place very slowly (Sakai 1979a). Thus, the survival temperature of primordia of whole buds is greatly affected by the cooling rate (Table 4.13). In very hardy species, such as *Abies sachalinensis*, no exotherm was detected even after cooling to −50 °C when cooling involved decrements of 5 K daily down to −20 °C, and a rate of a K h^{-1} from −20° to −50 °C. These primordia were all killed. If the buds were cooled by 2 K h^{-1} continuously from −3 °C, LTEs appeared at around −32 °C (Fig. 4.22). In shoot primordia of Alaskan spruces, *Picea glauca* and *P. mariana*, too,

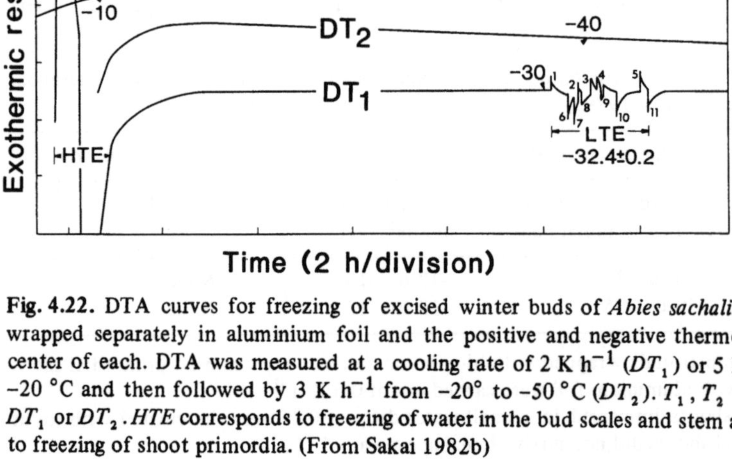

Fig. 4.22. DTA curves for freezing of excised winter buds of *Abies sachalinensis*. Six buds were wrapped separately in aluminium foil and the positive and negative thermocouple placed in the center of each. DTA was measured at a cooling rate of 2 K h^{-1} (DT_1) or 5 K decrements daily to −20 °C and then followed by 3 K h^{-1} from −20° to −50 °C (DT_2). T_1, T_2 temperature curves of DT_1 or DT_2. *HTE* corresponds to freezing of water in the bud scales and stem axis, *LTE* corresponds to freezing of shoot primordia. (From Sakai 1982b)

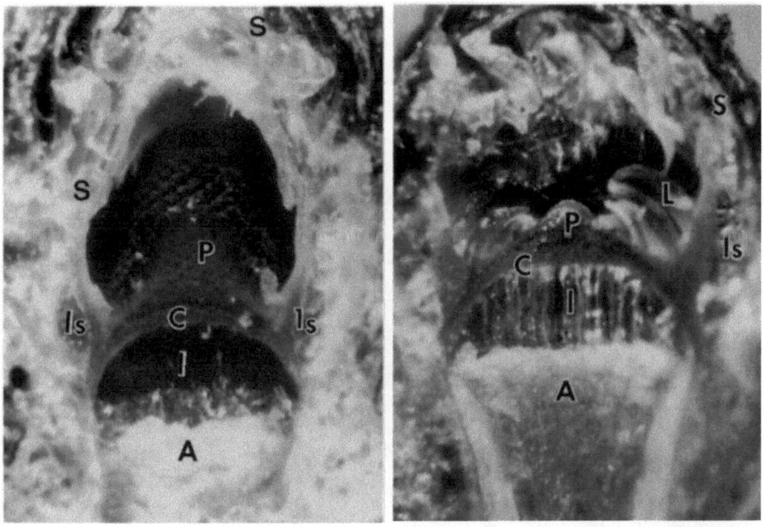

Fig. 4.23. Longitudinal sections of the vegetative buds of *Abies sachalinensis* (*left*) and *Larix sibirica* (*right*) held at −20 °C for 1 day following freezing at daily decrements of 5 K. The shoot primordia remained unfrozen. *I* ice underneath the crown; *Is* ice inside scales surrounding the basal part of the shoot primordia; *A* head of axial core pith; *C* crown; *L* leaf primordia; *P* shoot primordia; *S* scales. The crown is characteristically curved, resulting in an increased surface area. (From Sakai 1982a,b)

ice segregation was demonstrated: the freezable water in primordia of Alaskan spruces was fully translocated under slow cooling to −30° or −40 °C. These freeze-dehydrated buds survived immersion in LN_2 when subsequently rewarmed very slowly (Sakai 1979a). Extraorgan freezing occurs also in male and female flower buds of conifers (Sakai 1982b). It thus appears that both vegetative and floral primordia share the same freezing pattern in most coniferous genera.

Microscopic observation revealed that in buds exposed to −10 °C for 1 day, the bud tissues were frozen, while the shoot primordia remained greenish and soft. This kind of freezing evidently is extraorgan freezing. Much needle ice accumulates beneath the crown pushing down the pith and inside the bud scales surrounding the basal primordium (Fig. 4.23). This suggests that water moves out of the shoot primordia through the crown and to the scales via the vascular system. Ice segregation outside the shoot primordia increased with decreasing temperatures to about −30 °C or below (Dereuddre 1978; Sakai 1982a). As freeze-induced dehydration proceeds, the crown increases its curvature, forming a dome which facilitates freeze dehydration, especially in the smaller shoot primordia (Fig. 4.23).

Ice redistribution from primordia through the crown as a result of a water vapour pressure gradient between the primordium and the ice outside appears to be a similar phenomenon to the formation of needle ice near the soil surface, which is most likely to occur if the size of soil particles is 5–10 μm (Horiguchi 1979). It might reasonably be expected that microcapillaries of similar size (Fig. 4.24) would permit an efflux of water through the crown, but not internal ice propagation.

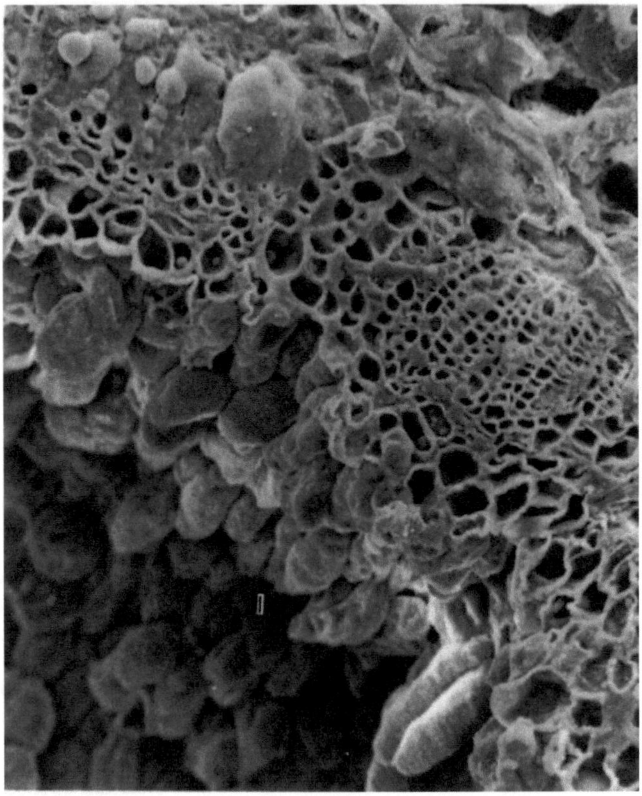

Fig. 4.24. Scanning electron micrography of a longitudinal section of the crown of *Abies sachalinensis*. *I* site of ice segregation; x 800 (Sakai and Niki, unpubl.)

In extraorgan freezing, there seems to be no low temperature limit, as long as shoot primordia can resist intensive freeze dehydration and its related stress. Thus, it appears that survival at the temperature of LN_2 is not an exclusive property of extracellular freezing. In contrast to extracellular freezing the cooling rate is decisive for survival or death during extraorgan freezing probably because winter bud primordia are much larger than cells.

Extraorgan freezing in the primordia explains why the hardiest boreal conifers survive winter cold in subarctic regions where the air temperatures drop to −50 °C or below. Primordia of these conifers survive intensive freeze dehydration at −60 °C or below, when cooled very slowly, although the constituent cells have little ability to tolerate "extracellular freezing" in the strict sense of the word. Winter buds of boreal conifers have enhanced hardiness by reducing their size, by increasing the rate of water translocation and by evolving the ability to tolerate intensive freeze dehydration.

4.4 Comparison and Classification of Frost Survival Mechanisms

In summarizing the characteristics of freezing avoidance in shoot and floral primordia of coniferous buds, floral primordia of angiosperms and hydrated seeds, it appears that the conventional term "supercooling" is not appropriate to describe cases of efficient freezing prevention, especially at naturally occurring cooling rates. Freezing avoidance by deep supercooling can be categorized into two types. One type is a "true supercooling" of overwintering xylem ray parenchyma in temperate woody plants and of large seeds (Ishikawa and Sakai 1982) which involves little or no water redistribution. The other type, which can be termed "extraorgan freezing", accompanies water migration from a supercooled organ to a specific space outside the organ, resulting in dehydration of the organ. This type is cooling rate-dependent: the slower the rate, the higher the ability to supercool. Extraorgan freezing can be further distinguished as being one of three types according to the limitation of survival (Table 4.14). In the first type, the organ which is dehydration-tolerant becomes fully dehydrated whilst cooling slowly, so that there is no freezable water within the organ; thus, the organ can tolerate even the temperature of LN_2. The shoot primordia of extremely hardy conifers native to Alaska, NW Canada and Siberia, hydrated seeds of lettuce and *Oenothera* are of this type. In the second type, the organ can be intensively dehydrated although it is not dehydration-tolerant: the limit for survival lies between $-35°$ and -50 °C. In the third type, even though the organ is dehydrated partially during slow cooling, freezable water remains in the organ and the organ is killed by the breakdown of supercooling. In this sense, the third type is close to supercooling: it could be termed "extraorgan freezing on account of ice segregation and tolerance to freeze dehydration", whereas it could be regarded as "supercooling" with respect to the state of the water remaining in the tissue. The floral primordia of conifers from warm temperate regions are killed between $-25°$ and -30 °C as soon as supercooling breaks down (Sakai 1979a). In warm climates, however, temperature below -25 °C seldom occur. Thus, it may be postulated that even warm temperate conifers tolerate extraorgan freezing under natural conditions.

There is a gradient in the rate of dehydration due to water migration from the cell, tissue or organ to an outer space (Table 4.15): rapid in extracellular freezing, slow in extraorgan freezing and little or none in supercooling. Extracellular freezing and extraorgan freezing are similar, in the sense, that the tissues and organs are more or less dehydrated provided that they are cooled slowly. The difference between the two seems to lie in the fact that in extraorgan freezing the cells in the intact organ do not have a barrier either at the cellular level or within the organ, but they do have barriers at the organ level or outside the organ. The cells in the organ may, of course, have the ability to freeze extracellularly to some degree, too.

One possible reason for the inability of shoot primordia of conifers and floral primordia of angiosperms to undergo extracellular freezing is that the plasma membranes and/or cell walls are somewhat different from those of tissues which are able to freeze extracellularly. Another possibility is that there is no space for extracellular ice to form in the tissues. In extracellular freezing the plasma membrane functions as a barrier, whereas in extraorgan freezing the crown of coniferous shoot primordia, the endosperm envelope in hydrated seeds and the vascular systems in floral primordia appear to be

Table 4.14. Classification of extraorgan freezing and deep supercooling. (From Ishikawa and Sakai 1982; Sakai 1982b)

Classification	Typical examples	Ice segregation (rate)	Degree of dehydration by slow cooling	Dehydration tolerance	Survival of immersion in LN_2 after prefreezing at −30° or −40 °C	Limit of survival temperature
I	Buds of extremely hardy boreal conifers, hydrated seeds of lettuce and *Oenothera*	Yes (fast)	Fully dehydrated	Tolerant	Yes	None (below −70 °C)
II	Buds of very hardy conifers, flower buds of *Prunus padus*	Yes (fast)	Considerably or fully dehydrated	Intolerant	No	−40° to −50 °C
III	Buds of temperate conifers, flower buds of rhododendrons, most of the hydrated seeds	Yes (slow)	Partially dehydrated	Intolerant	No	above −30 °C
Deep supercooling	Xylem parenchyma, large seeds	No	None or little	None or little	No	Homogeneous nucleation at −40° to −45 °C

Table 4.15. Characteristics of extracellular freezing, extraorgan freezing and deep supercooling. (From Ishikawa and Sakai 1982)

Freezing pattern	Typical cells, tissues, organs	Freezing dehydration (rate)	Hardiness of constituent cells	Cooling rate dependency	Barriers against external ice inoculation	Lower limit of survival temperature (°C)
Extracellular freezing	Cortical cells, leaf cells	Yes (rapid)	Hardy	Small	At cell level	None
Extraorgan freezing	Primordia, small seeds	Yes (slow)	Little or nonhardy	Great	At organ level (not at cellular level)	None
Deep supercooling	Xylem ray parenchyma	Little or none	Little or none	Little or none	Cell membrane and/or cell wall	−41 to ~−45

involved in preventing ice propagation and in the translocation of water from the supercooled organ to the ice sink.

The definitions of "freezing avoidance" and "freezing tolerance" have been complicated by the recognition of extraorgan freezing. Most cases of extraorgan freezing can be considered to be freezing tolerance in view of the dehydration tolerance observed at naturally occurring cooling rates. This is probably the mechanism of field survival, and it is this rather than the cause of death under laboratory conditions that is of interest in elucidating the adaptive mechanisms employed by plant tissues confronted with subzero temperatures.

Twenty years ago already, Idle (1966), in studying the formation and interpreting the pathophysiological significance of ice aggregates in plant tissues came to the conclusion that events observed in plants frozen rapidly would be unlikely to be identical with those occurring in nature, particularly in the early stages of ice formation. The rapid cooling rates used in studying the freezing mechanisms of seeds (20 K h^{-1}; Stushnoff and Junttila 1978), floral primordia (8.5 K h^{-1}; Rajashekar and Burke 1978) and coniferous primordia (60 K h^{-1}; Becwar et al. 1981) are probably unsuitable for distinguishing between extraorgan freezing and supercooling, especially when the samples are immediately cooled from room temperature. To interpret survival mechanisms under natural conditions, the freezing process should be observed at different cooling rates and the results compared. This holds not only for translocated ice formation, but equally for extracellular freezing as demonstrated with mountain plants by Beck et al. (1982) and Larcher and Wagner (1983).

It can be concluded that supercooling and extraorgan freezing are the survival mechanisms at subfreezing temperature for xylem ray parenchyma, flower of angiosperms buds, shoot and floral primordia of conifers and seeds. The other tissues of woody and herbaceous plants, such as the bark, leaves, vegetative buds of angiosperms etc., survive freezing temperatures lower than $-10°$ to -15 °C only by extracellular freezing.

4.5 Survival at Ultralow Temperatures

The question of whether there is a limit to the low-temperature tolerance of plants was approached long ago by a number of investigators. The search for the lowest survival temperatures mainly employed lower plants (Luyet and Gehenio 1938; see also Table 6.1), or higher plant organs in the dry dormant state, such as seeds (Lipman 1936), pollen grains (Davies and Dickinson 1971; Nath and Anderson 1975) and dormant tubers (Sakai 1960b). Becquerel (1954) demonstrated that dry spores and pollen grains were able to survive the lowest temperatures even to a fraction of a degree above absolute zero.

4.5.1 Desiccated Plant Tissues

In desiccated plant cells the small amount of moisture is matrically bound and, therefore, will not be converted to ice even when cooled to very low temperatures. Leaves

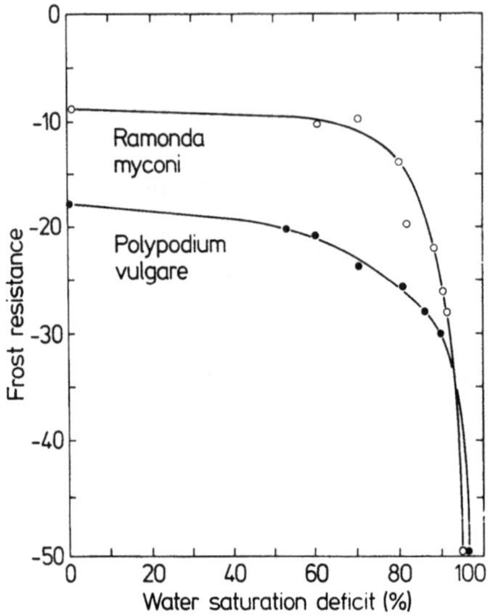

Fig. 4.25. Desiccation-enhanced freezing tolerance of leaves of drought-tolerant vascular plants in winter. (From Kappen 1966)

of certain ferns and phanerogams, which are damaged by frost at −10° to −20° when saturated with water, withstand −50 °C at 5% relative water content (Kappen 1966; Fig. 4.25). Critical water contents below which plant organs can survive the temperature of LN_2 are presented in Table 4.16; the effect of decreasing water content of seeds on frost survival is shown in Fig. 4.26. For soybean seed, where the critical moisture content is 19% (f.w.) or 0.235 g H_2O g^{-1} dry matter, Okamura (1973) distinguished three states of hydration: (1) the bound water state below 7% fresh weight, (2) the semibound water state at 7–19% and (3) the solvent water state over 19%. The semibound state can be referred to as the gel state according to Koga et al. (1966). Simatos

Table 4.16. Survival at very low temperatures of desiccated plants

Plant and plant part	Critical water content (% f.w.)	Rewarming	Exposure temperature (°C)	References
Anemone sp., tubers	20–25	Slow	−196	Sakai (1960b)
Tortula ruralis (moss)	21	Slow	−196	Malek and Bewley (1978)
Mnium affine (moss)	30	Slow	−190	Luyet and Gehenio (1938)
Lactuca sativa, achenes	16–18	Slow	−196	Stushnoff and Junttila (1978)
Triticum sativum, caryopses	18	Slow	−196	Ishikawa and Sakai (1978)
Cryptomeria japonica, pollen	17–20	Slow	−196	Ichikawa and Shidei (1971)

The critical water content for maintaining viability after direct immersion in LN_2 is greatly dependent on the subsequent rewarming rate. The above data correspond to a rewarming procedure at room temperature; when rewarmed rapidly in water at 30 °C, the critical water content may be somewhat higher.

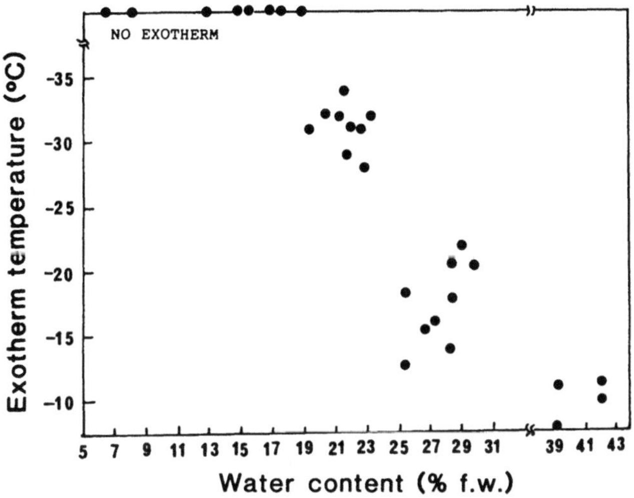

Fig. 4.26. Effect of water content on the exotherm temperature of winter wheat grains. *No exotherm* (upper end of the ordinate) was detected during slow cooling to −40 °C and rapid cooling to −50 °C. (From Ishikawa and Sakai 1978, partly revised)

et al. (1975) demonstrated that 0.25 g H_2O g^{-1} is the minimum water content above which the system acquires sufficient mobility to give rise to a detectable glass transition. The hydration binding between protein or solute and liquid water (12.5−33.5 kJ mol^{-1}) is greater than that of ice crystals (about 21 kJ mol^{-1}). Thus, some water bound to proteins can remain unfrozen at very low temperatures. On the basis of high-resolution NMR spectroscopy, Kuntz et al. (1969) suggested that the bound water is not "ice-like" in a strict sense, although it is less mobile than liquid or solvent water at the same temperature. When dehydrated seeds, pollen, tubers, etc. are allowed to imbibe sufficient water, most of them are killed by very slight freezes. Hydrated tubers of *Anemone* and *Spiloxene* (over 30% H_2O f.w.) were not able to tolerate freezing even at −5 °C (Sakai 1960a,b; Sakai and Yoshie 1984).

4.5.2 Hydrated Plant Tissues

In an experiment carried out by Scholander et al. (1953) twigs of woody plants wintering in N. Alaska were collected, transferred to the laboratory in the frozen state and immersed in liquid oxygen (−183 °C). All were killed, although the untreated control twigs produced buds normally in a greenhouse. From this result, it was concluded that even the hardiest plants wintering at severe climates cannot withstand freezing at the temperature of liquid oxygen.

In extracellularly frozen cells, the lower the temperature the greater the amount of ice, formed at equilibrium, until the temperature is reached at which all the freezable water has crystallized. Thus, equilibrium freezing is considered as an effective means for freeze dehydration of hardy cells. For the purpose of freeze dehydration to various degrees, twig pieces of *Morus bombycis* were slowly frozen in winter to different temperatures and held there for 16 h prior to immersion in LN_2. The survival gradually increased with decreasing temperature from −10° to −30 °C, reaching the maximum at −30 °C or below if rewarmed very slowly (Sakai 1956, 1960a). These twig pieces, and also twigs of *Salix koriyanagi* and *Populus sieboldii*, immersed in LN_2 after prefreezing at -30 °C, remained alive at least for 160 days (Sakai 1956, 1960a). In addition, twig pieces of mulberry were not injured even when cooled from 0 °C to -196 °C at a cooling rate of 0.5 K min^{-1}. From these results it appears that there is a definite temperature range at which the freezable water in cells is extracted by equilibrium freezing, and that the cells in this state are not injured even when exposed to an extremely low temperature. Below this temperature, the intensity of cold does not seem to exert any important effect upon these cells. Twigs of ***Salix koryanagi, S. sachalinensis*** and ***Populus sieboldii*** following prefreezing at -20° to -30 °C for 16 h, and shoots of *S. sachalinensis* sampled in Sapporo during winter when air temperatures were below -20 °C, remained alive after immersion in LN_2 and subsequent slow rewarming in air at -30 °C for 16 h and then at 0 °C.

We have three alternative models in mind: Completely hardy cells (1) withstand substantial dehydration during freezing such that at low temperature all remaining water is water of hydration (bound water); (2) form aqueous glasses intracellularly during freezing to low temperatures, glasses which remain amorphous solids at relatively high subzero temperatures; (3) are capable of withstanding considerable intracellular freez-

Fig. 4.27. A twig of *Salix sachalinensis*, cooled to the temperature of liquid helium following prefreezing at -30 °C in 1961, developed roots and grew to a tree of 8 m height by 1982. (Photo: A. Sakai)

Table 4.17. Prefreezing temperatures necessary for maintaining viability after immersion of woody twigs in liquid nitrogen. The twigs were cooled stepwise in hourly decrements of 5° to 10 °C and held at the prefreezing temperatures for 4 h prior to immersion in LN_2. After removal from LN_2 the twigs were transferred to air at −30 °C for 1 h before thawing in air at 0 °C. (From Sakai 1965, partly revised)

Species and date			Prefreezing temperature prior to immersion in liquid nitrogen (°C)								Hardiness at slow freezing[a]
			−10	−15	−20	−30	−40	−50	−60	−70	
Betula platyphylla											
var. *japonica*	(Nov. 20)		●	●	○	○	○	○	○	○	−90
	(Dec. 20)		●	○	○	○	○	○	○	○	−90
	(Feb. 20)		●	○	○	○	○	○	○	○	−90
	(Mar. 23)	(N)	●	●	●	○	○	○	○	○	−90
	(Mar. 23)	(H)	●	○	○	○	○	○	○	○	−90
Salix sachalinensis	(Dec. 20)		●	○	○	○	○	○	○	○	−90
Populus simonii	(Dec. 20)		●	○	○	○	○	○	○	○	−90
Malus pumila cv. Jonathan	(Jan. 20)		●	●	○[b]	○[b]	○[b]	○[b]	○[b]	○[b]	
Pinus pumila	(Dec. 20)		●	●	●	△	○	○	○	○	−90

N = untreated; H = hardened at −5 °C for 5 days; ○ no damages, △ injured, ● killed.
[a] Twigs survived freezing to −90 °C at least.
[b] Buds and cortex remained alive, but xylem was injured below −35 °C.

ing. In *Populus balsamifera* Hirsh et al. (1985) reported that the above model (2) is the most appropriate. The authors provided evidence that winter-hardened poplar resists the stresses of freezing below −28 °C by glass formation of the bulk of the intracellular solution during slow cooling. This implies that the formation of the solid intracellular glass prevents further water loss to extracellular ice at lower temperatures. During slow warming of the samples after slow cooling to −70 °C at 3 °C/h, a sudden endotherm was detected by differential scanning calorimetory at −28 °C, indicating an equilibrium glass transition without devitrification.

The prefreezing temperature necessary to maintain viability depends on the relative hardiness of the plant, it varies from −15° to −40 °C, the hardier samples requiring less prefreezing (Table 4.17; Sakai 1965). Table 4.18 contains a list of plant species which survived freezing to superlow temperatures. Twigs of *Salix sachalinensis, Populus maximowiczii* and *Betula platyphylla* even survived freezing to the temperature of liquid helium (about −269 °C) following prefreezing at −20° or −30 °C. The treated twigs of *S. sachalinensis* rooted and grew up to trees of 8 m height in 1982 (Fig. 4.27). These results indicate that extremely hardy plants can tolerate freezing near the temperature of absolute zero.

Table 4.18. Survival at very low temperatures of plants in the hydrated state (all plants were rewarmed slowly in air at 0 °C)

Plant and plant part	Exposure temperature (°C)	Freezing method	Reference
Morus bombycis (cortical tissues)	−196	Prefreezing[a] at −30 °C	Sakai (1956, 1960a)
Morus bombycis (cortical tissues)	−120	Continuous freezing at 0.5 K min^{-1}	Sakai (1956, 1960a)
Salix koriyanagi (twigs)	−196	Prefreezing at −30 °C	Sakai (1956, 1960a)
Salix koriyanagi (twigs)	−120	Continuous freezing at 0.5 K min^{-1}	Sakai (1956, 1960a)
Pinus strobus (needles)	−189	Continuous freezing at 0.07 K min^{-1}	Parker (1960a)
Pinus pumila (twigs)	−196	Prefreezing at −40 °C	Sakai (1965)
Polystichum retrosopaleaceum (gametophytes)	−196	Prefreezing at −30 °C	Sato and Sakai (1980a)
Betula pendula (twigs)	−253	Prefreezing at −70 °C	Tumanov et al. (1959)
Ribes nigrum (twigs)	−253	Prefreezing at −70 °C	Tumanov et al. (1959)
Ribes uva-crispa (twigs)	−253	Prefreezing at −70 °C	Tumanov et al. (1959)
Salix sachalinensis (twigs)	−269	Prefreezing at −30 °C	Sakai (1962b)
Populus maximowiczii (twigs)	−269	Prefreezing at −30 °C	Sakai (1962b)
Betula platyphylla (twigs)	−269	Prefreezing at −30 °C	Sakai (1962b)

[a] Freezing temperature prior to immersion in liquid nitrogen (−196 °C), hydrogen (−253 °C) and helium (−269 °C) respectively.

5. Cold Acclimation in Plants

5.1 The Annual Course of Frost Resistance

In zones with a seasonal climate, plants undergo periodic transition from a lower to a higher level of resistance. All categories of frost resistance are affected, the tolerance to equilibrium freezing (Fig. 5.1) as well as the ability to undergo deep supercooling (Fig. 5.2). Two types of seasonal acclimation can be distinguished, each being an expression of the different mechanisms by which they achieve their resistance (see Table 4.3):

1. Species Responding Directly to Low Winter Temperatures. Plants growing in regions with a mild winter or in habitats shielded from severe frost acquire greater resistance in the cold season as a *direct* consequence of the progressive drop in temperature. This enhanced resistance may be due simply to a lowering of T_f and T_{sc}, or to the acquisition and further development of freezing tolerance. However, if the plants are protected from cold they can remain at a low level of resistance throughout the year. This is the situation in the rhizomes of the cosmopolitan fern *Pteris aquilinum*, although not in other ferns of the temperate zone (Kappen 1964). A further example is provided by the observation of Larcher (1954) that the leaves of olive trees cultivated in glasshouses

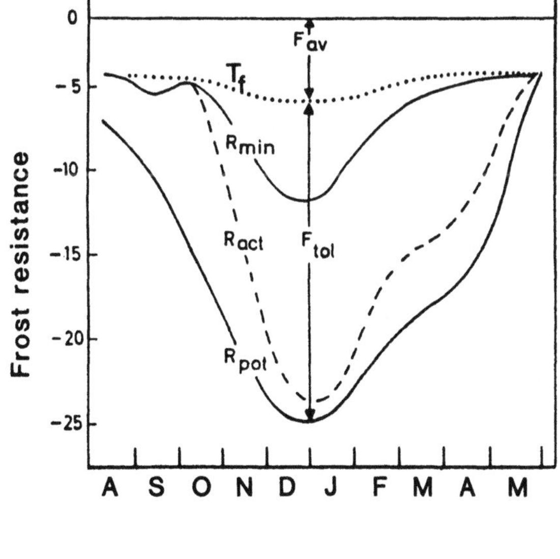

Fig. 5.1. Annual variation of freezing tolerance (F_{tol}) and freezing avoidance (F_{av}) of leaves of *Rhododendron ferrugineum*. T_f threshold freezing temperature; R_{min} frost resistance after dehardening at 15 °C; R_{act} frost resistance at the habitat; R_{pot} highest frost resistance after hardening at −6 °C for 5 days. (From Larcher 1987a, after Pisek and Schiessl 1947)

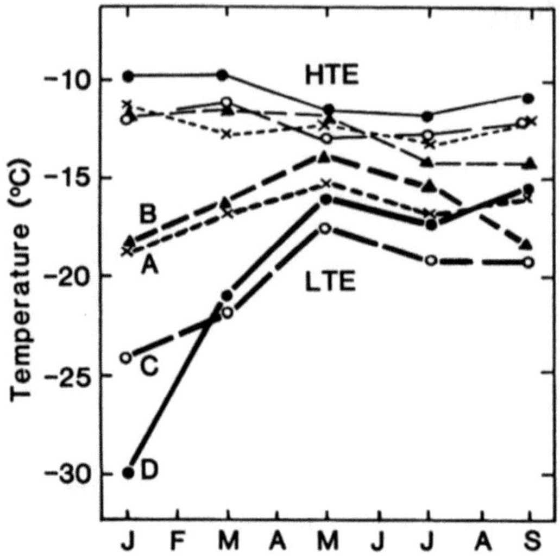

Fig. 5.2. Trends of seasonal variations in high (*HTE*) and low temperature exotherms (*LTE*) of xylem samples from: *A* evergreen trees, e.g. *Camellia japonica, Cinnamomum camphora, Quercus glauca; B* warm-temperate deciduous trees, e.g. *Albizzia julibrissin, Melia azedarach, Rhus toxicodendron; C* temperate deciduous trees, e.g. *Castanea crenata, Quercus serrata, Zelkova serrata; D* cool-temperate deciduous trees, e.g. *Alnus hirsuta, Cornus controversa, Magnolia obovata.* (From Kaku and Iwaya 1978)

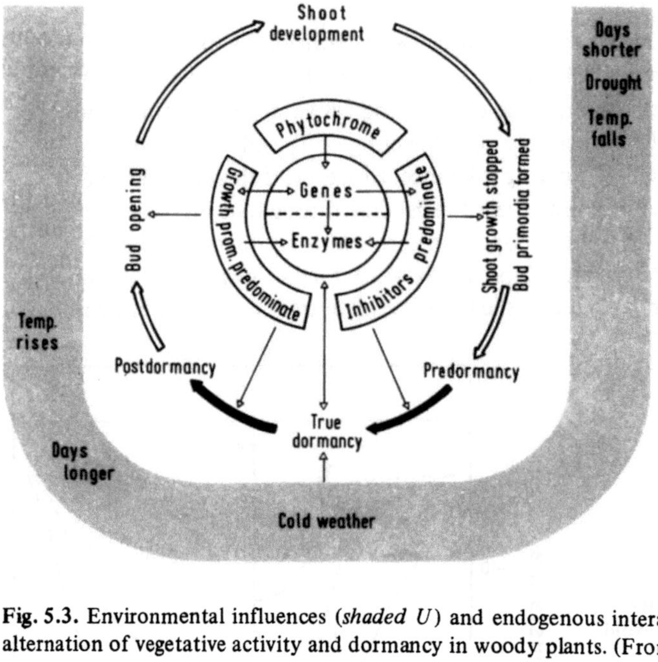

Fig. 5.3. Environmental influences (*shaded U*) and endogenous interactions affecting the seasonal alternation of vegetative activity and dormancy in woody plants. (From Larcher 1983a)

exhibit damage at any time of year at −6 °C, whereas under Mediterranean winter conditions they tolerate temperatures down to −10 °C. A similar kind of behaviour is seen in woody plants of the subtropics (e.g. *Salix safsaf, S. tetrasperma:* Sakai 1978, and *Pinus caribaea:* Oohata et al. 1981) and in many herbs.

2. *Species with Inherent Periodicity in Resistance.* Woody plants of the temperate zone and perennial herbs that overwinter with storage organs adjust their development and metabolic activities to the changing season well in advance. A gradual transition to developmental arrest begins in autumn, and in late winter and spring the plants slowly begin to resume activity (Fig. 5.3). This tendency to a seasonal alternation between activity and rest has a genetic background (Bünning 1953; Villiers 1975), the degree to which it is expressed varying according to species and provenance. Species whose periodicity in resistance is dependent upon true dormancy attain a basic level of increased resistance in winter even if protected from frost in a glasshouse (Tyurina et al. 1978) or beneath a covering of snow. True dormancy also plays an important role in preventing loss of hardiness in mid-winter during periods of unseasonable warm weather.

5.2 The Seasonal Cold Acclimation Process

Seasonal cold acclimation involves a sequence of processes which are mutually interdependent, each stage preparing the way for the next. The typical stage pattern of cold acclimation in woody plants from temperate zones is shown in Fig. 5.4.

The *first stage* of cold acclimation, after growth has ceased, appears to be induced by short days and proceeds at 10° to 20 °C in fall, during which abundant organic substances are stored. The most important synthetic processes during the first stage of acclimation are accumulation of reserve starch and neutral lipids. These storage materials are essential substrates and energy source for the metabolic changes occurring during the second stage.

The *second stage* of cold acclimation is induced by low temperature below about 5 °C, especially subzero temperatures. In this stage of hardening, proteins and membrane lipids are neosynthesized and/or undergo structural changes, ultimately leading to maximal hardiness.

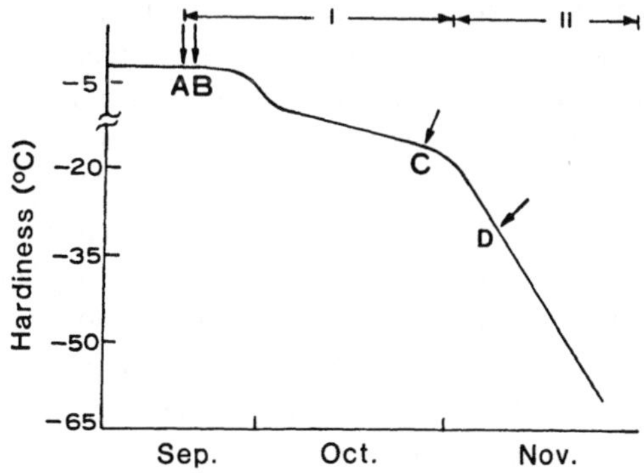

Fig. 5.4. Model of the stage pattern of cold acclimation of woody plants in mid-latitudinal habitats. *A* growth cessation; *B* induction of bud dormancy; *C* defoliation; *D* release from deep dormancy. *I* and *II* indicate the first and the second acclimation stage (A. Sakai, original)

5.2.1 Seasonality Timing

With the exception of the humid equatorial regions, seasonal periodicity of day length, temperature and precipitation means that there is regular alternation between periods in which the climatic conditions are favourable to plant growth and those in which they are not. Thus, a regular alternation between developmental activity and cessation of growth is imposed on the plants. In regions with climatic rhythmicity, the process can, to a considerable degree, be programmed and occur spontaneously, especially in the temperate and cold climate zones (Bünning 1953). A plant species is well acclimated if the growing season is utilized without risk of injury as the unfavourable season approaches (Langlet 1937; see also Larcher 1983a). In general, this is ensured by coupling of the acquisition of resistance to the developmental processes. A lack of synchronization between periodic plant activity and the rhythmicity of the climate restricts the spread of a species; this limitation has been overcome by the differentiation of rhythmo-ecotypes during evolutionary adaptation. The physiological background of seasonal periodicity in plants is dealt with in detail by Kozlowski (1971), Perry (1971), Wareing and Saunders (1971), Villiers (1975), Vintejoux and Dereuddre (1981), Fuchigami et al. (1982) and Saure (1985).

The Onset of Dormancy. In an uniform environmental study of clones of a widely ranging species, *Cornus stolonifera*, collected from 21 locations in N. America and grown at St. Paul, Minnesota, the timing of cold acclimation differed significantly: northern clones ceased growth 50 days earlier than those from southern or coastal regions (Smithberg and Weiser 1968). Variations in the time of height growth cessation were also observed between clonal lines of *Populus trichocarpa* native to the west coast of N. America between 32° and 65° latitude (Pauley and Perry 1954). When grown in the same day-length regime, and otherwise uniform environment, the time of growth cessation among clones varied from mid-July to late October depending on provenance. The timing of developmental arrest assumes a role of critical survival value: through the selective pressure exerted by the first killing frosts of autumn, only those genotypes that terminate shoot growth sufficiently early are capable of survival within any uniform day-length zone in which topography or other factors may cause considerable variations in the length of the growing season. The group of clones originating between 40° and 50° N is actually composed of a broad longitudinal sampling of *P. trichocarpa* ecotypes, extending from the Pacific coast to western Montana, and with a vertical dispersion from near sea level to about 1500 m altitude. The length of the frost-free season in this latitudinal zone is, in fact, known to vary widely within comparatively short distances, due to elevation, topography, etc. Figure 5.5 shows the scatter diagram resulting when the length of the average growing season for those clones native only to the narrow latitudinal zone 45° to 47° N is plotted against the date of growth cessation at Weston (near Boston). It can be seen that *P. trichocarpa* populations are capable of adaptation to growing seasons of varying length within uniform day-length zones by selection of those types with suitable photoperiodic response. The time at which growth ceases is directly correlated with the length of the frost-free season prevailing in the native habitat of each clone, reflecting, of course, the modifying influence of the individual genotype, tree age and other environmental influences.

Fig. 5.5. Correlation between the average growing season in Weston and the date of cessation of height growth at Weston of *Populus trichocarpa* clones native only to the latitudinal zone 45–47° N. (From Pauley and Perry 1954)

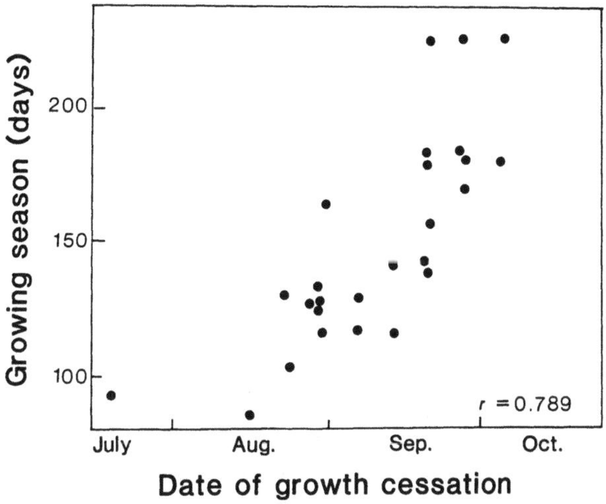

Chilling Requirement to Break Dormancy. Many temperate woody plants have a distinct chilling requirement for breaking dormancy, usually 3 to 4 weeks at 5 °C (Vegis 1973). In some woody plants, subzero temperatures between −5° and −10 °C are effective in breaking bud dormancy (Fig. 5.6). Breakdown of deep dormancy of woody plants grown in Sapporo usually occurs at subzero temperatures prevailing in November. Chilling requirement appears to be an important factor setting the southern growth limit of woody plants. In some cases, application of gibberellic acid will break bud dormancy without cold treatment, and it seems likely that, as in seeds, a change in the balance of

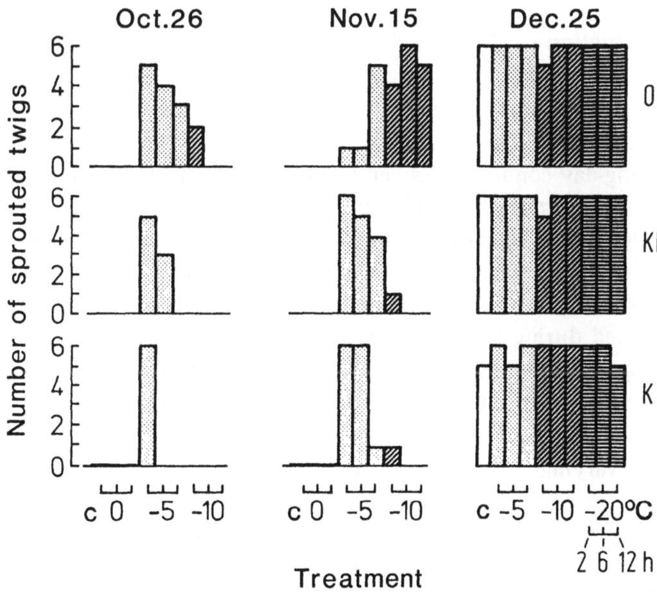

Fig. 5.6. Breaking of true dormancy of mulberry twigs by exposure to subzero temperatures. Twigs frozen at −5 °C were cooled slowly to −10° and −20 °C and held there for 2, 6 and 12 h respectively. *C* untreated control; *O, KR, K* are different varieties. (From Hasegawa and Tsuboi 1960)

growth-regulating substances is the trigger for bud break (Wareing and Phillips 1978). In cereals, bud dormancy is broken by an adequate exposure to temperatures between +2° to +6 °C and by increasing day length (Vasilyev 1934, 1961; Rudorf 1938; Goujon et al. 1968).

5.2.2 Induction of Cold Acclimation and the First Stage of the Hardening Process

The major events at the onset of the hardening process were summarized by Weiser (1970) as follows: (1) growth cessation is a prerequisite to cold acclimation in woody plants; (2) leaves are the site of perception of the short-day stimulus which initiates the first stage of acclimation; (3) short-day induced leaves are the source of a translocatable factor which promotes acclimation; (4) the hardiness-promoting factor (Irving and Lanphear 1967; Fuchigami et al. 1970) moves from the leaves to overwintering stems; (5) plants exposed to long-day and cold-night temperatures will eventually become fully hardy, but with delay and the risk of being damaged by early frosts; (6) plants severely depleted of photosynthates cannot acclimate.

Only a very low degree of hardening can be achieved by diseased plants, by individuals deficient in mineral nutrients (for references, see Chandler 1954; Biebl 1962a; Alden and Hermann 1971; Larcher 1985a), or by plants that have been unable to accumulate sufficient carbohydrate reserves either due to loss of foliage or as a result of curtailed period of growth (for examples, see Kramer and Wetmore 1943; Tumanov et al. 1972; Tumanov and Krasavtsev 1975; Stergios and Howell 1977; Nissila and Fuchigami 1978).

Although the increase in hardiness during the first stage of cold acclimation is relatively small (- 10° to - 15 °C) it is nevertheless significant.

Induction of Cold Acclimation in Northern Trees. Cold acclimation of many northern trees appears to be induced by short days (Tumanov et al. 1965, 1972, 1973; Irving and Lanphear 1967; Weiser 1970; Williams et al. 1972). In some cases, the timely autumn cold acclimation can be prevented or retarded if the plants are kept under long-day conditions as can be seen in trees near street lights (Kramer 1937). In Sapporo, *Populus nigra* normally ceases growth in September and defoliates in mid- to late October. The twigs of poplars close to street lights still retain green leaves in mid-November, whereas shaded sides of the same tree, and trees at a distance of as much as 20 m had already defoliated (Fig. 5.7). Winter acclimation of twigs which were illuminated during the night was markedly retarded in late autumn, although they had full hardiness by mid-winter (Table 5.1). The influence of day length on the progress of frost resistance becomes obvious in plants that are exposed, under constant temperature conditions, either to the natural photoperiod or to an artificially selected programme of varying day length. An experiment of this type revealed the dominant role of day length in *Pinus cembra* and *Rhododendron ferrugineum* (Schwarz 1970). At a constant temperature of +15 °C and under natural photoperiod the annual course taken by frost resistance was similar to that seen in plants growing in the open. Long-day treatment throughout the year resulted in a lower resistance in winter, whereas with year-round short day, resistance remained high even in the summer. Howell and Weiser (1970a) found an increase in frost resistance in apple trees under short-day conditions in con-

Fig. 5.7. *Populus nigra* retaining green leaves in mid-November near a street light in Sapporo. (Photo: A. Sakai)

Table 5.1. Frost resistance (°C) of buds and twigs of *Populus nigra* depending on exposure to street lights (A. Sakai, unpubl. results)

Exposure to light	November 10th		January 10th
	Unhardened	Hardened[a]	
Twigs close to the street light retaining green leaves	– 5	–10	–70
Shaded twigs of the same tree with yellow leaves	–15	–40	–70
Twigs of a defoliated tree at 20 m distance from street light	–30	–70	–70

[a] Hardening at 0 °C for 10 days.

junction with temperatures above 15 °C. Treatment of plants in the open with artificially prolonged daylight delayed the increase in resistance until the onset of the first frost; from then on, cold compensated for the long-day effect (Fig. 5.8). In experiments performed by Bervaes et al. (1978) 50% damage to pine seedlings in a non-hardened state was observed at –8 °C, but not until –19 °C following short-day treatment for 3 weeks and cooling to 2 °C, at –13 °C following short-day and warmth (20 °C), and at –16 °C following long-day and cold (2 °C). Thus, even at warm temperatures, short-day treatment induced higher frost resistance, although cold was more effective than reduction of the photoperiod. In spruce, the differences observed were smaller and the short-day effect was stronger. There are also cases in which short-day alone is ineffective, but in combination with cold results in hardening, e.g. the acquisition of resistance by roots of *Potentilla fruticosa* (Johnson and Havis 1977) and *Rubus*

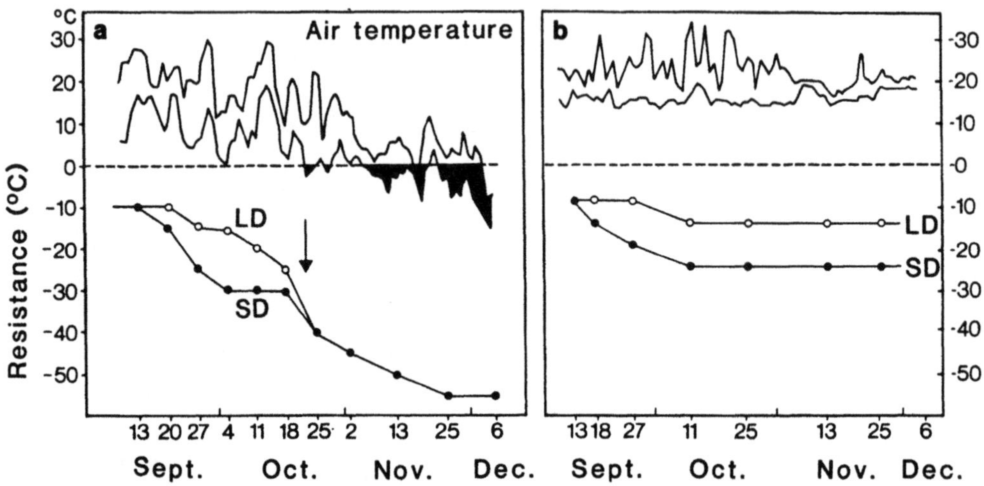

Fig. 5.8a,b. Seasonal hardening patterns of young apple trees at different photoperiods in (a) the field and (b) in a warm greenhouse. *SD* natural short days in autumn; *LD* long-day treatment (photoperiod 18 h using additional incandescent light). Air temperatures are daily maxima and minima. *Arrow:* First leaf-killing frost. (From Howell and Weiser 1970a)

chamaemorus (Kaurin et al. 1982). The roots of *Pinus glauca*, however, react to short-day without cold (Johnson and Havis 1977).

Induction of Cold Acclimation in Trees of the Warm Temperate Zone. The woody evergreen plants of the subtropics and regions with mild winters temporarily arrest growth during the coldest period (Cooper and Peynado 1959). The process of cold acclimation in such plants stops at the first stage. A *characteristic of one-phase hardening* seems to be that it can begin very suddenly and be complete within a very short time, even 1 or 2 days, and that it can be lost again equally quickly. Plants of this type are capable of adapting to acute frost stress without any great reduction in metabolism or development (Larcher 1981c).

Pinus species can be distinguished into two phenological groups (Oohata 1979). One group of pines is characterized by a single period of bud flushing and shoot extension. This uninodal type, native to temperate and boreal climates, develops winter dormancy imposed by short days. The multinodal type which is characterized by successive flushes of shoot growth during the same year, responds only slightly or not at all to the photoperiodic stimulus and continues to grow intermittently throughout the year without dormancy. In these species (e.g. *Pinus clausa, P. elliottii, P. taeda* and *P. radiata* from N. America; *P. greggii* and *P. patula* from Mexico; and *P. insularis* and *P. merkusii* from SE Asia) falling temperature is the decisive signal for hardening. Broad-leaved evergreen trees distributed in warm temperate Japan, such as *Myrica, Cinnamomum* and *Quercus* show periodic growth under long-day and short-day (8 to 12 h) conditions at 18 °C or higher temperatures. Bud dormancy is imposed in autumn by lowering of temperatures to 10 °C; to break dormancy chilling is required (Yurugi and Nagata 1981). *Eucalyptus* and other woody angiosperms of the Southern Hemisphere, except subalpine or cool-temperate *Nothofagus* species (Wardle and Campbell 1976b), have

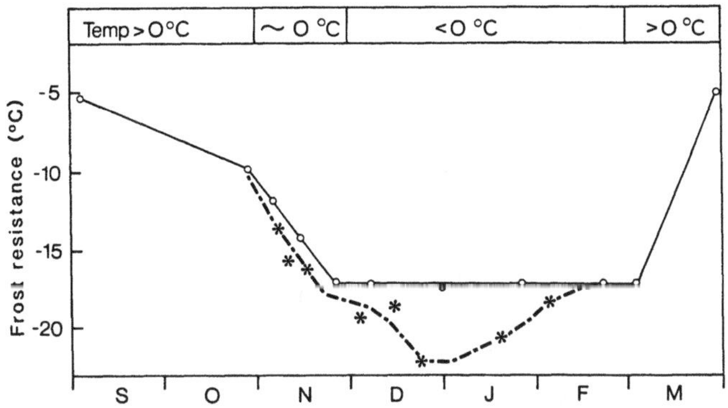

Fig. 5.9. Development of frost hardiness in rape during winter. Plants with (o) high and (*) low water content during the periods with subzero mean air temperatures. (From Kacperska-Palacz 1978)

been found to be insensitive to photoperiod with respect to cold acclimation (Paton 1978, 1982). In the hardiest *Eucalyptus* species the ability to respond readily to hardening and dehardening temperatures (Paton et al. 1979) appears to compensate for the absence of a short-day imposed winter dormancy.

Induction of Acclimation in Herbaceous Plants. Herbs differ from woody plants in their less complicated mode of transition to the freezing-tolerant state and in the greater lability of the acquired resistance (Siminovitch 1981). Nevertheless, hardening in overwintering species of cereals (Tumanov 1979; Olien and Smith 1981; Siminovitch 1981), alfalfa (Paquin and Pelletier 1980) and winter rape (Kacperska-Palacz 1978) also proceeds in recognizable phases (Fig. 5.9). Similarly, the biochemical processes involved correspond roughly, in the first phase at least, with those familiar from woody plants. In winter cereals, alfalfa and rape, hardening is *initiated* by temperatures of +5 °C to +2 °C, apparently without the transition to a state of readiness to harden seen in woody plants. The day length is only of subsidiary importance, if at all. Accumulation of carbohydrates is also essential for herbaceous plants to acquire higher resistance. Since winter cereals and winter rape do not accumulate starch in autumn, the necessary carbohydrates have to be provided by photosynthesis. For these plants, therefore, light as a source of energy is essential at the beginning of the hardening process. If photosynthesis functions sufficiently and if the consumption of photosynthates is reduced by interruption of growth, carbohydrates can accumulate (Tumanov 1979; Levitt 1980; Klimov et al. 1981). The cessation of growth is induced by low temperatures or drought (Tumanov 1931; Fuchs 1933; Siminovitch 1981), and mediated by phytohormones (Kacperska-Palacz 1978; Spomer 1979; Whightman 1979; Carter and Brenner 1985).

5.2.3 The Second Stage of Frost Hardening

The process of cold acclimation is promoted by subfreezing temperatures. The temperature most effective in increasing hardiness depends on species, tissues and developmental stage of the plant. Twigs of various woody plants, partially acclimated in fall,

Table 5.2. Effect of hardening temperatures upon the increase of frost resistance of twigs of *Populus simonii* on October 30th. (From Sakai 1966a)

Freezing temperature (°C)	Before hardening	Effect of hardening temperatures[a] 0	−3	−5	−10	0_{14}	0_7 plus -3_7	0_7 plus -5_7	0_7 plus -10_7
		7 days				14 days in total			
−20	○	○	○	○	○				
−25	△	○	○	○	○				
−30	●	○	○	○	△	○	○	○	○
−70	●	△	○		●	○	○	○	○
−30 LN[b]	●	●	○		●	○	○	○	○
−20 LN[b]	●	●	●	●	●	●	○	○	●
−15 LN[b]						●	△	●	●

[a] ○ uninjured, △ injured, ● dead or seriously injured.
[b] −30 LN, −20 LN, −15 LN indicate prefreezing temperatures prior to immersion in LN_2. For rewarming the twigs were transferred from LN_2 to −10 °C for 1 h and then in air at 0 °C.

survive freezing to −10° or −15 °C: hardening at −3 °C appeared to be the most effective in increasing resistance (Table 5.2). Hardiness was also greatly enhanced by treatments at −3 °C for 14 days, −5 °C for 3 days and −10 °C for 1 day (Pogosyan and Sakai 1969). In nature, the hardiness of trees increases remarkably when the daily minimum temperature falls to subzero for a week (Sakai 1959; Tumanov and Krasavtsev 1959; Howell and Weiser 1970a). However, the question arises as to whether or not subfreezing temperature is a triggering stimulus or a prerequisite for enhancing hardiness to extremely high levels (to −70 °C or below), as proposed by Tumanov (1979). *Salix dasyclados* and *S. ledebouriana*, after hardening at 0 °C for 14 days in mid-October, 1 week before the first frost, became hardy to −70° or −196 °C; the same result was obtained with *S. sachalinensis* (Sakai 1966a). Potted specimens of *Populus* x *euramericana* cv. *'gelrica'*, which were kept at 15 °C for 2 weeks in early October and then hardened at 0 °C for 2 weeks, survived −196 °C after continuous freezing or after prefreezing at −20 °C (Sakai and Yoshida 1968). These results indicate that a subzero temperature is not always a prerequisite for enhancing hardiness to the level of −70 °C or below, especially in very hardy trees.

Tumanov and his co-workers remark that three hardening steps are necessary for full development of frost hardiness: (1) a temperature slightly above 0 °C, (2) a temperature slightly below zero (−3° to −5 °C), (3) slow cooling for about 1 month or more at temperatures from −10° to −60 °C. Prolonged exposure to temperatures in this range causes twigs to acquire a state of hardiness which may not commonly be attained in nature (Tumanov et al. 1959). Willows, white birches, pines and black currants were artificially hardened by this method and finally became hardy even at −196 °C or −253 °C (Tumanov 1979). However, experiments have not been made to determine whether this hardening at very low temperatures for long periods of time is essential for resistance to temperatures of LN_2. The experiments shown in Table 5.3 and other results (Sakai 1960a, 1965) demonstrate that twigs (bark and vegetative buds) of very hardy trees can be acclimated to absolute frost tolerance by prefreezing

Table 5.3. Effect of the environmental temperature on mid-winter frost resistance of *Salix sachalinensis* grown in different localities. (From Sakai 1970a)

Locality and year		Degree of freezing resistance (°C)[a]	Mean air temperature (°C)[b]	Mean minimum air temperature (°C)[c]
Hachijo Island	1967	−30 .. −40	7.8	5.5
	1969	−10 .. −15	12.4	10.0
Kumamoto	1966	−30 .. −40	4.7	0
	1969	−50 .. −70	2.7	− 2.2
Shimizu	1966	−40 .. −50	5.7	0
	1968	−40 .. −50	5.4	− 0.7
	1969	−25 .. −30	8.6	4.5
Sapporo	1967	below −150	− 6.3	−11.1
	1969	below −150	− 6.9	− 9.5

[a] Tested at Sapporo.
[b] Mean air temperature during 20 days before the twigs were collected.
[c] Mean minimum air temperature during 10 days before the twigs were collected.

at −20° or −30 °C without gradually lowering the temperatures to −60 °C. However, in buds which survive by extraorgan freezing, and in xylem which avoids freezing by supercooling, the survival temperature becomes considerably lower if hardening is achieved by daily decrements of 5 °C to −30 °C or −40 °C (Sakai 1979a, 1982b; Gusta et al. 1983).

Under natural conditions plants are exposed to fluctuating temperatures. Some reports indicate that alternating temperatures are at least as effective for acclimation as constant low temperatures (Angelo et al. 1939). Others have failed to obtain hardy plants with alternating temperatures (Day and Peace 1937). Suneson and Peltier (1934) seem to have resolved these differences, for they obtained maximum hardiness by exposure to alternating temperatures during November and December, followed by sustained low temperature for 3 weeks. In young *Cryptomeria* plants the daily temperature fluctuation at the basal stem (between 23° or 19° and 0 °C) caused a decrease in hardiness. An increase in hardiness was observed in *Cryptomeria* plants subjected to a daily alternation between 10° and 0 °C (Sakai 1968a; Horiuchi and Sakai 1978), and an even greater improvement under temperatures alternating between 13° and −1 °C. Subsequent hardening at 0° or −3 °C further increased the hardiness. In young trees the hardiness of the stem base near the ground is lower than that higher up the stem where the temperature is lower than at the base.

5.3 Environmental Control of the Level of Frost Resistance

The level of frost resistance at any particular time, the actual frost resistance (R_{act}), corresponds to the state of the hardening process at that time and the degree of hardening (see Fig. 5.1). The latter depends directly on climatic conditions, such as warm

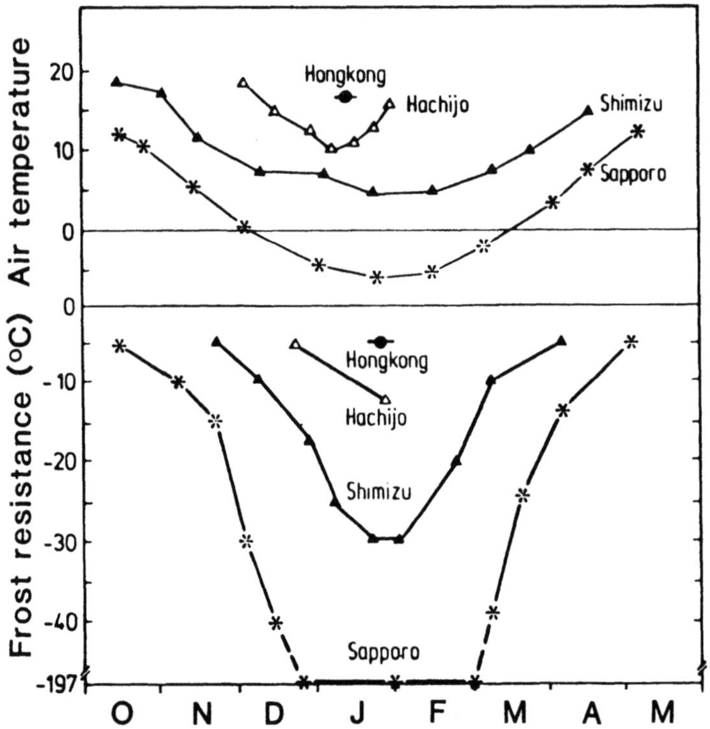

Fig. 5.10. Frost resistance (LT_i) of *Salix babylonica* growing in different climates. Cuttings from the same tree growing in Shimizu (ca. 35° N, warm-temperate climate) were planted in Hong Kong (ca. 20° N, subtropical), Hachijo Island (ca. 33° N, warm-temperate) and Sapporo (ca. 43° N, cool-temperate). Air temperature: monthly averages. (From Sakai 1970a)

and cool weather, availability of light and humidity. Due to the complexity of the external factors involved, plants of one and the same species may exhibit unequal degrees of hardening at different localities, or at one and the same site in successive winters.

Twigs from the same clones of *Salix babylonica* were planted at various locations with different winter temperatures. The winter hardiness differed considerably depending upon the temperature regime of the planting site and on the prevailing temperature in different winters (Fig. 5.10; Table 5.3). The temperature most effective in producing maximum hardiness was nearly the same for temperate and subtropical or tropical willows (Sakai 1970a). These results demonstrate that only by exposure to stressful low temperatures does the whole genotypic capacity of frost resistance become manifest.

5.3.1 Fluctuation of Frost Resistance During Winter

Frost resistance fluctuates during winter, increasing as the temperature drops to subzero, decreasing as it rises above zero. The extent and speed of modulation of R_{act} depend on the type of plant (woody plants, graminoids or dicotyledonous herbs), on the predominant survival mechanism (supercooling or tolerance of equilibrium freezing) and on the specific hardening capacity. During winter, the degree to which frost resistance can be influenced by temperature varies with the depth of dormancy (Pisek and Schiessl 1947; Scheumann and Schönbach 1968; Howell and Weiser 1970b; Tumanov et al. 1973; Fuchigami et al. 1982). At the onset of winter, woody plants react more strongly to hardening stimuli than to dehardening temperatures. During the period of deep dormancy the range of variation is reduced in both directions. In the late winter, at the post-dormancy stage, the plants are more easily dehardened and less easily rehardened.

The short-term changes of R_{act} show that modulations in resistance are in good agreement with temperature fluctuations. Examples of the response of woody plants can be found in Pisek and Schiessl (1947), Pisek (1958), Proebsting (1959, 1963), Proebsting et al. (1980), Howell and Weiser (1970a,b), Ketchie and Beeman (1973), for wheat in Andrews et al. (1974a) and for alfalfa in Paquin and Pelletier (1980). Proebsting (1963) demonstrated that the winter hardiness of peach flower buds, which survive frost by extraorgan freezing, varied between -18° and -26 °C depending on environmental temperature within a short period (Fig. 5.11). He also noted that there was a temperature level (-18 °C) above which peach bud hardiness did not decrease until the end of the post-dormancy period, in spite of a warm spell of 2 days at 13 °C. Andrews and Proebsting (1985) obtained nearly the same results with sweet cherry flower buds.

During the post-dormant period the effect of air temperature on the changes of resistance during winter was investigated on poplar twigs which had attained the highest level of hardiness. Continuously frozen twigs showed no decrease in hardiness at -3°, -5° and -10 °C even if kept for 1 month at least, whereas twigs after thawing at 0 °C showed with time a gradual decrease in hardiness from -70 °C to -30 °C (Sakai 1966a). However, the dehardened twigs were subsequently fully rehardened at subzero temperatures, especially at -5° or -10 °C. Twigs of *Betula pubescens* survived slow freezing to -120 °C or immersion in LN_2 throughout the winter (Table 5.4). Twigs, collected on March 25, still retained the maximal hardiness in spite of a warm spell during which the average daily temperature remained above zero point for 7 days. Even in the twigs exposed to temperatures of 2°, 10° and 18 °C outdoors for 3 days, little or no decrease in hardiness was observed during the winter dehydrated state. Similar results, were obtained with twigs of *Populus* x *euramericana* cv. *'gelrica'* and some boreal willow twigs.

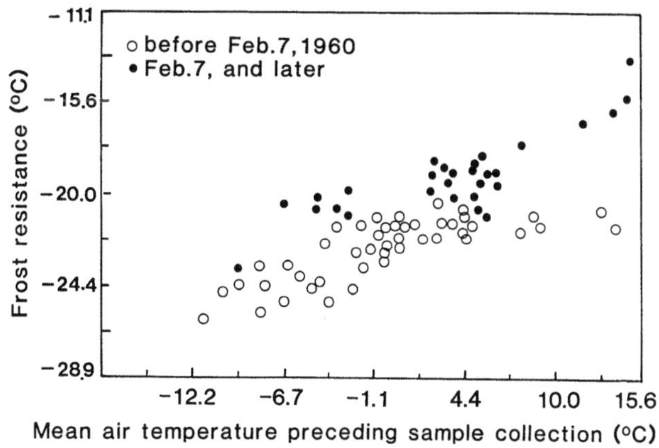

Fig. 5.11. Relation between the frost resistance (LT_{50}) of peach flower buds and the mean temperature of the 2 days preceding collection of the samples. (From Proebsting 1963)

Table 5.4. Changes in winter hardiness of twigs of *Betula pubescens* after exposure to temperatures above 0 °C[a]. (From Sakai 1973)

Freezing conditions	Feb 8–12		Feb 5–8		Feb 24–27			March 4–10			March 25
	Field −6.4 °C[b]	2 °C[c] (O)	−7 °C[b]	10 °C[c] (O)	−5 °C[b]	18 °C[c] (O)	18 °C[c] (I)	−3.2 °C[b]	8 °C[c] (I)	18 °C[c] (I)	1.8 °C[d]
− 70	○	○	○	○	○	○	○	○	○	○	○
−110	○	○	○	○	○	○	○	○	○	△	○
− 70 → −150	○	○	○	○	○	○	○	○	○	△	○
− 30 LN	○	○	○	○	○	○	○	○	○	△	○
− 20 → −150	○	○	○	○	○	○	○	○	○	●	○
− 20 LN	○	○	○	○	○	△	△	○	○	●	○

○ Normal; △ Injured; ● Killed.

[a] Twig pieces were frozen at −10 °C for 30 min, then cooled in 10 °C steps at 30 min intervals to −30 °C. These twigs were further cooled to −70 °C (from −30° to −70 °C: 0.8 K min^{-1}) or −110 °C (from −70° to −110 °C: 0.5 K min^{-1}). Some twigs frozen to −70 °C were transferred to air at −150 °C. Some twigs frozen to −20° or −30 °C were held for 1 h at each temperature prior to immersion in liquid nitrogen (−20 LN_2 or −30 LN_2) or were transferred to air at −150 °C (60 K min^{-1}).

[b] Mean air temperature in the field for 3 or 6 days during which intact twigs were exposed to temperatures above 0 °C.

[c] Temperatures above 0 °C, to which intact or excised twigs were exposed outdoors (O) or indoors (I) for 3 or 6 days.

[d] Average air temperature for 1 week preceding the twig collection.

The killing temperature for xylem and bark from thawed (21 °C for 30 min) and nonthawed winter twigs of 14 deciduous species was compared in a controlled freeze test by Gusta et al. (1983): bark hardiness was unaffected by thawing, whereas 4 of the 14 taxa lost 2 to 7 K of xylem hardiness after thawing.

Tumanov and Krasavtsev (1959) reported that maximum hardiness produced by means of a gradual cold acclimation was quickly lost after thawing. For example, very hardy twigs thawed for as little as 6 h decreased in hardiness from -196 °C to -45°. Thus, there is a discrepancy between these results and those of Sakai (1973) which might be explained by the cooling methods used for evaluating hardiness. In Tumanov and Krasavtsev's experiments, nonthawed twigs hardened below -30 °C and thawed twigs after hardening were cooled by direct transfer to the various test temperatures (Krasavtsev 1960). By this method thawed twigs were cooled at much faster rates than frozen twigs at -30 °C or below. In Sakai's method, twig pieces were frozen at -10 °C for 30 min, then cooled in 10 K steps at intervals of 30 min to lower temperatures, in order to minimize the hardening effect which might be produced during the test. From -30 °C on, the twigs were directly cooled to -70 °C or below.

Fluctuations in frost resistance of winter cereals were investigated by Gusta and Fowler (1976). Winter wheat and rye collected in the fall and stored at -2.5 °C maintained the same level of hardiness for 17 weeks. Upon exposure to warm temperatures, they readily deacclimated, e.g. crowns of 'Kharkov' winter wheat and 'Frontier' rye dehardened from -19° to -11 °C, and from -24° to -14 °C respectively, after 6 days at 15 °C. Dehardened plants were capable of rehardening within a short period upon exposure to cold-acclimating conditions. Tuber-bearing *Solanum* species from the Andes with hardening ability (e.g. *S. acaule* or *S. commersonii*) require about 15 days of cold acclimation to attain the maximum level of frost resistance; deacclimation occurs much faster (Chen and Li 1980a,b; Fig. 5.12).

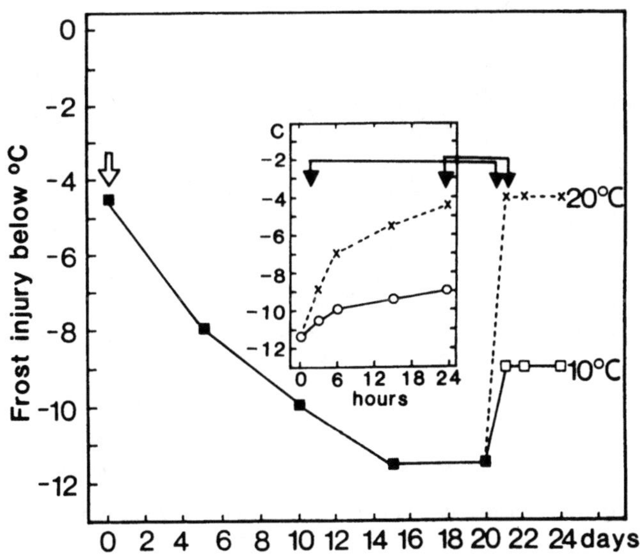

Fig. 5.12. Time course of cold acclimation and deacclimation of *Solanum commersonii*. Plants were treated at 2 °C for 20 days and then transferred to either 10° or 20 °C. *White arrow:* beginning of cold acclimation; *black arrows:* deacclimation. The *inset* demonstrates the dehardening rate at hourly intervals. (From Chen and Li 1980a)

5.3.2 Water Content of the Plant and Frost Resistance

The water content of a plant influences its frost resistance via the cell sap concentration and the degree of hydration of the protoplasm. Prolonged drought, which results in starvation, may weaken the plant and hence lead to a decrease in its resistance (Fraser and Farrar 1957). Gradual desiccation, on the other hand, stimulates an increase of the general stress tolerance (Levitt 1969). Mulberry twigs desiccated in autumn at 15 °C for 2 days showed nearly the same increase in frost resistance as twigs hardened at 0 °C for 10 days; during desiccation starch was converted into sucrose (Sakai 1962). With shoot apices of cereals Cloutier and Siminovitch (1982) observed an increase in frost resistance after drying for 24 h which was comparable to the hardening effect after exposure to 3 °C for 2 weeks. *Opuntia humifusa* survives low winter temperatures better if the shoots contain only 65% water (Koch and Kennedy 1980).

Drying is especially effective if it is accompanied by low temperatures or frost (see Fig. 5.9; Kacperska-Palacz 1978; Paquin and Mehuys 1980). Yelenosky (1979) found that small citrus trees showed an increase in frost resistance after drying, similar to that seen after cold hardening; however, the hardening effect was not cumulative. Apparently, in plants of the freezing-sensitive type, like citrus, the levels of resistance can only be slightly improved.

An abrupt decrease in hardiness due to a warm spell in winter or in early spring seems to be associated with the increase of water content. As long as a low water content is maintained, warm weather during winter is unlikely to cause any great decrease in hardiness. In grapes wintering in Sapporo, the basal part at 20 cm depth below the snow surface remained at temperatures between -1° and -5 °C until mid-March; the

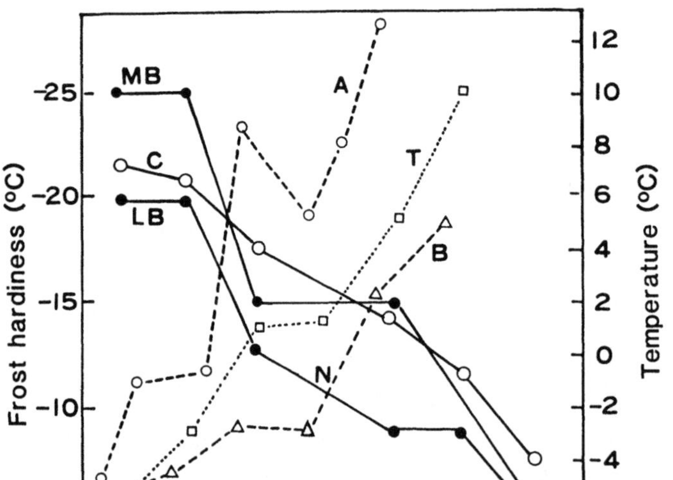

Fig. 5.13. Sudden decrease in frost hardiness of grape canes in early spring in Sapporo. *Solid lines:* hardiness; *MB* main buds; *LB* lateral buds; *C* stem tissue. *Broken lines:* temperatures; *T* mean air temperature; *A* mean maximum and *B* mean minimum temperature of the canes (at 20 cm below surface of snow cover until mid-March). (From Pogosyan and Sakai 1969)

water ascent in the shoot from root was completely blocked, though the soil remained unfrozen (Pogosyan and Sakai 1969). Little or no increase in water content was observed in the buds and upper parts of the shoots in mid-March. Their hardiness still remained near maximum as in winter, though the temperatures of the upper vines often rose to 10° to 15 °C during the day. A warm spell lasting for 3 days from March 18th increased the mean air temperature to 2.5 °C for the first time, and the snow melted. Subsequently, the basal stem remained almost completely unfrozen during day time, though freezing often occurred at night. The water content of the buds and upper stems increased as much as 5% (f.w.) and hardiness decreased abruptly within a week (Fig. 5.13).

5.4 Biochemical and Structural Changes During Cold Acclimation

5.4.1 Metabolic Pathway Shifts from the Growing State to the Dormant State

The metabolic pathway shifts associated with cold acclimation of dormancy-developing and non dormant tissues have been the subject of much research. Sagisaka (1969, 1972, 1974) revealed a shift of glucose-6-phosphate metabolism from glycolysis to the pentose phosphate cycle in poplar twigs in early autumn and vice versa in spring. The pentose phosphate cycle supplies various substrates and $NADPH_2$ for important biological reactions in plant cells. In early September, activities of glucose-6-phosphate and 6-phosphogluconate dehydrogenases began to rise, while transketolase, transaldolase and isocitrate dehydrogenase activities decreased (Fig. 5.14). Glucose-6-phosphate and 6-phosphogluconate dehydrogenase activities were highest in the wintering twigs. During the transition to dormancy a number of enzymes were synthesized de novo; in spring

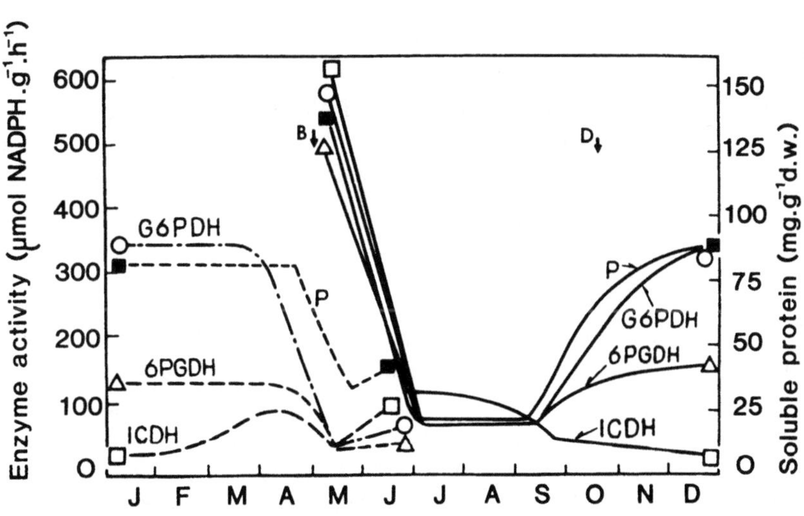

Fig. 5.14. Changes in the activity of glucose-6-phosphate dehydrogenase (G6PDH), 6-phosphate gluconate dehydrogenase (6PGDH) and isocitrate dehydrogenase (1CDH) in poplar bark tissues. *P* soluble protein; *B* bud opening; *D* defoliation. *Broken lines*: 2-year-old twigs. (From Sagisaka 1974, partially modified)

they were degraded within a short time (Sagisaka 1972, 1974; Sagisaka and Asada 1983). The increased activity of glucose-6-phosphate dehydrogenase and important changes in catalytic properties of pyruvate kinase (a key enzyme in the control of glycolysis) were also observed in leaves of winter rape exposed to cold (Sobczyk and Kacperska-Palacz 1980; Sobczyk et al. 1984). In *Chlorella ellipsoidea* cells, which become frost hardy when exposed to 3 °C for 2 days (during the intermediate stage in the ripening phase of their life cycle), glucose-6-phosphate dehydrogenase was found to be highly activated during the first 6 h of hardening (Sadakane et al. 1980; Sadakane and Hatano 1982; Hatano and Kabata 1982): de novo synthesis of glucose-6-phosphate dehydrogenase isoenzymes was induced by low temperature, and even at 5 °C the activity was half of that measured at 25 °C. It seems likely that new isoenzymes are directly involved in the development of frost hardening. The main hardening process in *Chlorella* seems to be similar to that in higher plants (Hatano et al. 1976, 1978).

In addition to data indicating the important shift in glucose catabolism there is ample evidence that in cold-acclimated cells the reducing power utilizing reactions are favoured. The content of ascorbic acid (Futrell et al. 1962; Andrews and Pomeroy 1978) and of the reduced form of glutathione (Esterbauer and Grill 1978; Guy and Carter 1982) were found to increase during the acclimation of different plants. Glutathione reductase appears to be an ubiquitous enzyme in wintering perennials (Sagisaka 1982). The activities of ascorbate free radical reductase, ascorbate peroxidase and dehydroascorbate reductase increased after growth cessation and remained high during winter (Nakagawa and Sagisaka 1984). The authors suggest that, in winter, ascorbate not only detoxifies the peroxides produced in the tissues, but also serves for regulating the level of $NADPH_2$. Some increase of the $NADPH_2$ level and an increased anabolic reduction charge were actually observed in winter rape leaves subjected to cold treatment (Kacperska, pers. commun.). Studies on [^{14}C]-incorporation into certain metabolites revealed that cold promoted those metabolic pathways which depend on the availability of the reduced pyridine nucleotides (Kacperska 1985).

In cold-acclimated winter rape leaves, a pronounced increase of ATP and of adenylate energy charge have been also reported (Sobczyk and Kacperska-Palacz 1978; Sobczyk et al. 1985). Therefore, a high availability of ATP and $NADPH_2$ seems to be a necessary requirement for the cold acclimation process (Kacperska-Palacz 1978); it allows for synthesis of RNA, protein phospholipid and other substances at low temperature.

5.4.2 Sugars and Related Compounds

There is ample evidence that the major changes in total osmotic potential which accompany the seasonal course of freezing tolerance, are due to changes in the concentrations of sugars and polyhydric alcohols. Measurements made in herbaceous and woody plants over the years clearly show that soluble carbohydrates increase from fall to winter when plants are subjected to low temperatures, and decrease in spring as they deharden (Fig. 5.15; for references, see Levitt 1956, 1980; Biebl 1962a; Sakai 1962a; Parker 1963; Jeremias 1964; Alden and Hermann 1971; Tumanov 1979; Siminovitch 1981). The sugar content increases also during artificial hardening (Fig. 5.16). In midsummer, exposure to 0 °C fails to induce a rise in frost resistance and an increase in

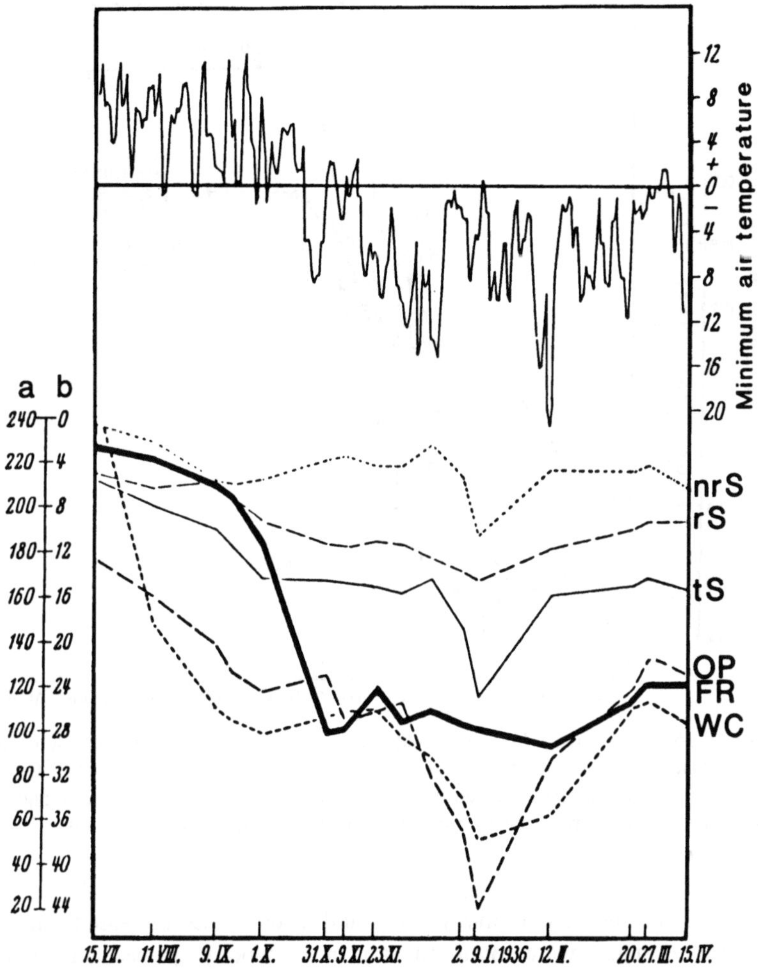

Fig. 5.15. Seasonal variation of frost resistance (*FR, heavy line*) and cell sap parameters in leaves of *Rhododendron ferrugineum* under field conditions (cf. air temperatures) at the alpine timberline. *WC* leaf water content (%, dry weight basis); *OP* osmotic potential (bar); *tS* total soluble carbohydrates (expressed as osmotic pressure equivalent, bar); *rS* reducing sugars (bar); *nrS* nonreducing sugars (bar). *Left scales: a* leaf water content (%); *b* frost resistance (LT_{10} at −°C), osmotic potential (bar) and sugar concentration (bar). (From Ulmer 1937)

sugar content in mulberry twigs. The sugar increase after hardening in the fall or in early spring is paralleled by hydrolysis of accumulated starch in woody plants (Sakai 1962a, 1966a; Reuther 1971). The rate of starch to sugar conversion at low temperature differs remarkably in different twig tissues in many woody plants, the starch in xylem parenchyma being more slowly converted in sugar than that in the cortex (Sakai 1966a). On the other hand, many wheat varieties and also winter rape fail to show a parallel relationship between sugar content and freezing tolerance. It is even possible to find an inverse relationship between hardiness and sugar content if a small number of varieties is compared (Green and Ratzlaff 1975). Parker (1959) found that raffinose and stachyose increased markedly in the bark and leaves of six conifer species during late

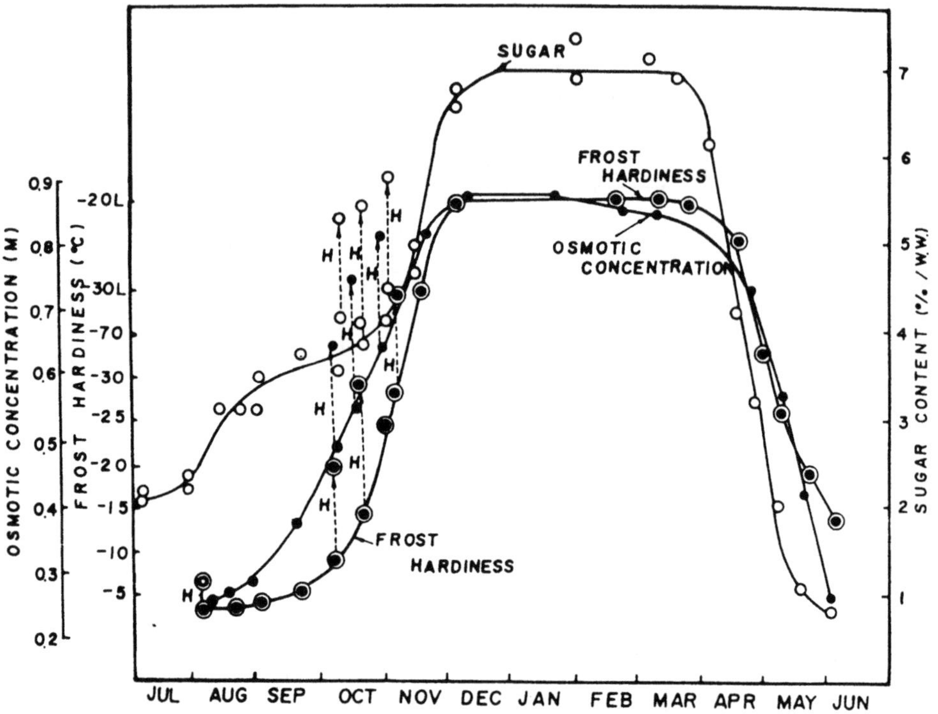

Fig. 5.16. Relationship between frost hardiness and sugar content of stem cortex from *Robinia pseudoacacia*. *Broken lines with arrows:* changes after hardening (*H*) at 0 °C for 2 weeks. Osmotic concentration is indicated as equivalent of molar solutions of NaCl; −20 L and −30 L represent prefreezing temperatures prior to immersion in LN_2. (From Yoshida and Sakai 1968)

fall; sucrose and sometimes glucose and fructose also increased. The proportion of the different sugars were found to vary in 18 woody species, but no specific sugar was consistently correlated with hardiness (Sakai 1962a). In certain grass leaves, the major accumulation at low temperatures is due to fructosans (D. Smith 1968a). The kind of carbohydrate accumulated in winter is, of course, dependent on the normal carbohydrate metabolism of the particular plant. In several species (*Gardenia jasminoides, Punica granatum, Sorbus aucuparia, Malus* sp.) polyhydric alcohols, such as sorbitol or mannitol, amount to about 40% of the total soluble carbohydrate content and may therefore play some role in hardiness (Sakai 1966d; Raese et al. 1977; Ichiki and Yamaya 1982). An accumulation of glycerol, which has an antifreeze effect in insects, was not found in woody plants (Sakai 1961).

The ability of soluble carbohydrates to increase frost hardiness was shown in sugar feeding experiments (Sakai 1962a). The importance of metabolism of the sugars in long-term absorption experiments was demonstrated by Tumanov and Trunova (1963). Tumanov and his co-workers assert that the maximum hardiness potential of winter cereals as well as of callus tissue from woody plants cannot be attained unless, in addition to exposure to the low temperature, they are fed with sucrose (Ogolevets 1976). In *Chlorella*, where the hardening effect in the dark is very limited, the addition of glucose causes a remarkable increase in frost resistance (Hatano 1978). This suggests that the chloroplasts serve as substrate donors in the light during hardening. Such results lead on the ask whether sugar plays a role in cold acclimation besides being a source of energy and substrates.

The study of sugar localization in the cells of hardy plants is of great importance for explaining their role at low temperatures and for understanding the mechanism of their protective effect (Heber 1959). Sugars were shown to accumulate in chloroplasts at low temperatures in cabbage, spinach (Krause et al. 1982) and wheat (Trunova and Zvereva 1974). Much sugar is accumulated in the vacuole which enhances the water-retaining capacity of the cells and decreases the degree of their dehydration by extracellular freezing. If sugar feeding leads to an increase in cell sap concentration, the sugars must accumulate in the vacuole. If, however, the sugars are metabolized, they must enter the protoplasm. Sugiyama and Simura (1967, 1968) attempted to decide this question by immersing shoots of the tea plant into 0.1 and 0.3 M [^{14}C]-labelled sucrose solution at 14°–18 °C for 2 days. This led to an increase in freezing tolerance, osmotic concentration and content of total sugars. Radioautographs indicated that [^{14}C] was found only in the thin cytoplasmic layer adjacent to the cell wall. Nevertheless, the osmotic concentration of the cell sap increased by 25 to 30% and the total sugar content by about 30%.

The many reports of a correlation between accumulation of soluble carbohydrates and freezing tolerance indicate that, at least in some plants, sugars must play an important role in the mechanism of tolerance in three ways: (1) The osmotic effect: accumulation of sugars can decrease the amount of ice formed, and therefore improve the avoidance of freeze-induced dehydration. (2) The metabolic effect: the metabolization of sugars in the protoplasm at low, hardening temperatures produces other protective substances or energy. (3) The cryoprotective effect: the protection of cells and biomembranes by sugars and sugar alcohols has been repeatedly demonstrated (Sakai 1962; Heber and Santarius 1973; Steponkus et al. 1977; Heber et al. 1979; Krause et al. 1982a; Santarius 1982). The molecular nature of noncolligative protection by sugar and sugar alcohols has not yet been elucidated, but data obtained suggest the possibility that specific interactions between cell structure and sugar could protect or alleviate the deleterious effects caused in biomembranes during freezing.

5.4.3 Amino Acids

Many investigators have measured the amino acid content of plants in an attempt to find a correlation with frost hardiness. In some cases, the amino acid content may increase with hardiness, in others, it may fail to show any relationship; where a relationship exists, it seems to reflect a general accumulation of organic nitrogen during fall. In the case of forage plants, arginine and alanine were predominant among the several amino acids that increased more in the hardy than in the nonhardy varieties (D. Smith 1968b). Among the free amino acids that accumulated during hardening of winter rape, proline showed by far the greatest augmentation, although asparagine also increased (Kacperska-Palacz and Wcislinska 1972). In poplar twigs arginine was the major amino acid during winter, while glutamine and glutamate became dominant at the time of budding (Sagisaka 1974). Free amino acids in 31 plant species at the stage of wintering and the beginning of growth were analyzed by Sagisaka and Araki (1983). Three types could be distinguished: a group which accumulated arginine alone (11 of 31 species); a group which accumulated proline alone (4 of 31) and a group which accumulated arginine and proline (15 of 31). A comparison of the amino acid composition in winter and at the beginning of budding indicated that the rise in the proline level was related to the onset of growth. Synthesis of arginine and proline and translocation of proline would seem to indicate that these substances play the role of filling-in materials in any particular tissue. At the time of active growth, an excessive lowering of the level of metabolic key intermediates would be avoided by storage of proline as a source of soluble nitrogen as well as a translocation material (Sagisaka and Araki 1983).

Proline acts as a cryoprotectant (Withers and King 1979). The photophosphorylation activity of washed thylakoid membranes was protected from freezing inactivation by adding proline, arginine, threonine, D-amino butyrate or lysine (dependent on the counter ion), at a concentration of about 0.15 M or even less (Heber et al. 1971). Although a large amplitude in the level of arginine and proline was observed by Sagisaka and Araki (1983), the toxic amino acids for survival during freezing, e.g. valine, leucine, isoleucine, phenylalanine (Heber et al. 1971), always remained at lower levels during winter. These results suggest that in wintering plants, arginine and proline play a role not only in the storage of nitrogen, but also, in combination with sugars, in the protection from freezing injury.

5.4.4 Nucleic Acids and Proteins

Increase in RNA accompanies the augmentation of proteins during the fall hardening in black locust trees (Siminovitch et al. 1968; Fig. 5.17) and in dogwood (Li and Weiser 1969). In 1-year-old apple twigs, RNA began to increase 1 week prior to the sudden enhancement of freezing tolerance (Li and Weiser 1969). The soluble RNA (sRNA) increased by 38% in 1 week, and light and heavy ribosomal RNA (rRNA) increased by

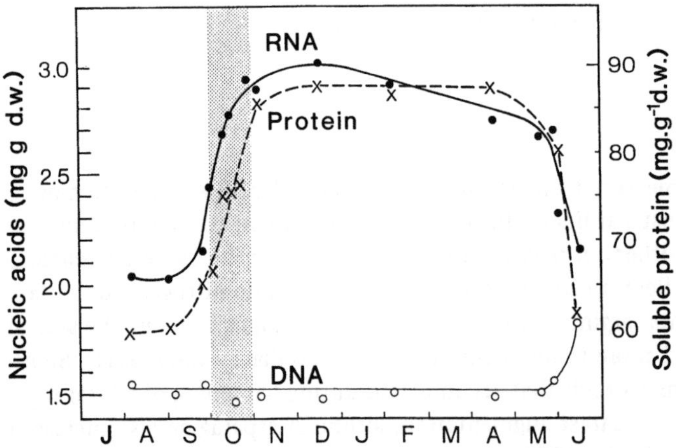

Fig. 5.17. Seasonal variation in DNA, RNA and water soluble protein in cortical tissues of *Robinia pseudoacacia*. *Shaded area:* hardening period. (From Siminovitch et al. 1968)

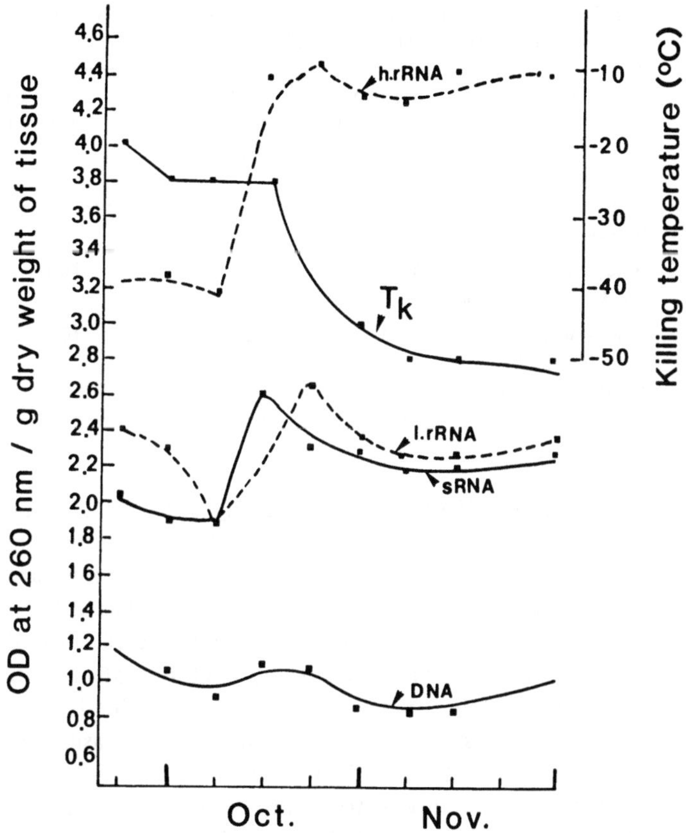

Fig. 5.18. Changes in nucleic acids of apple twigs during cold acclimation. *sRNA* soluble RNA; *l.rRNA* light ribosomal RNA; *h.rRNA* heavy ribosomal RNA; T_k killing temperature; *OD* optical density. (From Li and Weiser 1969)

42% in 2 weeks just prior to and during the stage of rapid hardening (Fig. 5.18). In leaves of ***Buxus microphylla*** var. ***koreana*** there was an increase in RNA, mainly rRNA, during cold acclimation which was closely paralleled by an accumulation of water-soluble and membrane-bound proteins (Gusta and Weiser 1972). DNA increased slightly during hardening; the most dramatic change was in the heavy rRNA (Fig. 5.18). Low temperatures nearly doubled the rRNA synthesis in winter wheat, but not in non-resistant spring wheat (Devay and Paldi 1977) and in potato species (Chen and Li 1980b); DNA failed to increase in quantity or in the percent hybridized to RNA. The lack of an augmentation of DNA is presumably due to the absence of meristems.

The quantitative variation in RNA content at low temperatures is probably due to a higher rate of RNA synthesis rather than to a lower rate of ribonuclease activity (Rochat and Therrien 1976a,b; Li and Palta 1978). The difference in rRNA metabolism between cold-acclimating and nonacclimating potato species, and experiments with specific RNA synthesis inhibitors during cold acclimation, indicate that the process of hardening is initiated at the level of translation. Probably, the 'hardiness message' presumably mRNA, exists in the plant system and functions only under cold-acclimating conditions

Table 5.5. Effect of temperature on the contents of sugars, soluble proteins and frost hardiness of winter wheat plants. (From Trunova 1982)

Treatment	Sugar per dry weight (%)	Soluble protein, mg g^{-1} dry weight	Frost hardiness, % of plants that survived			
			−10°C	−13°C	−16°C	−18°C
1. Before hardening	18.9	11.2	90	0	0	0
2. After 2 days at 18 °C in 10% sucrose	50.1	12.5	100	0	0	0
3. The same as No. 2 + 7 days at 2 °C in the dark in water	47.2	38.9	100	100	100	85

(Chen and Gusta 1983). In *Chlorella* (L_2 cell stage), it was observed that RNA and protein increased greatly in connection with a raise in freezing tolerance. Hatano et al. (1976) demonstrated that cycloheximide inhibits protein accumulation and the development of frost hardiness, while chloramphenicol has little or no effect. The inhibitors of RNA synthesis, actinomycin D and 5-fluorouracil, also inhibit the development of hardiness. Chloramphenicol specifically inhibits protein synthesis on chloroplastic ribosomes, while cycloheximide suppresses protein synthesis on the cytoplasmic ribosomes. These results suggest that protein synthesis on cytoplasmic ribosomes of *Chlorella* may be involved in the induction of cold acclimation.

Evidence for the importance of protein synthesis in the process of cold acclimation has been presented for wheat (Trunova 1982) and winter rape (Kacperska-Palacz et al. 1977a,b). In wheat plants kept in 10% sucrose solution at 18 °C for 2 days in the light, sugars accumulated to as much as 50% dry weight, but soluble protein increased only about 1% in the tillering nodes, resulting in no raise in hardiness (Table 5.5). When the plants were then placed in the dark at 2 °C for 7 days, soluble protein accumulated by almost 3.5 times with concomitant increase in hardiness. De novo protein synthesis, which was essential for hardening, was induced by low temperatures. During cold acclimation of potato plants, sugar accumulation occurred on the second and the third day prior to any measurable change in frost hardiness (Chen and Li 1982). Potato plants which do not develop freezing tolerance also accumulated sugar when exposed to similar environmental conditions. An augmentation of leaf-soluble protein was observed only in those species which were able to cold-acclimate (Fig. 5.19). The net increase of soluble protein was significantly and positively correlated with a higher frost hardiness in those species (Chen and Li 1982). The question arises as to how the augmentation of soluble protein confers protection against freezing stress, and whether or not functional proteins necessary to advance cold acclimation are induced.

The amount of soluble proteins may sometimes fail to parallel changes in hardiness (Levitt 1980). In spring the water-soluble proteins remained at a high level, while frost hardiness declined markedly. Siminovitch et al. (1968) also demonstrated that dehardening in spring was not accompanied by loss of soluble protein, but was intimately related to the decrease in membrane bound proteins (Fig. 5.20). Siminovitch et al.

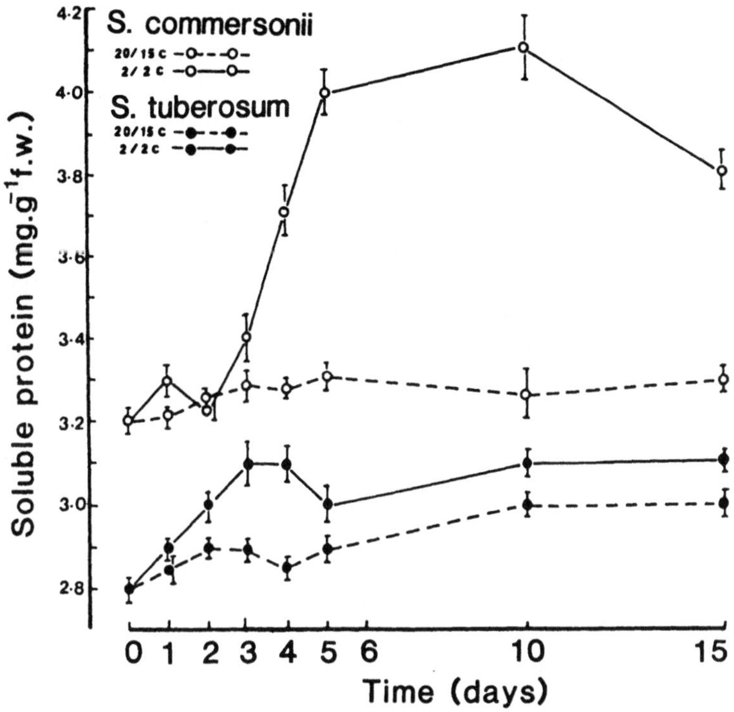

Fig. 5.19. Soluble protein content of leaves of tuber-bearing *Solanum* species at high and cold-acclimating temperatures. *S. commersonii* begins to acclimate after 3 days of cold treatment and becomes frost hardy to −12 °C by the 15th day. *S. tuberosum* is unable to be hardened. (From Chen and Li 1982)

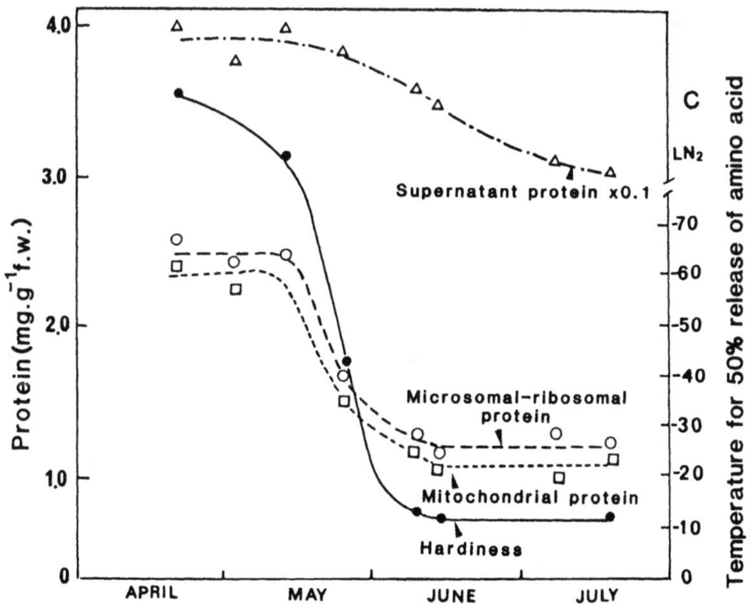

Fig. 5.20. Changes in mitochondrial protein, microsomal-ribosomal protein and supernatant protein of cortical tissues from *Robinia pseudoacacia* during spring dehardening. (From Siminovitch et al. 1968)

(1968) succeeded in hardening bark tissue of black locust trees to -45 °C, under conditions preventing an increase in soluble proteins. These results seem to indicate that membrane-bound proteins may account for the cold acclimation.

Qualitative changes in proteins (presumably enzymes) during cold acclimation were reported for many plants (Levitt 1980). There is evidence of alterations in isoenzyme compositions during cold acclimation (Table 5.6). The effect of cold-acclimating temperatures on the structure and function of soluble proteins has been studied in 'Puma' rye by Huner et al. (1982). Results with RuBPCase (ribulose bisphosphate carboxylase oxygenase) indicate a change in electrophoretic properties during cold acclimation (Huner and MacDowall 1976a,b). RuBPCase from acclimated sources is more stable at low temperature and has a higher affinity for CO_2 than enzyme from nonacclimated sources (Huner and MacDowall 1979a,b). The large subunits of RuBPCase from the

Table 5.6. Characteristics of enzymes during cold acclimation of various plants. (From Li 1984)

Plant	Enzymes	Observations	Reference
Cabbage	RuBPCase	Structural changes	Shomer-Ilan and Waisel (1975)
Spinach	Phosphatase	Changes in isozymes	Heber (1968)
	Glutathione reductase	Activity increased	Guy (1983)
Tobacco	RuBPCase	Conformational changes	Chollet and Anderson (1977)
Winter rape	Glucose-6-PO_4 dehydrogenase	More stable to freezing stress	Sobczyk et al. (1980)
	Glyceraldehyde-3-PO_4 dehydrogenase	More stable to freezing stress	Sobczyk et al. (1980)
	NADP-isocitrate dehydrogenase	More stable to freezing stress	Sobczyk et al. (1980)
	Pyruvate kinase	More stable to freezing stress	Sobczyk et al. (1980)
	Phosphoenolpyruvate carboxylase	Activity increased	Sosinska and Kacperska (1979)
	RuBPCase	Activity decreased	Sosinska and Kacperska (1979)
Winter rye	RuBPCase	Electrophoretic property changes	Huner and MacDowall (1976a,b)
	RuBPCase	Conformational changes in vivo	Huner and MacDowall (1976b)
	RuBPCase	Less S-S aggregaton	Huner and MacDowall (1979a)
	RuBPCase	Functional changes	Huner and MacDowall (1979b)
Winter wheat	ATPase	Adaptation changes	Jian et al. (1982)
	Invertase	Changes in isozymes	Roberts (1975)
	Peroxidase	Changes in isozymes	Roberts (1969a)
	RuBPCase	Electrophoretic property changes	Huner and MacDowall (1976a)
Chlorella	Glucose-6-PO_4 dehydrogenase	Activity increased at 0 °C	Sadakane and Hatano (1982)

acclimated sources are less susceptible to sulfhydryl-disulfide aggregation than those from nonacclimated materials. These data indicate that enzymes from acclimated plants retain their function at low temperatures and may be more stable to freezing stress.

A decrease in intramembrane protein particles on the freeze-fracturing face was observed with chloroplast thylakoid membranes by Garber and Steponkus (1976) and Steponkus et al. (1977). In addition, they observed a change of particle size distribution on this face, which might be based on the structural rearrangement of the hydrophobic matrix of the chloroplast thylakoids. Sugawara and Sakai (1978) reported that the number of inner membrane particles of the plasma membrane, especially on the fracture face E (face to the cell wall), was reduced markedly after hardening of calli of Jerusalem artichoke tubers and were restored to the initial level after dehardening (Fig. 5.21). These changes in particle population were intimately associated with changes in hardiness. This conclusion is supported by results of Pearce (1985) with cell membranes of wheat adapted to extracellular freezing. The alteration in the protein concentration on the plasma membrane may be explained by the significant increase of the phospholipid to protein ratio as hardiness proceeds (Yoshida 1984a).

Changes in quality and quantity of plasma membrane proteins of mulberry bark cells, particularly the glycoproteins, were observed during the early fall to winter season (Yoshida 1984b). Most of the polypeptide changes occurred near the period of growth cessation. Since the plasma membrane is involved in numerous physiological functions, these changes would explain why growth cessation is a prerequisite for the development of cold hardiness in woody species. Substantial changes in plasma membrane proteins during cold acclimation have also been observed in herbaceous plants (Uemura and Yoshida 1984; Ishikawa and Yoshida 1985).

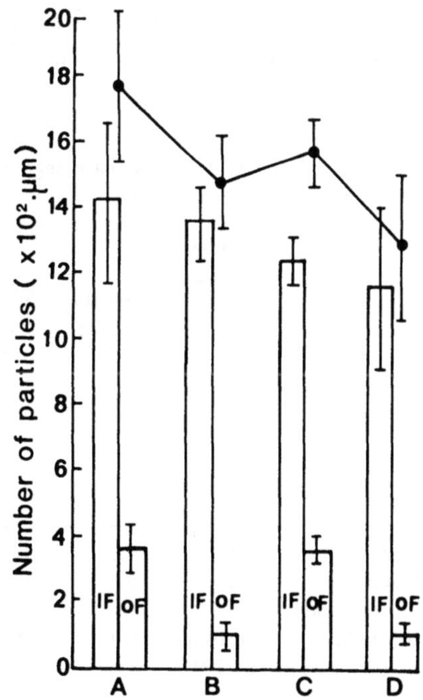

Fig. 5.21. Change in particle concentration on the fracture faces of plasma membranes of *Helianthus tuberosus* during hardening and dehardening. *A* untreated; *B* hardened at 0 °C for 14 days; *C* dehardened at 20 °C for 4 days; *D* rehardened at 0 °C for 14 days. *Vertical lines:* standard deviation. *IF* fracture face P; *OF* fracture face E; • total particle number. (From Sugawara and Sakai 1978)

5.4.5 Lipid Changes

Lipids are essential components of the cell membranes and are the only substances in the plant, that, like water, are present in the liquid state at normal temperatures. Temperature reduction initiates a physical change of state in the hydrophobic matrix of the membrane and produces a phase transition from liquid crystalline to the solid (gel) state which increases the susceptibility of the membrane to stress. Thus, it may be conceivable that cell survival is a function of the integrity or fluidity of membranes. Recent attempts to explain tolerance to stresses imposed by low and subfreezing winter temperatures have focussed on analyzing the structural components of membranes (Lyons et al. 1979).

Phospholipid Changes. Siminovitch et al. (1968) first demonstrated that phospholipid changes in starved tissues appeared to be related to hardiness. A double girdling was performed in a trunk of *Robinia pseudoacacia* in early August when carbohydrate had already accumulated in the stem tissues. Tissues depleted of organic nutrients showed none of the characteristic activities of normal trees in autumn, such as synthesis of RNA, soluble protein and cytoplasmic substances. Yet the tissues of the starved segment increased their hardiness considerably from summer to winter (Table 5.7). They did show a slight increase in [^{14}C]-leucine incorporation, but the largest increase was in phospholipids. In potted trees, Yoshida and Sakai (1973) observed that the development of hardiness was always accompanied by an increase in phospholipids (Fig. 5.22), especially phosphatidyl choline and phosphatyl ethanolamine (Yoshida 1974). The quantities of these substances were inversely related to the environmental temperatures and showed a remarkable augmentation at 0 °C (Fig. 5.23). In many herbaceous overwintering plants too, phospholipids increased with hardiness (de la Roche et al. 1972; Grenier and Willemot 1975; Sikorska and Kacperska-Palacz 1979, 1982). All results indicate that low temperature significantly affects lipid metabolism and hardiness.

Table 5.7. Changes in starved bark segments, encircled on *Robinia pseudoacacia* trees in August. (From Siminovitch et al. 1968)

Time	Killing temperature (°C)	Content of					
		Sugars (% dry wt)	Soluble proteins (mg g^{-1} dry wt)	RNA (mg g^{-1} dry wt)	Total lipid (mg g^{-1} dry wt)	Lipid phosphorus (μg g^{-1} dry wt)	[1-^{14}C]-Leucine incorporated (cpm/400 mg fresh wt and 4 h)
August	–16	7.2	37	2.18	29.3	113	5,650
November (encircled)	–45	2.1	39	2.21	28.3	187	6,150
November (control)	Not killed by LN_2	9.1	75	3.12	32.0	210	15,025

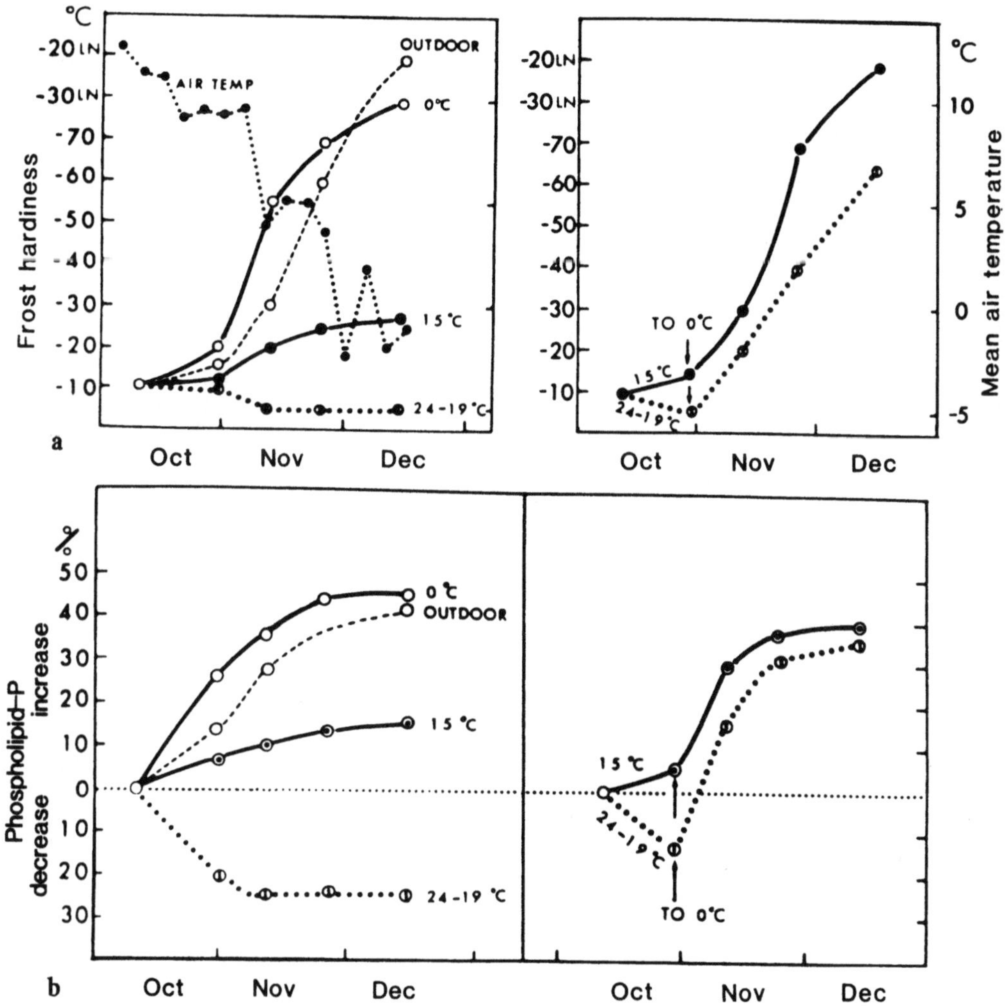

Fig. 5.22a,b. Changes in hardiness (**a**) and phospholipid content (**b**) of poplar bark at different temperature regimes: 24 °C day/19 °C night, 15°/15 °C, both under natural day length, and at 0°/0 °C under illumination of 10 klx for 8 h day^{-1}. *Arrows* indicate the time when potted plants were transferred from 24°/19 °C or 15 °C to the hardening chamber at 0 °C. On October 11th, 2 weeks before defoliation, all leaves were removed prior to the experiment to exclude temperature influences on the leaves. (From Yoshida and Sakai 1973)

Phosphatidyl choline is one of the major phospholipids in nonphotosynthesizing membranes in wheat seedlings. An attempt was made to modify the phospholipid composition by increasing the proportion of phosphatidyl choline and to observe whether this affects freezing resistance (Horváth and Farkas 1981). Application of 15 mM choline chloride at 25 °C to wheat seedlings was sufficient to develop a frost hardiness similar to that acquired by the untreated seedlings subjected to cold hardening for 48 h (Table 5.8). These data indicate that the polar-head group composition of membrane phospholipids in plants can be easily manipulated and point to the importance of phosphatidyl choline in cold acclimation.

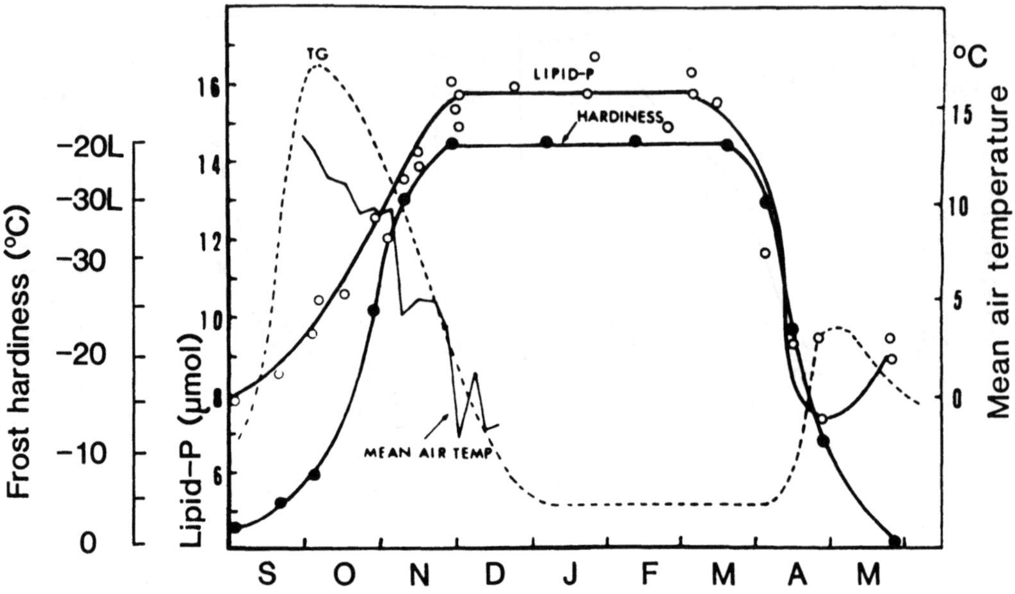

Fig. 5.23. Seasonal changes in hardiness and total amount of phospholipids in poplar cortex; *TG* triglycerides. (From Yoshida 1974)

Table 5.8. Frost resistance of winter wheat seedlings grown in the presence or absence of choline chloride. (From Horváth et al. 1981)

Choline chloride (mM)	Duration (days)	Temperature (°C)	LT_{50}[a] (°C)
0	8	25	− 2.2
15	8	25	−12.3
0	48	b	−12.5
15	8 + 8	25 + 2	−13.7
0	8 + 8	25 + 2	− 7.3
15 + 0	8 + 8	25 + 2	− 9.7

[a] Frost injury was monitored by following the electric conductance of the leaves. Leaves, immersed in liquid nitrogen served as controls for the totally damaged state. Each mean value is based on replicate experiments made on 25 leaves. The deviation from the mean was no more than ± 10%.
[b] Totally hardened seedlings.

Other lipids have also been found to increase during the hardening of some plants; for example digalactosyl diglyceride (DGDG) at the expense of monogalactosyl diglyceride (MGDG) in the case of poplar bark (Yoshida 1974) and in pine chloroplasts (Bervaes et al. 1972; de Yoe and Brown 1979). In chloroplast lamellae of *Pinus strobus*, DGDG and linolenic acid (18:3) increased between mid-November and December. Unsaturation was achieved by preferential incorporation of glycerolipids containing linolenic acid. From their results, de Yoe and Brown (1979) considered that hardy pine

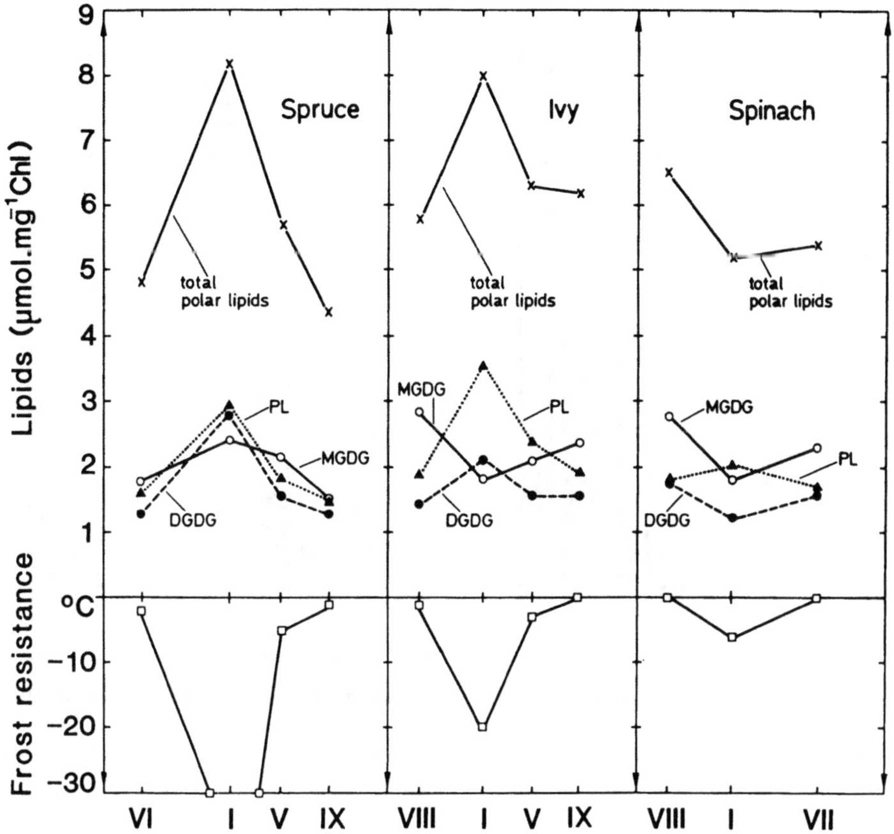

Fig. 5.24. Seasonal changes in the polar lipid content and in the contents of monogalactosyl diglyceride (*MGDG*), digalactosyl diglyceride (*DGDG*) and phospholipids (*PL*: phosphatidylcholine, phosphatidylinositol, phosphatidylethanolamine, phosphatidylglycerol) of 1-year-old spruce needles and of ivy and spinach leaves. (From Senser and Beck 1984, extended by M. Senser)

chloroplasts maintain lamellar viscosity by increasing lipid unsaturation and tolerate freeze dehydration by increasing the interfacial water-binding ability of the lamellae.

Comparative studies carried out by Senser and Beck (1984) revealed characteristic differences in the membrane lipid composition of the moderately hardy leaves of spinach, the more hardy leaves of *Hedera helix* and the very hardy needles of *Picea abies* upon frost hardening (Fig. 5.24). In comparison, in spruce, the membrane lipid content after hardening in winter was almost twice that of the freezing-sensitive summer state. Phospholipids and the two galactolipids MGDG and DGDG contributed to this winter maximum (Senser 1982). The membrane lipid content of the ivy leaves increased in a similar manner; however, the increase in DGDG occurred mostly at the expense of MGDG. Since, in the ivy leaves, the phospholipids accumulated concomitantly, the phospholipid to galactolipid ratio increased, whereas in spruce needles this ratio did not change throughout the year. The elevated phospholipid to galactolipid ratio in the moderately hardy leaves of spinach was due to a decrease in DGDG and especially in MGDG with only minor increase in the level of phospholipids. The isolated mem-

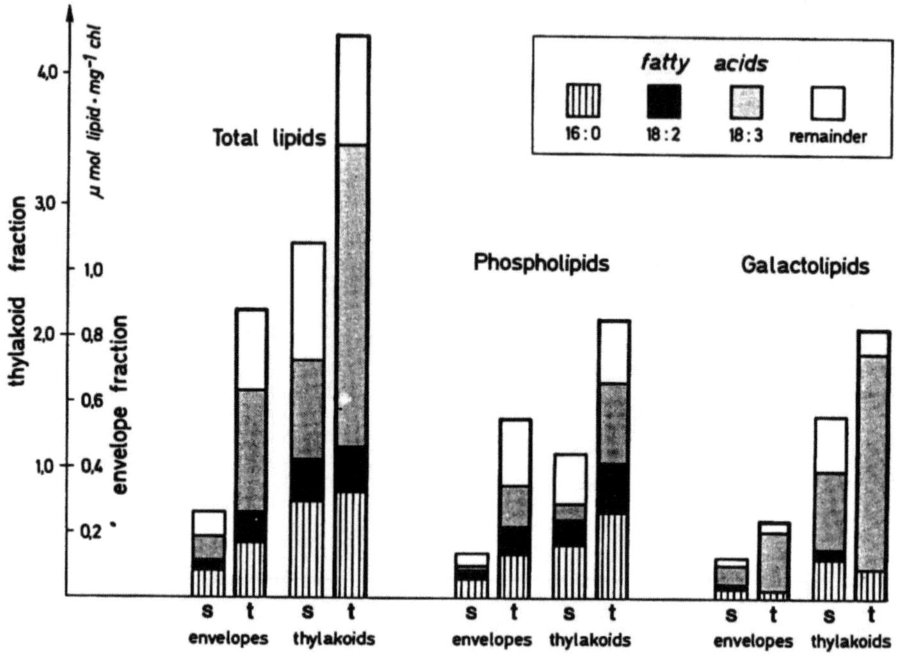

Fig. 5.25. Changes in the content of the main fatty acids of total polar lipids and of phospho- and galactolipids of the thylakoid and envelope fraction of chloroplasts isolated from (*s*) freezing-sensitive and (*t*) freezing-tolerant 1-year-old spruce needles. (Compiled by M. Senser based on data from Senser and Beck 1984)

branes of spruce chloroplasts showed alterations quite similar to those observed with intact needles (Senser and Beck 1982a; Fig. 5.25). The envelope fraction showed a remarkable shift in the direction of phospholipid accumulation together with a four-fold increase in the content of linolenic acid; the linolenic acid content of the thylakoid fraction was 2.5-fold that of the chloroplasts of frost-sensitive needles. It was assumed that the higher lipid to protein ratio in the envelope of frost hardened organelles renders a high elasticity to the membrane.

Investigations performed by Senser and Beck (1982b) on spruce trees which had either been cultivated in long-term experiments (at least 3 years) under different artificial combinations of photoperiod and temperature or had been submitted to out-of-season hardening and dehardening experiments revealed the presence of a strong correlation between the development of frost resistance, on the one hand, and of an augmentation of the membrane lipids, on the other. It was postulated that changes in the lipid metabolism during the course of these experiments had taken place in an at least two-step process, corresponding to successive phases of frost hardening. A first, short-day dependent step was characterized by an augmentation of membrane lipids, in particular of phospholipids. A second step, induced by subzero temperatures, featured a preferential incorporation of polyunsaturated fatty acids into the membrane lipids. The regulation of the progressive changes in the membrane are discussed by the authors on the basis of phytochrome balance.

Many investigators have measured the change in unsaturation of the fatty acids in the attempt to find a connection with frost hardiness. In some reports, an increase in unsaturation parallels hardiness (Gerloff et al. 1966; Farkas et al. 1975; de la Roche et al. 1975; Willemot 1977; Senser and Beck 1984: spruce), in others, any relationship fails (Yoshida 1974; Siminovitch et al. 1975; Smolenska and Kuiper 1977; Senser and Beck 1984: spinach). In many of the studies relating unsaturation to hardening, total lipids extracts or total membrane preparations were investigated rather than isolated, purified membranes. This has been primarily due to the lack of methods for isolating sufficient quantities of purified plasma membranes from tissues. Recently, Yoshida et al. (1983) and Uemura and Yoshida (1983) developed a method of isolating plasma membranes from plant tissues utilizing an aqueous polyethylene glycol-dextran two-polymer phase system containing NaCl.

Changes in the Plasma Membrane During Cold Acclimation. An enrichment of total phospholipid content, on a protein basis, with cold acclimation has been found in plasma membranes isolated from cortical cells of *Morus bombycis* (Yoshida 1984b; Fig. 5.26) and from herbaceous plants (Uemura and Yoshida 1984; Ishikawa and Yoshida 1985). The plasma membranes of mulberry bark were highly enriched in sterol compounds, particularly during the active growing season. Free sterols and glycoside sterols fluctuated similarly. Sterols, expressed on a protein basis, were relatively stabile until mid-

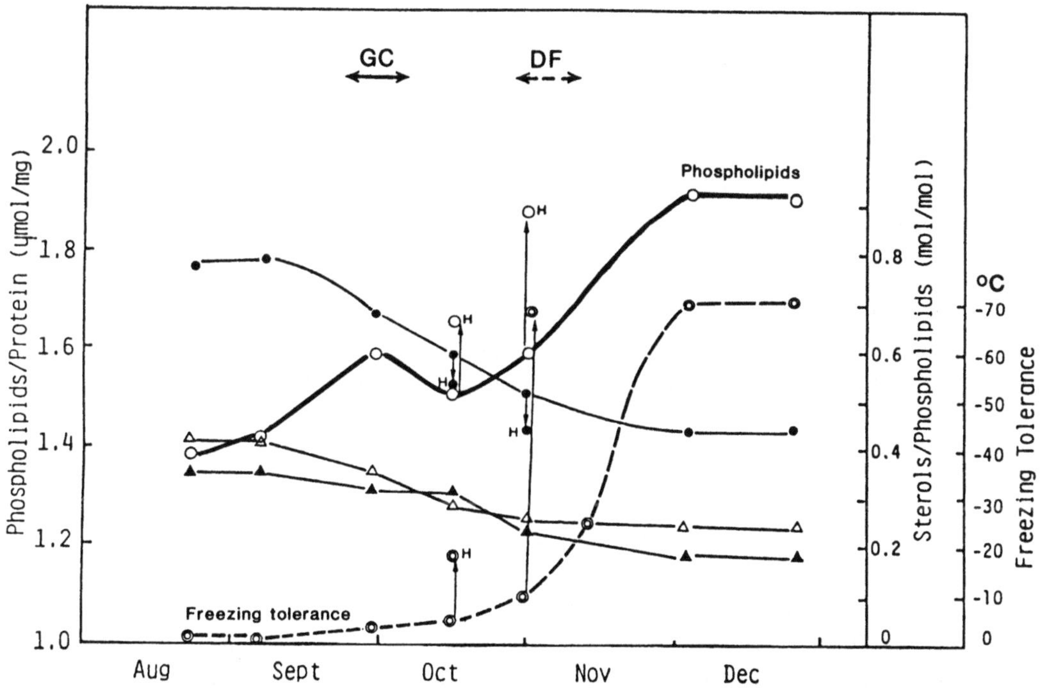

Fig. 5.26. Seasonal changes in phospholipids and sterols of plasma membranes from mulberry cortex. ○ Phospholipids/protein; ▲ molar ratio of free sterols to phospholipids; △ molar ratio of sterol glycosides to phospholipid; ● molar ratio of total sterols to phospholipids; ◎ freezing tolerance. *GC* growth cessation; *DF* defoliation period. *Lines with arrows:* changes after hardening (*H*) of excised twigs at 0 °C for 2 weeks. (From Yoshida 1984b)

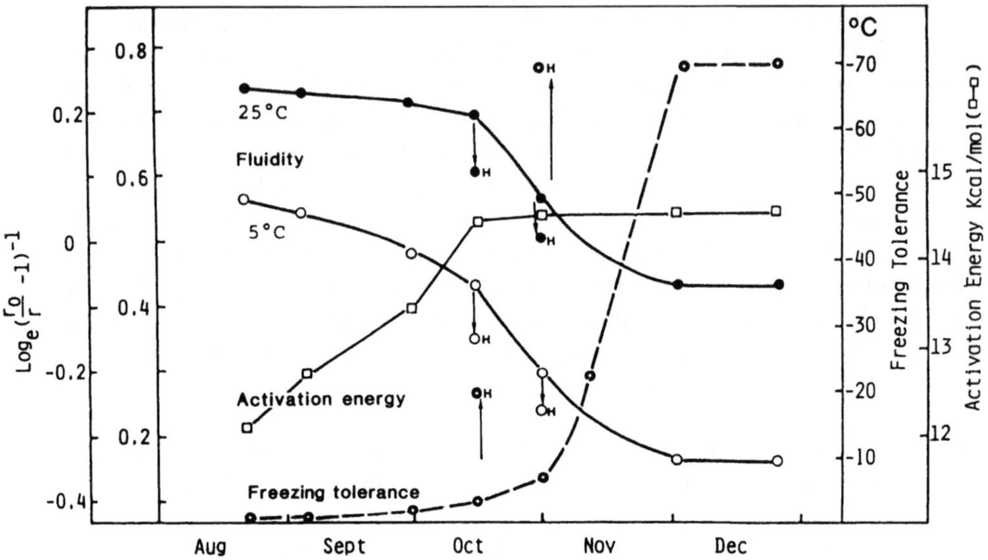

Fig. 5.27. Seasonal changes in relative fluidity, activation energy of anisotropy parameter values and freezing tolerance of plasma membranes of mulberry cortex; cf. also Fig. 5.26. (From Yoshida 1984b)

September, but had decreased approximately 20% by November and December. Due to a large augmentation of phospholipids from October to November, the sterol to phospholipid ratio diminished with time (see Fig. 5.26). Using fluorescent polarization techniques, Yoshida (1984b) demonstrated that the fluidity of the plasma membranes increased from mid-October to December in correspondence to the increase in freezing tolerance (Fig. 5.27). Vigh et al. (1979a), using electron spin resonance techniques, reported that the plasma membranes in protoplasts isolated from winter wheat leaves become more fluid with cold acclimation. The change in fluidity from summer to mid-October was small, but enough to indicate that qualitative alterations affecting membrane fluidity also occurred during this period. From August to the beginning of October, there was a marked increase in linoleate and a decrease in linolenate with little change in cold hardiness (Fig. 5.28). During this period, palmitate remained relatively constant. Since these are the three major fatty acids found in the plasma membranes of mulberry bark cells, the ratio of unsaturated to saturated fatty acids did not change during this period. After growth cessation at the end of September the augmentation of linoleate continued while the content of palmitate started to decrease gradually; the content of linolenate diminished from summer to the middle of October. The ratio of unsaturated to saturated fatty acids increased as cold hardiness increased (see Fig. 5.28). An analogous situation was found in the endomembranes.

Fatty acid composition of total lipids and of individual phospholipids from purified plasma membranes of needles of *Pinus sylvestris* was determined by Hellergren et al. (1984) who found that cold acclimation was not connected with a decreased saturation level. In the plasma membranes of some of the herbaceous plants tested by Yoshida et al. (1983), only minor changes in fatty acid unsaturation of phospholipids and sterol content were observed during cold acclimation, though phospholipids always augment-

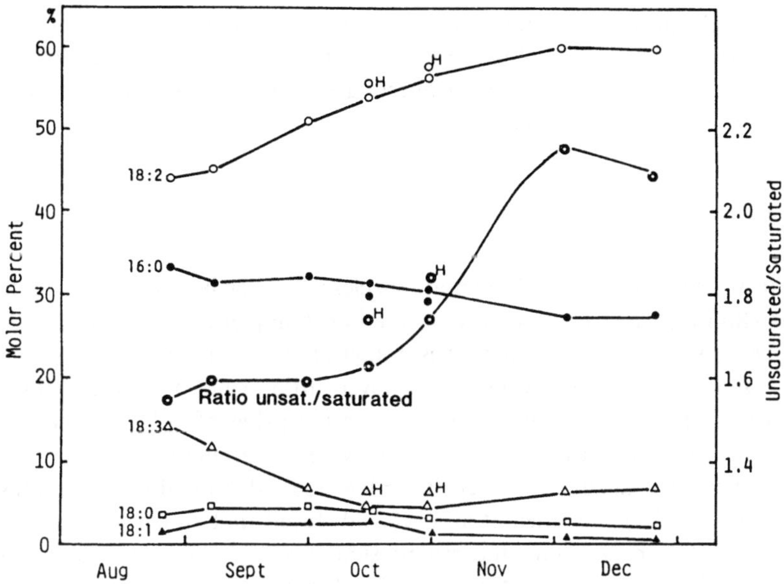

Fig. 5.28. Seasonal changes in fatty acid composition of plasma membranes of mulberry cortex. 16:0 palmitate; 18:0 stearate; 18:1 oleate; 18:2 linoleate; 18:3 linolenate; *H* values after hardening at 0 °C for 2 weeks. (From Yoshida 1984b)

Table 5.9. Changes in fatty acid composition of phospholipids of different membrane fractions from *Dactylis glomerata* after cold acclimatation. (From Yoshida and Uemura 1984)

Membrane fraction	Sampling date	Fatty acids							Unsaturated/ saturated
		16:0	16:1	18:0	18:1	18:2	18:3	20:0	
					Mol %				
Plasma	Oct. 2	26.1	Trace	0.7	7.5	40.6	23.1	1.2	2.54
membrane	Dec. 8	26.1	Trace	0.6	6.1	41.5	23.8	1.0	2.57
Endomem-	Oct. 2	23.9	0.4	0.8	9.6	34.4	30.2	0.7	2.92
branes	Dec. 8	21.0	0.7	0.4	6.1	38.6	32.4	0.7	3.49

Membrane systems were isolated and purified using an aqueous two polymer phase system. Endomembranes include ER, Golgi membranes and mitochondria.

ed. In the endomembranes, however, significant fatty acid unsaturation was observed during cold acclimation (Table 5.9).

The fatty acid composition of the cell membranes, especially the incorporation of branched-chain, more unsaturated or short-chain fatty acids seems to be an important determinant in the membrane fluidity and phase transition of lipids. Phospholipid and sterol contents, and the ratio of sterols to phospholipids, also play an important role in maintaining an optimal range of membrane fluidity ensuring structural and functional stability. It is known that sterol glycosides and β-hydroxy sterols affect the physical properties of membranes at low temperatures (Chapman 1975; Kleemann and MacConnel 1976). Thermotropic properties of the plasma membranes may also be compensated through quantitative and qualitative changes in the membrane proteins and in turn changes in the protein-lipid interactions (Yoshida and Uemura 1984). Thus, comprehensive studies on membrane property are necessary to elucidate changes in the plasma membrane related to hardiness.

5.4.6 Abscisic Acid

In *Solanum commersonii*, a potato species which can be frost-hardened, free abscisic acid (ABA) increased considerably on the fourth day of cold treatment at 2 °C (Fig. 5.29). The peak of ABA content was followed by accumulation of soluble protein (see Fig. 5.19) and more hardiness. However, little or no increase in ABA was observed in leaves of *S. tuberosum*, a plant that is incapable of cold acclimation. By supplying exogenous ABA, Chen and Li (1982) were able to induce frost hardiness in *S. tuberosum* which is unable to trigger the endogenous ABA build up. From these results, Chen and Li (1982) and Chen et al. (1983) deduced that an elevation of ABA during cold acclimation may induce the protein synthesis responsible for the development of frost hardiness. Lalk and Dörffling (1985) observed that hardening increases the level of ABA in wheat. In agreement with the findings of Chen and Li (1982), in their study the elevated level of ABA was preceded by a rise in osmolarity. On the other hand, the pressure potential remained constant or was slightly higher during the hardening period. Thus, they concluded that accumulation of ABA was due to the hardening process and not to simple water stress caused by low temperature inhibition of water uptake by the roots. Furthermore, they were able to promote hardiness in wheat by spraying the plants with 10^{-4} M ABA 24 h before the freezing test. Chen et al. (1983) treated potato plants with high concentrations of ABA at 25 °C; these plants attained hardiness to nearly the same level as untrated plants held at 2 °C. Similar results were ob-

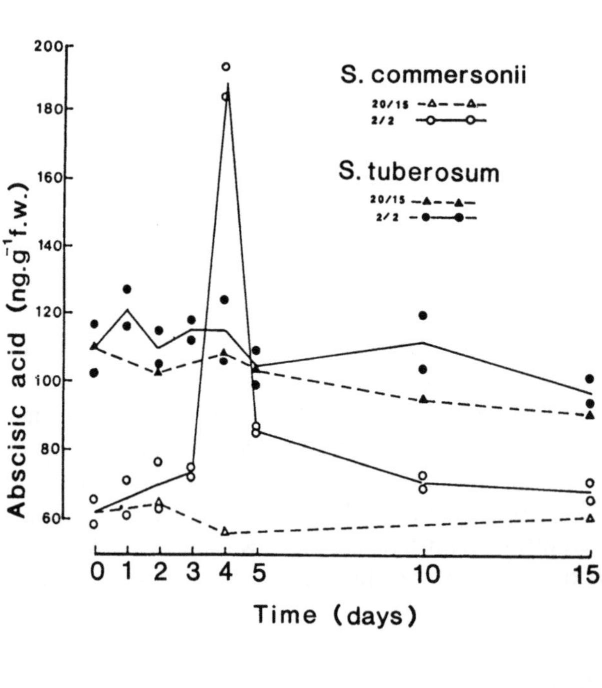

Fig. 5.29. Changes in concentration of endogenous abscisic acid in *Solanum* leaves during cold acclimation; cf. also Fig. 5.19. (From Chen and Li 1982)

tained with suspension cultured cells (Chen and Gusta 1983) which suggests that ABA can substitute for low temperature by triggering the genetic system responsible for inducing the hardening process. Orr et al. (1985) showed that 5×10^{-5} M ABA, added to alfalfa suspension cultures, enhanced hardiness by almost 10 K only in combination with cold exposure and in the absence of kinetin. In vitro translation products of polyadenylated RNA from 0 to 14 days of non-acclimating (20 °C), cold-acclimating (5 °C), and ABA-treated (at 20 °C) stem cultured plantlets of *Solanum commersonii* were analyzed by electrophoresis by Tseng and Li (1986). The levels of translatable mRNA for at least five translation products are increased by cold acclimation and ABA treatment. The comparison of the results with protein synthesis in vitro revealed that those proteins whose synthesis was stimulated in the presence of ABA or by cold treatment are good candidates for a role in the induction of cold hardiness. Johnson-Flanagan and Singh (1986) also reported that microspore-derived suspension cultures of winter rape *(Brassica napus)* developed freezing tolerance within 8 days in the presence of ABA. Pulse-chase labeling of the suspension cultures with ^{35}C-methionine during hardening and subsequent fluography indicated that a 20-kD and a 17-kD polypeptide were preferentially synthesized. In vitro translation of the isolated poly-A RNA in wheat germ or rabbit reticulocyte lysates showed that both the 17 and 20 kD bands were not products of a general stress reaction, but rather a genuine to the induction of freezing tolerance. It is also well known that exposure of chilling-sensitive plants to cool temperatures raises the endogenous level of ABA (Rikin et al. 1976; Daie and Campbell 1981; Eamus and Wilson 1983).

The molecular mechanism by which ABA affects freezing or chilling resistance is not clear. Recently, Farkas et al. (1985) showed that ABA inhibits the incorporation of mevalonate into free sterol and stimulates the incorporation of choline chloride into phosphatidylcholine. They hypothesized that ABA is acting via enzymes in lipid metabolism and brings about a modification of the physical state of biomembranes. Although the general evidence points to effects on the translation of proteins, the key problems are to elucidate the functions of these proteins and the molecular mechanisms involved in enhancing hardiness.

5.4.7 Cytological Changes During Cold Acclimation

Much attention has been focussed on biochemical changes in the membranes, but few studies have been reported on cytological manifestations (Srivastava and O'Brien 1966a,b; Siminovitch et al. 1968; Krasavtsev and Tutkevich 1971; Podbielkowska and Kacperska-Palacz 1971; Pomeroy and Siminovitch 1971; Otsuka 1972; Senser et al. 1975; Senser and Beck 1979, 1982a, 1984; Niki and Sakai 1981, 1983; O'Neill et al. 1981; Mittelstädt and Müller-Stoll 1984). In electron microscope studies, Pomeroy and Siminovitch (1971) were first to observe cytological changes in connection with the seasonal cycle of frost hardiness. Unhardy bark cells of *Robinia pseudoacacia* in summer show a spindle- or lens-shaped nucleus situated close to the cell wall. Most of the cell space is occupied by a large vacuole, and the cytoplasm appears to be confined to a thin peripheral layer lining the cell wall (Fig. 5.30). In contrast, in late October the cells show a relatively large, spherical nucleus, usually situated in the centre of the cell,

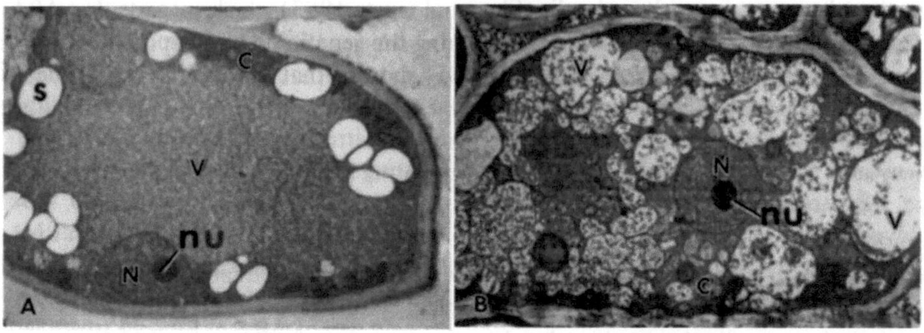

Fig. 5.30A,B. Stem parenchyma cells of *Robinia pseudoacacia* on (**A**) September 5th and (**B**) October 24th. *C* cytoplasm; *N* nucleus; *nu* nucleolus; *V* vacuole; *S* starch; x 16,000. (From Pomeroy and Siminovitch 1971)

and a maze of radiating strands of cytoplasm traversing the vacuole. As a result of augmentation of cytoplasm, a dense and extensive cytoplasm containing numerous small vacuoles is characteristic for winter hardy cells (Fig. 5.31a). Cell vacuoles contain a variety of lytic enzymes, such as protease, acid phosphatase, ribonuclease, carboxypeptidase, aminopeptidase, invertase, hydrolase, ATPase, etc. (Matile and Winkenbach 1971; Matile 1975). Autophagic activity of vacuoles was observed in many plant cells, which results in the digestion of cytoplasmic structures and reorganization of distinct cytoplasmic organelles (Matile 1968; Matile and Moor 1968; Otsuka 1972). In mulberry cortical cells, seasonal changes in the vacuole from winter to spring consist in an engulfment of the tonoplast, fusion and inflation of small vacuoles and coalescence into larger vacuoles (Fig. 5.31b). It appears that such intracellular digestion from winter

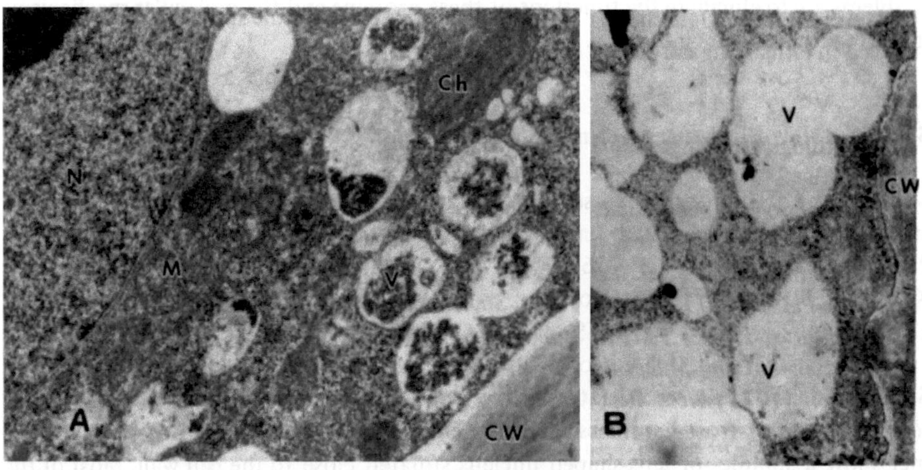

Fig. 5.31A,B. Ultrastructure of cortical cells of *Morus bombycis*. **A** Twig collected on January 29th; small vacuoles are dispersed throughout the dense cytoplasm; x 12,000. **B** May 20th; vacuoles have fused, coalescing into larger vacuoles; x 13,500. *CW* cell wall; *Ch* chloroplast; *M* mitochondrion; *N* nucleus; *V* vacuoles. (From Niki and Sakai 1981)

to spring plays an important role in the adaptation to changing environmental conditions. Vacuoles have other important functions in addition to those concerned with lytic processes, one of which concerns the storage of metabolic intermediates, such as organic acids, amino acids and sugars.

Pomeroy and Siminovitch (1971) observed a seasonal transition in the plasma membrane from a physical state of relative smoothness and regularly in summer to a highly folded state in winter. They considered that a highly folded membrane state would facilitate water flow and alleviate the stresses of contraction and expansion during freeze-thaw cycles. However, as shown in Fig. 5.31a, the plasma membrane of cortical cells of mulberry twigs in winter was relatively smooth, and highly folded states were not observed. Only after cold acclimation in October at 0 °C for 20 days (Niki and Sakai 1981) or -3 °C for 7 days (Niki and Sakai 1983), when hardiness increased from -15° to -70 °C, the plasma membrane was highly folded (Fig. 5.32) and microvesicles with a double lipid layer membrane appeared in the peripheral cytoplasm (Fig. 5.33). These microvesicles originate from the endoplasmic reticulum (ER) (Niki, unpublished data). A very similar ultrastructure was seen in the cold-acclimated cells collected at the first decade of November. On April 10, at a decreased hardiness of -15 °C, the plasma membrane was already smooth and regular. When these dehardened cells were rehardened at 0 °C for 10 to 15 days, the hardiness increased, the plasma membrane became folded and microvesicles reappeared near the periphery of the cytoplasm (Niki and Sakai 1981). From these results, it appears that a highly folded state of the plasma membrane and the formation of numerous microvesicles represents a transition as-

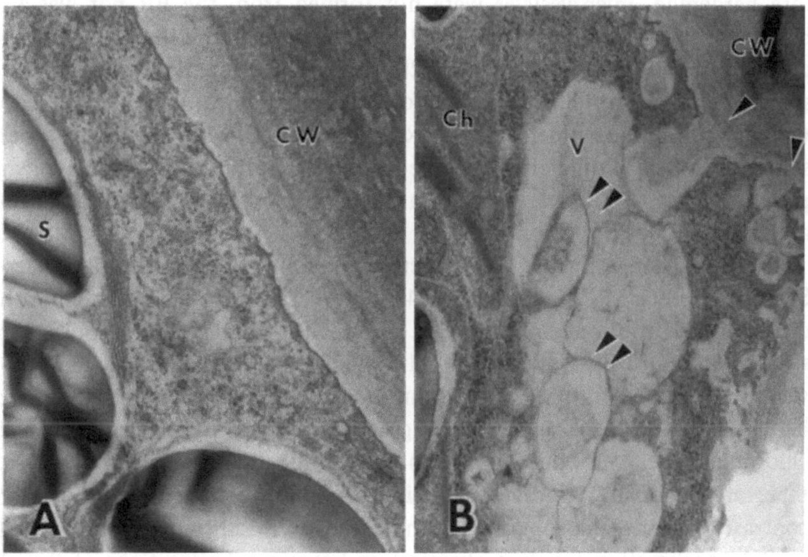

Fig. 5.32A,B. Ultrastructure of cortical cells of *Morus bombycis*. **A** Twig collected on October 16th; plasma membranes smooth and regular, starch grains fill the amyloplast. **B** Twig hardened at 0 °C for 20 days; the highly folded plasma membrane invaginates into the cytoplasm; *double arrows* indicate the engulfment of vesicles. *CW* cell wall; *Ch* chloroplast; *V* vacuole; *S* starch. x 16,000. (From Niki and Sakai 1981)

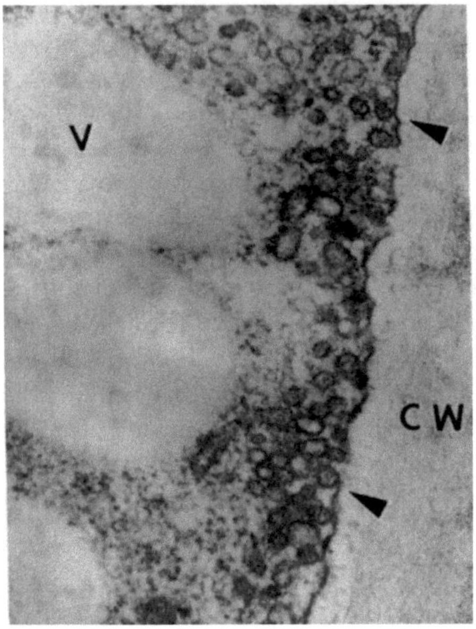

Fig. 5.33. Ultrastructure of a bark cell from a hardened mulberry twig collected on October 27th and hardened at 0 °C for 13 days. Numerous microvesicles with double lipid layer membranes appear in the peripheral cytoplasm. Some microvesicles have fused with the plasma membrane (*arrows*). *CW* cell wall; *V* vacuole. x 20,000. (From Niki and Sakai 1981)

sociated with higher freezing tolerance, rather representing a special membrane structure characteristic for extremely hardy cells in the winter state.

When mulberry cortical cells were treated with cycloheximide in late October before hardening, microvesicles were not formed from the ER nor was an increase in hardiness observed (Niki and Sakai 1983). However, suppression was not effective in cells which were already hardy to -20 °C (twigs collected on November 4). Evidently, some functional proteins necessary for full hardiness to develop must have been synthesized before November 4. The ER is considered to be the locus of membrane biosynthesis and is intimately concerned with the turnover of plasma membrane components. Thus, ER plays an important role in membrane transformation during cold acclimation.

Seasonal changes in the ultrastructure of spruce chloroplasts have been studied by Senser and Beck (1979, 1982a, 1984). Spring, summer and winter chloroplasts can be distinguished from one another (Fig. 5.34). Frost-hardened chloroplasts in winter are swollen and deformed, the thylakoid system is reduced and disorganized, and the envelope is enlarged forming protrusions and invaginations. From the onset of the frost-hardening process, the membrane augmentation leads to an increase in the number of chloroplasts per cell which is evidenced also by the observation of division stages of these organelles. In ivy leaves a division of chloroplasts could not be detected, but, in contrast to spruce chloroplasts, the thylakoids were substantially multiplied.

Cold acclimation involves chemical and structural alterations of the plasma membrane to resist freeze dehydration, mechanical stress, molecular packing and other events caused by extracellular freezing. Cytological changes associated with an abrupt increase in hardiness occur at 0° or -3 °C within 7 to 10 days. However, these cytological changes may be indirect. For a better understanding of the cytological changes during hardening comprehensive studies on a molecular basis are required.

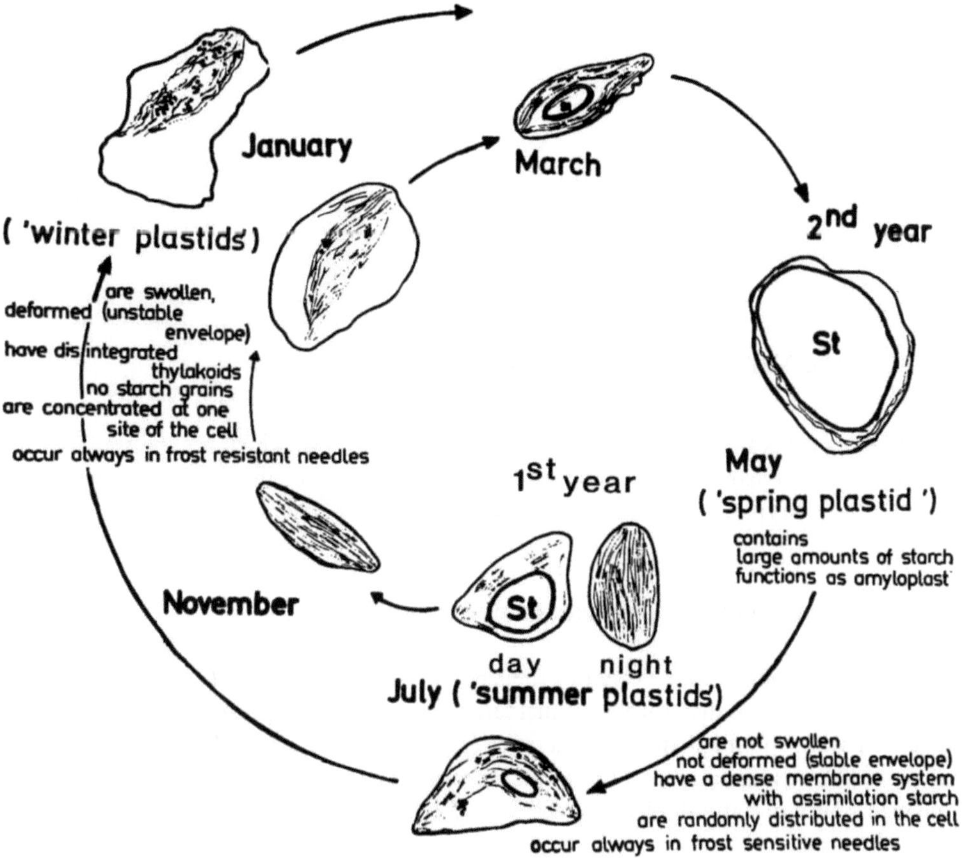

Fig. 5.34. Seasonal changes in the ultrastructure of spruce chloroplasts. *St* starch. (From Senser and Beck 1979, modified by M. Senser)

6. Frost Resistance in Plants

6.1 Genetic Variation in Frost Resistance

Frost resistance is a genetically determined physiological trait which is expressed under environmental constraints. The ability to harden is the basis for the differences between individuals, ecotypes, varieties and species with respect to potential frost resistance.

6.1.1 Differences in Frost Resistance Between Plant Taxa

If plants are arranged in the order of their potential frost resistance (i.e. their maximal attainable resistance) they form a gradual series from extremely freezing-sensitive species to plants with unlimited freezing tolerance (Larcher and Bauer 1981). This wide range of potential frost resistance is the result of evolutionary adaptation to cold stress via the mechanism of natural selection (see Sect. 7.6).

Microorganisms and Thallophytes. In an air-dry anabiotic state microorganisms and thallophytes are able to survive immersion in liquid nitrogen (see Sect. 4.1.3).

In the hydrated state, gram-negative *bacteria*, especially Enterobacteriaceae and Pseudomonadaceae, are killed by exposure to -15° to -30 °C for periods of between several days and several weeks, whereas gram-positive bacteria, among them also many soil bacteria, are tolerant to freezing at very low temperatures (Christophersen 1973; Mazur 1966). The high freezing tolerance of prokaryonic organisms (Cyanobacteria: Holm-Hansen 1963) seems to be unrelated to ecological factors.

Among the *fungi*, certain highly resistant species are capable of surviving temperatures of -40° to -192 °C in the vegetative state (Kärcher 1931; Hwang 1968). Wood-dwelling fungi, fungi that fructify either very early or very late in the year, and mycorrhizal fungi from high altitudes are especially resistant (Lindegren 1933; Moser 1958). Some species of ectomycorrhizal fungi, such as *Suillus punctipes* and *Pisolithus tinctorius* are more sensitive to cold and their mycelia are damaged at -10 °C. Others, for example *Xerocomus badius* and *Paccaria laccata*, survive but suffer appreciable disturbances in development (France et al. 1979). Gelatinous basidiomycetes, such as *Auricularia auricula-judae, A. mesenterica* and *Stereum hirsutum* survived freezing at -18 °C for 5 days: within a few days after thawing they returned to about the original level of spore discharge (Ingold 1982). The agarics *Flammula velutipes* and *Tubaria furfuracea* have sporophores that are active throughout winter (Ingold 1981).

Table 6.1. Potential cold resistance of algae, lichens and mosses in the hydrated state. (From Larcher and Bauer 1981, revised)

Plant and habitat	Low temperature injury below °C[a]	References
Algae		
Arctic seas		
Intertidal	−10 to −60	Biebl (1968, 1970)
Sublittoral	−2 to −4	Biebl (1968, 1970)
Temperate seas		
Intertidal	−8 to −40	Biebl (1939, 1958, 1972), Parker (1960), Terumoto (1964)
Sublittoral	−2 to +4	Biebl (1958, 1970)
Tropical seas		
Intertidal	−2 to +11	Biebl (1962b)
Sublittoral	+3 to +14	Biebl (1962b)
Arctic and Antarctic freshwaters		
Cyanophyceae	−70 to −196	Holm-Hansen (1963)
Phycophyta	−15 to −30	Holm-Hansen (1963), Biebl (1969)
Temperate lakes		Holm-Hansen (1963)
Cyanobacteria	(−25) −70 to −196	Terumoto (1964), Duthie (1964),
Phycophyta	−2 to −20	Biebl (1967a), Schölm (1968)
Hot springs	[+15 to +20]	Soeder and Stengel (1974), Brock (1978)
Epiphytic and epipetric algae		Kärcher (1931), Edlich (1936)
Soil algae	−70 to −196	Holm-Hansen (1963)
	−25 to −196	
Lichens		
Arctic, Antarctic	−80 to −196	Kappen and Lange (1972), Riedmüller-Schölm (1974)
Desert, high mountains	−78	Kappen and Lange (1970, 1972)
Temperate zone	−50 and below	Kappen and Lange (1972)
Mosses		
Arctic	−50 to −80	Riedmüller-Schölm (1974)
Temperate zone		
Marchantiales	−5 to −10	Clausen (1964), Dircksen (1964)
Jungermanniales	−10 to −15	Clausen (1964), Dircksen (1964)
Musci (hydrophytes)	−8 to −20	Irmscher (1912), Dircksen (1964)
Musci (bogs)	−7 to −15	Dircksen (1964)
Musci (forest floor)	(−10) −15 to −35 (−55)	Irmscher (1912), Dircksen (1964), Hudson and Brustkern (1965), Antropova (1974)
Musci (epiphytic, epipetric)	−15 to −30	Irmscher (1912)
Humid tropics	<0 to −7	Biebl (1964, 1967b)

[a] ( ) exceptional values, [ ] estimated from growth limitations.

Among the *algae*, every grade of cold resistance is encountered (Table 6.1) from chilling susceptibility in the thermophilic algae and in tropical seaweeds, to absolute freezing tolerance in soil and freshwater algae, sometimes in good agreement with the thermal characteristics of the environment (Biebl 1962b, 1970; Lüning 1985). If they are exposed to large seasonal temperature variations, algae adjust their frost resistance by direct response to cold (Parker 1960b; Terumoto 1959, 1965; Alexandrov et al. 1970).

Lichens, with very few exceptions *(Roccella fucoides, Umbilicaria vellea)*, are notable for an extremely high resistance to cold, regardless of their distribution, whereby the phycobiont is rather more sensitive than the mycobiont (Kappen and Lange 1970).

Mosses can, when in an active state and well provided with water, be either freezing-sensitive (e.g. certain tropical rain forest mosses: Biebl 1964) or freezing-tolerant in winter (Irmscher 1912; Hudson and Brustkern 1965). It is remarkable that of the tropical mosses and Hymenophyllaceae, so far investigated, not a single chilling-sensitive species has been found (Biebl 1964, 1967b).

Cormophytes. Especially susceptible to frost are the nonhardening species whose freezing point lies between -1° and $-3\ ^\circ C$, such as tropical and subtropical herbaceous plants and trees of the humid tropics (Biebl 1964; Rowley et al. 1975; Ludlow 1980; Earnshaw et al. 1987). In some families all species are freezing-sensitive, although to a different degree; examples are the palms (see Table 7.4) and those plant families whose distribution is confined to regions without frost below about $-10\ ^\circ C$. The transition from the freezing-sensitive to the freezing-tolerant category of frost resistance can be traced within families (Table 6.2: Poaceae; Table 6.3: Cactaceae), within tribes (Table 6.4: Sempervivoideae; Table 6.5: Aurantieae) and even within genera (Table 6.6: *Erica*; Table 6.7: *Solanum*). Among the cormophytes that become freezing-tolerant in the hardened state, various levels of resistance are achieved among closely related taxa, e.g. within the genera *Magnolia, Deutzia, Viburnum* (Bialobok 1974), *Salix, Populus* (Sakai 1978d), *Camellia* (Sakai and Hakoda 1979), *Rhododendron* (Bialobok 1974; Sakai and Malla 1981; Sakai 1982c; Iwaya-Inoue and Kaku 1983; Sakai et al. 1986); for further examples, see Tables 6.8, 7.8, 7.12 and Fig. 7.25.

Table 6.2. Potential frost survival (LT_{50} at °C) and resistance mechanisms of Poaceae. (Data from Cloutier and Andrews 1984; Dvorak and Fowler 1978; Fowler et al. 1977; Fuller and Eagles 1978; Gusta et al. 1980; Ivory and Whiteman 1978; Johnston and Dickens 1976; A. Larsen 1979; Miller 1976; Noshiro and Sakai 1979; Rajashekar et al. 1983; Rogers et al. 1977; Rowley et al. 1975; Rowley 1976; M. Holzner, unpubl.; M. Schwienbacher-Mascotti, unpubl.)

C_4-Grasses			
Cenchrus ciliaris	− 2	fs	
Setaria anceps	− 2 ... − 4	fs	a
Chloris gayana	− 3	fs	
Pennisetum clandestinum	− 3	fs	
Saccharum officinarum	− 3	(cs), fs	b
Saccharum spontaneum	− 4	fs	
Eremochloa ophiuroides	− 6		
Eragrostis curvula	− 7	(fs)	
Cynodon dactylon	− 3 ... − 7	(fs)	a, b
Paspalum dilatatum	− 8 ... −10	(fs)	b
Bothriochloa ischaemum			

Table 6.2 (continued)

C_3-Grasses			
Dactylis glomerata	−10		b
Festuca arundinacea	− 7 ... −13		
Lolium perenne	−12 ... −19	ft	a, b
Phragmites communis	−13 ... −15	ft	
Phleum pratense	−15 ... −17	ft	
Festuca rubra	−17 ... −27	ft	a, b
Poa pratensis	−17 ... −30	ft	a, b
Elymus mollis	−20	ft	
Stipa pennata	−26	ft	
Puccinellia distícha	−27	ft	
Stipa capillata	−30	ft	
Bromus inermis	−30	ft	
Agrostis palustris	−35	ft	
Cereals			
Zea mays (C_4)	− 2 ... − 4	(cs), fs	a, b
Avena sativa	− 9 ... −13	ft	b
Hordeum vulgare	−16 ... −20	ft	a, b
Triticum boeticum	− 5 ... − 9	(fs) ft ?	a, b
Triticum urartu	− 5 ... −11	ft ?	a, b
Triticum turgidum	− 6 ... −13	ft ?	a, b
Triticum aestivum	−14 ... −22	ft	a, b
Triticale	−16 ... −22	ft	
Secale cereale	−19 ... −30	ft	b

Abbreviations: a = depending on provenance; b = depending on variety; (cs) = chilling-sensitive stages or varieties; fs = freezing-sensitive; (fs) = freezing-sensitive with ability to be cold-acclimated; ft = freezing-tolerant.

Table 6.3. Frost resistance of Cactaceae (°C). LT_{50} = 50% of chlorenchyma killed. [Nobel (1982) and unpubl. data of M. Ishikawa and L. Gusta (*Opuntia fragilis*)]

Species	Location	Altitude	LT_{50}
Opuntia fragilis[a]	Saskatoon, Canada	600 m	−50[b]
Coryphanta vivipara	Wyoming	2240 m	−20
Coryphanta vivipara var. rosea	Nevada	1970 m	−22
Pediocactus simpsonii	Wyoming	2250 m	−18
Opuntia polyacantha	Wyoming	2230 m	−17
Denmoza rhodacantha	N. Argentina	2260 m	−10
Eriosyce ceratistes	Central Chile	1940 m	−10
Stenocereus thurberi	N. Mexico	350 m	− 9
Ferocactus acanthodes	S. California	840 m	− 8
Ferocactus wislizenii	S. Arizona	850 m	− 8
Trichocereus chilensis	Central Chile	1760 m	− 8
Trichocereus candicans	N. Argentina	1860 m	− 7
Opuntia bigelovii	S. California	850 m	− 7
Lophocereus schottii	S. Arizona	410 m	− 7
Ferocactus covillei	S. Arizona	1160 m	− 7
Ferocactus viridescens	S. California	40 m	− 6
Opuntia ramosissima	S. California	220 m	− 4

[a] On sunny dunes at 52° 10′, latitude at the northern limit; water content about 60% f.w.
[b] At −50 °C without any damage, at −70 °C slight injury.

Table 6.4. Killing temperatures (LT_{10} at °C) and threshold freezing temperatures (T_f) of Semperivivoideae from the Canary Islands. (From Lösch and Kappen 1981).[a]

Species	LT_{10}	T_f
Aeonium ciliatum	– 4	–4.2
Aeonium glutinosum × *glandulosum*	– 4.5	–3.5
Aeonium cuneatum	– 5	–4.6
Aeonium nobile	– 5	–3.6
Aeonium palmense	– 5	–3.5
Aeonium holochrysum	– 5	–4
Greenovia diplocycla	– 5	–3.2
Aichryson punctatum	– 5.5	–5
Aeonium haworthii	– 6	–6.1
Monanthes laxiflora	– 6	–2
Monanthes polyphylla	– 6	–4.2
Aeonium canariense	– 6	–3.4
Aeonium glutinosum	– 6	–6.6
Aeonium tabulaeforme	– 6.5	–2.3
Aeonium urbicum	– 7	–3
Aeonium glandulosum	– 7	–5.5
Aichryson laxum	– 7	–6.7
Greenovia aurea	– 7	–4.2
Aichryson pachycaulon	– 7	–7.3
Monanthes anagensis	– 7.5	–6.1
Monanthes muralis	– 8	–2.4
Aeonium sedifolium	– 8	–8
Aeonium goochiae	– 8	–8.3
Aichryson bollei	– 8	–5.3
Aeonium lindleyi	– 9	–4.1
Aichryson palmense	– 9	–8
Aeonium spathulatum	–10	–9.3

[a] Species with LT_{10} well lower than T_f are freezing-tolerant. As a rule, species of the Sections Aichryson, Holochrysa (primitive) and Urbica of the genus *Aeonium* are freezing-avoiders, of the Section Canariensia and of the genera *Greenovia* and *Monanthes* are tolerators.

6.1.2 Differences in Frost Resistance Within Populations

The gene pool of every population exhibits a range of variation enabling it to survive a certain degree of change in the environment. The variability of frost resistance within a population is essential for the survival of a species following frost of unusual severity or untimely occurrence, as well as for adaptation to long-term fluctuations in climate. The degree of scatter of resistance within the progeny is characteristic for a species and provides a measure of its scope for adaptation and selection. Intraspecific difference in freezing resistance among ecotypes and climatic races of widely ranging species have been reported, e.g. by Minckler (1951), Parker (1963), Schönbach and Bellmann (1967), Smithberg and Weiser (1968), Green (1969), Flint (1972), Maronek and Flint (1974), Rehfeldt (1979, 1980), Harwood (1980) and Clausen (1982). Agricultural and horti-

Table 6.5. Frost resistance (°C) of Aurantieae in the hardened state; LT_i initial damages; LT_{50} 50% injury. (After data of Yoshimura 1967; Larcher 1971; Konakahara 1975; Yelenosky 1977; Yelenosky et al. 1978; from Larcher 1981b)

Species	Leaves		Buds	Twigs (LT_{50})	
	LT_i	LT_{50}	LT_{50}	Cambium	Xylem
Citrus medica	− 4	− 5	− 7 ... −8	− 8 ... − 9	− 9 ... −10
C. limon	5	− 6 ... −7	− 9	−10 ... −11	−11 ... −12
C. paradisi	− 4 ... −5	− 7	− 9	−10 ... −11	−12
C. sinensis	− 4 ... −5	− 7	−10	−10 ... −11	−12
C. reticulata	− 6	− 8	−10	−13	−14
C. natsudaidai	− 6 ... −7				
C. unshiu	− 8 ... −9	−10			
C. aurantium	− 6	− 8 ... −9	−11	−14	−13
Eremocitrus glauca				− 9	
Fortunella margarita	− 7	− 9 ... −10	−13	−17	−16
Poncirus trifoliata	(−15)[a]		−23	−32	−25

[a] Deciduous.

Table 6.6. Frost resistance (LT_i at °C) of *Erica* species after prehardening at −1° to −3 °C for 15 days. (From Sakai and Miwa 1979)

Species	Leaves	Shoot bud	Cortex	Xylem
Species native to South Africa				
Erica baccans	−5	−	− 5	− 5
E. bauera	−	−8	− 8	− 8
E. bicolor	−7	−	− 7	− 7
E. caffra	−8	−	− 8	− 8
E. cerinthoides	−8	−	−10	−10
E. chamissonis	−8	−	− 8	− 8
E. chloroloma	−5	−	− 5	− 5
E. diaphana	−5 ... −7	−	− 5 ... −7	− 5 ... −7
E. formosa	−8	−	− 8	− 8
E. glandulosa	−8	−	− 8	− 8
E. gracilis	−5	−	− 8	− 8
E. x hiemalis	−5	−5	− 5	− 5
E. mauritanica	−8	−5	− 8	− 8
E. mammosa	−8	−	− 8	− 8
E. canaliculata	−8	−8	− 8	−10
E. patersonia	−6	−5	− 5	− 5
E. peziza	−8	−	− 8	− 8
E. quadrangularis	−5	−	− 5	− 5
E. sessiliflora	−5	−	− 5	− 5
E. speciosa	−8	−	− 8	−10
E. sparsa	−8	−	− 8	− 8
E. taxifolia	−5	−	− 5	− 5
E. versicolor	−8	−5	− 8	− 8
E. verticillata	−8	−	− 8	− 8
E. viridescens	−6	−	− 8	− 8
E. x willmorei	−5	−4	− 5	− 5

Table 6.6 (continued)

Species	Leaves	Shoot bud	Cortex	Xylem
Species native to Europe				
E. arborea	−15	–	−15	−20
E. cinerea	−15	–	−15 ... −17	−23 ... −25
E. herbacea[a]	−18 ... −23	<−20	−20 ... −25	−20 ... −40
E. lusitanica	−13	–	−15	−20
E. tetralix	−15 ... −18	–	−15	−20
E. vagans	−15	−15	−20	−25
E. watsonii (ciliaris x tetralix)	−18	–	−15 ... −18	−25
E. williamsii (tetralix x vagans)	−15	–	−15	−20

[a] Includes *E. carnea* and *E. mediterranea*.

Table 6.7. Classification of tuber-bearing *Solanum* species in terms of cold resistance and hardening ability. (From Chen and Li 1980a)

Categories	Species of *Solanum*	Killing temperature (°C)	
		Before acclimation[a]	After acclimation[b]
Frost resistant and able to cold harden	*acaule* (Oka 3885)[c]	−6.0	− 9.0
	commersonii (Oka 5040)	−4.5	−11.5
	multidissectum (PI 210042)	−4.0	− 8.5
	chomatophilum (PI 266387)	−5.0	− 8.5
Frost resistant, but unable to cold harden	*bolivense* (PI 265860)	−4.5	− 4.5
	megistacrolobum (Oka 3914)	−5.0	− 5.0
	sanctae-rosae (Oka 5697)	−5.5	− 5.5
Frost sensitive, but able to cold harden	*oplocense* (Oka 4500)	−3.0	− 8.0
	polytrichon (PI 184773)	−3.0	− 6.0
Frost sensitive and unable to cold harden	*brachistotrichum* (PI 320265)	−3.5	− 3.5
	cardiophyllum (PI 186548)	−3.0	− 3.0
	fendleri (PI 275163)	−3.0	− 3.0
	jamesii (PI 275163)	−3.0	− 3.0
	kurtzianum (Oka 4940)	−3.5	− 3.5
	microdontum (Oka 5623)	−3.0	− 3.0
	pinnatisectum (PI 275232)	−2.5	− 2.5
	stenotomum (PI 195188)	−3.5	− 3.5
	stoloniferum (PI 161770)	−3.0	− 3.0
	sucrense (EBS 1791)	−3.0	− 3.0
	tuberosum (cv. Red Pontiac)	−3.0	− 3.0
	venturii (Oka 498)	−3.5	− 3.5
	vernei (Oka 4476)	−3.5	− 3.5
	verrucosum (PI 195170)	−3.0	− 3.0
Chilling sensitive	*trifidum* (PI 255541)	−3.5	dead[d]

[a] Plants were grown in a regime of 20/15 °C day/night, 14 h light.
[b] Plants were grown in a regime of 2 °C, 14 h light for 20 days.
[c] Identification number at the Potato Introduction Station, Sturgeon Bay, Wisconsin.
[d] Plants were dead after 20 days of 2 °C treatment.

Table 6.8. Frost resistance (LT_i at °C) of roses in the hardened state. (From Rajashekar and Burke 1978, Sakai 1982c)

Species	Provenance	Leaves	Buds	Twig cortex	Xylem
Rosa gigantea	S. China		−10 ... −13	−10 ... −13	−10 ... −13
R. odorata	China		−15	−15	−15
R. laevigata	S.W. Japan, S. China	−10	−17	−15	−15
R. chinensis	China		−17	−17	−17
R. moschata	Himalaya, S. China		−17 ... −20	−17 ... −20	−17 ... −20
R. foetida	W. Asia		−17	−17	−17
R. wichuraiana	Japan	−13	−25	−23	−23
R. multiflora	Japan		−25	−23	−23
R. damascena	Asia Minor		−30	−25	−25
R. gallica	Europe		−30	−30	−30
R. canina	Europe		−60	−60	−30 ... −40
R. setigera	N. America		−60	−60	−30
R. villosa	Europe, W. Asia		−60	−60	−30
R. glauca	Europe		−60	−60	−30
R. rugosa	N. Japan		−70	−70	−35
R. acicularis	N.E. Asia, Alaska		−70	−70	−35 ... −40

cultural plants have been particularly well investigated in this respect (for references, see Larcher 1985a).

Altitudinal Variation in Frost Resistance. In mountain areas, the drop in temperature associated with rising altitude brings with it an increase in frequency and degree of frost to which the plants are exposed. The altitude-dependent differentiation of ecotypes and varieties within a single species is an expression of the increasing adaptive significance of the greater frost risk as well as the shorter growing period in higher elevations. In the mountains of the tropical, subtropical and warm temperate regions frost only occurs above a certain altitude. The severity of the frost rises with altitude so that, of the freezing-sensitive plants of lower and intermediate altitudes those populations survive that are best able to harden. Ecotype differentiation has been demonstrated in C_4 grasses (Table 6.9). Among genotypes of *Solanum acaule* the frost resistance was closely related to the altitudinal origin (Fig. 6.1). Intraspecific variation in the freezing resistance of *Eucalyptus pauciflora* has been reported by Pryor (1956) and

Table 6.9. Altitudinal gradient of frost resistance (°C) of ecotypes of the C_4-grass *Setaria anceps* in Kenya. (From Ivory and Whiteman 1978)

Provenance	LT_{50}
2100 m a.s.l.	−2.1
2200 m	−2.6
2400 m	−3.0
3100 m	−4.3

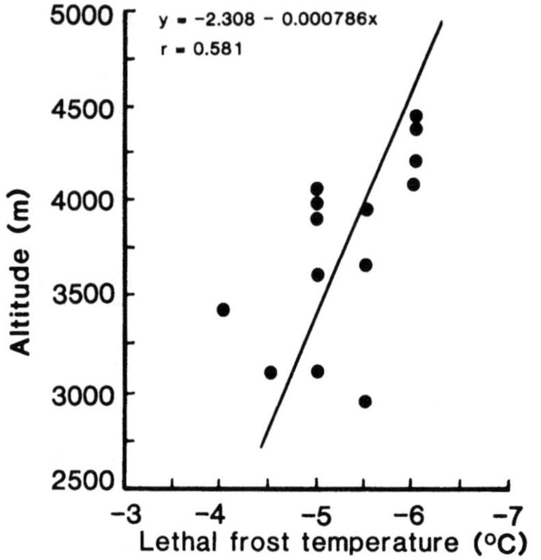

Fig. 6.1. Interrelationship between frost resistance and altitude of the genotype origin of *Solanum acaule*. (From Li et al. 1980)

Green (1969), who found a trend of increasing freezing resistance with increasing altitude of its seed source in Australia. Similar results were obtained by Sakai and Wardle (1978) with *Nothofagus solandri*, and by Alberdi et al. (1985) with *Nothofagus antarctica* and *Embothrium coccineum* from S. Chile. Depending on topography, an inconsistency can be found in this trend, e.g. if frost is less severe on slopes than on valley floors (Fig. 6.2). Such observations imply that under the influence of freezing events, natural selection may have been operative in bringing about increased hardiness in the population.

In the mountains of intermediate and higher latitudes the absolute temperature minima do not sink much below those in the valleys, but the periods of severe cold are longer. Thus, especially in trees, adaptation to altitudinal freezing stress is achieved not only by the development of a high winter frost hardiness, but primarily by the development of rhythmo-ecotypes which characteristically remain in a state of dormancy for a long time and which are able to promptly improve their resistance in autumn. This is shown by frost resistance tests carried out by Scheumann and Schönbach (1968) on *Larix leptolepis* and by J. Bo Larsen (1978a,b) on *Abies grandis* and *Pseudotsuga menziesii*.

In order to assess the altitudinal variation in freezing resistance among populations of *Abies sachalinensis* in the central mountainous area in Hokkaido, seeds were collected from 51 mother trees, and twigs from 80 selected clones from natural stands at different altitudes (Eiga and Sakai 1984). These seedlings and grafts were grown on the same plot. The distribution of freezing injury in winter buds of the open-pollinated progeny from mother trees at different altitudes from 250 m to 1200 m in two localities is shown in Fig. 6.3. The frequency of freezing injury to seedlings from the mother trees at 600 m showed a nearly normal distribution, whereas in those from 1200 m (near the forest limit), slight to moderate injuries were most frequent. Thus, the hardiness increased with increasing altitude. A highly significant difference in the freezing injury index of open-pollinated progenies was found for altitude and stand, but not for two provenances. The altitudinal variation in the freezing resistance and its variance

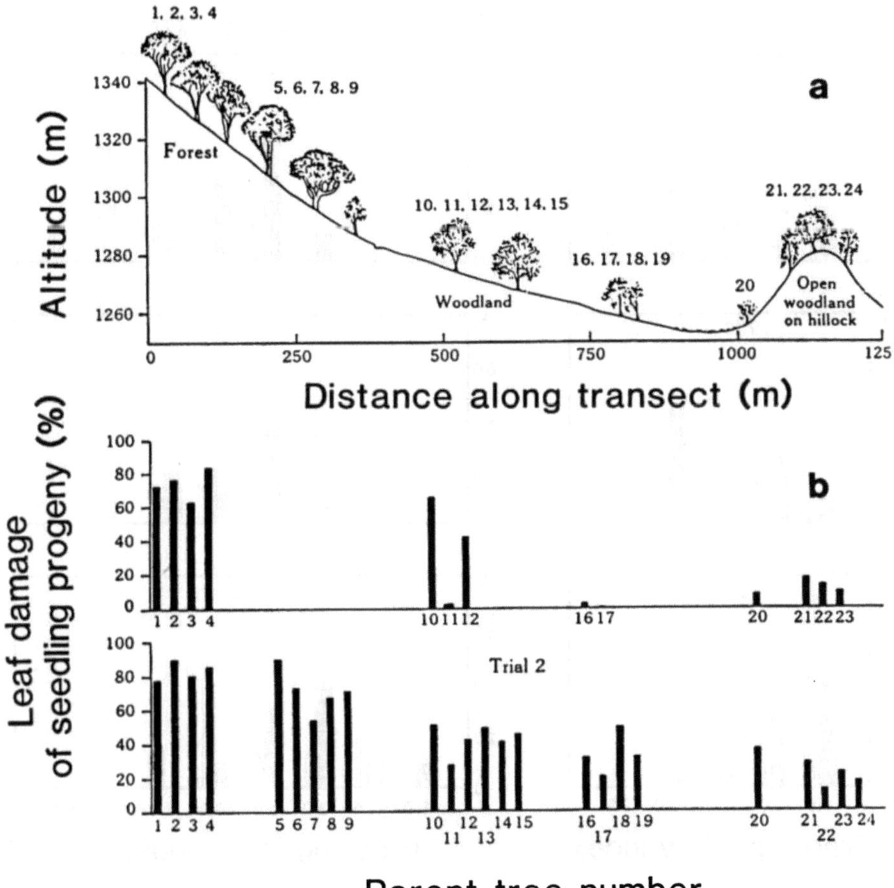

Fig. 6.2a,b. Relative frost resistance of seedlings of *Eucalyptus pauciflora* from (a) parent trees at and above valley-floor tree line; (b) leaf damage score after freezing at −14 °C for 2 h. The population from the slope with unimpeded air drainage achieved less hardiness than the population from depressions. (From Harwood 1980)

were also investigated with ramets from 80 clones ranging at different altitudes in a limited area within a 50 km radius (Fig. 6.4). The variance of the freezing injury index decreased with increasing hardiness from about 400 m to about 700 m. These facts indicate that the marked increase in hardiness with increasing altitude is caused by the increase in more hardy individuals and a decrease in less hardy individuals within each population.

Seedlings of *Abies sachalinensis* from mother trees at altitudes from 300 m to 1200 m showed remarkably little variability in the phenological phenomena, bud opening in spring and formation of terminal buds (Kurahashi and Hamaya 1981). In spite of such a remarkable uniformity of the growth period, the height growth in the early stages of the seedlings varied considerably according to seed source and planting site. The height growth of the seedlings depended on the weight of the seed which decreased with in-

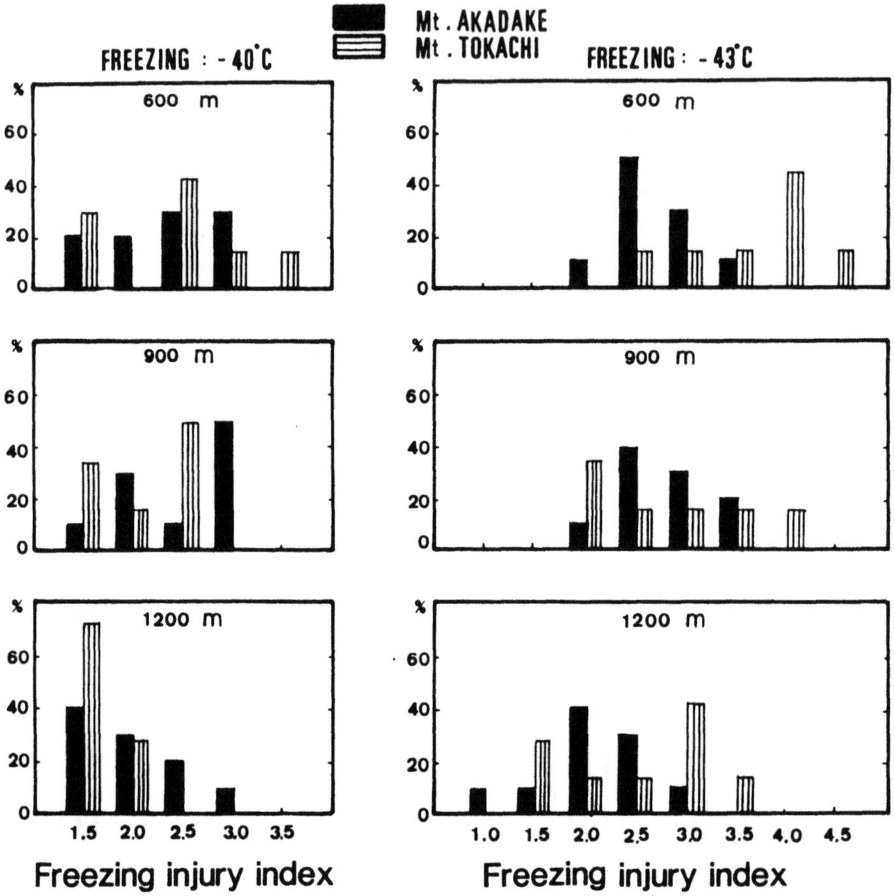

Fig. 6.3. Degree of injury of winter buds of the progeny of open-pollinated trees of *Abies sachalinensis* at different altitudes and from two provenances, after freezing at the indicated temperatures. A higher freezing injury index corresponds to increasing damage. *Ordinate:* percentage of mother trees whose lineage was damaged to the extent shown by the freezing injury index. (From Eiga and Sakai 1984)

creasing altitude of seed provenance. The results of reciprocal crossings between the sample trees chosen from one seed source of the lower group (530 m) and another of the higher group (1100 to 1200 m) revealed that variation of height growth of the seedlings along the altitudinal gradients is genetic rather than epharmonic. The offsprings from the highest altitudes (1100 and 1200 m) were characterized by high mortality in the sapling stage (within 7 years) in nearly every planting site at different altitudes. On the basis of the results obtained, Kurahashi and Hamaya (1981) concluded that the fittest genotypes for afforestation are found in the 400 to 700 m zone, where the Sakhalin fir is dominant. Above 1200 m altitude, the young trees are damaged and stunted to such a degree that normal stands with crown closure can only develop on particularly favourable sites. At such high altitudes, male flowers are often injured by late frost. This, and the small population density, lead to sexual isolation or inbreeding, and hence to very low rates of germination and high mortality in the early seedling

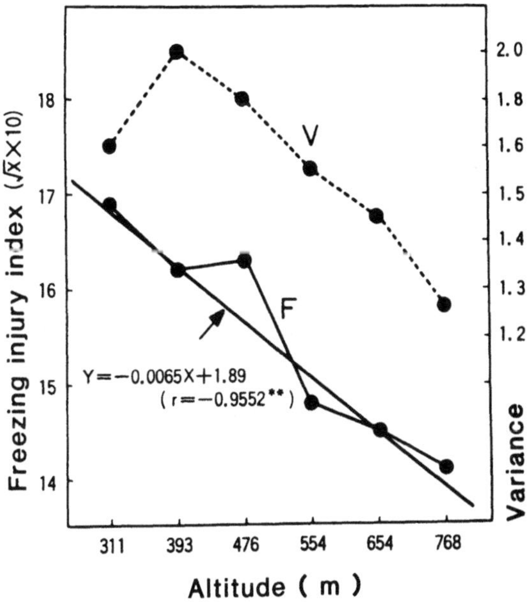

Fig. 6.4

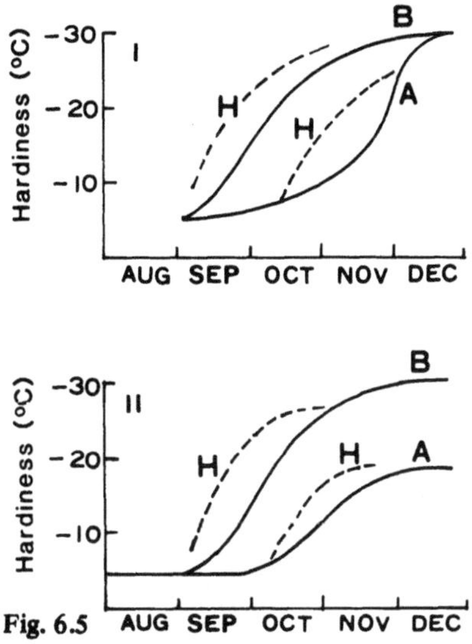

Fig. 6.5

Fig. 6.4. Freezing injury index (*F*) and its variance (*V*) of ramets originating at different altitudes. The test temperature was −43 °C. Statistical significance at the 1% level. (From Eiga and Sakai 1984)

Fig. 6.5. Patterns of cold acclimation among varieties or ecotypes. *Type I:* Difference in the onset of acclimation. *Type II:* Difference in both onset of acclimation and insufficient ability to harden. *H* hardening at 0 °C for 2 weeks. *A, B* are different varieties or ecotypes. (From Sakai 1959)

stage. The populations of *Abies sachalinensis* at 1200 m have evolved the highest winter hardiness, but seem to have lost their genetic variability and plasticity in the face of the increase of natural selection pressures with increasing altitude. Thus, they have become highly specialized to high altitudinal climate.

In *Picea abies* the individual range of variation in resistance at all altitudes is so large that an unequivocal classification is impossible (Holzer 1970). In any case, on account of the persistent low temperatures, the trees growing at the forest limits remain dormant far enough into the spring that even early sprouting spruces are scarcely affected by late frosts, in contrast to the situation in the valleys (Tranquillini 1979a).

Regional Variation in Frost Resistance. Differences in resistance from one population to another may be due either to genotypic differences in hardening ability or to differences in the length of their period of active growth (Sakai 1959a; Hamaya et al. 1968; Fig. 6.5). The contribution of these two possibilities is recognizable if the resistance is correlated with phenological phases which serve as indicators of the state of activity (Larcher 1968; Fig. 6.6).

A common method of studying intraspecies variability is the uniform environment plot. In an uniform environment study of clones of *Cornus stolonifera*, collected from 21 locations throughout N. America and grown at St. Paul, Minnesota, significant variations were found in plant form, growth rate and autumnal phenological events

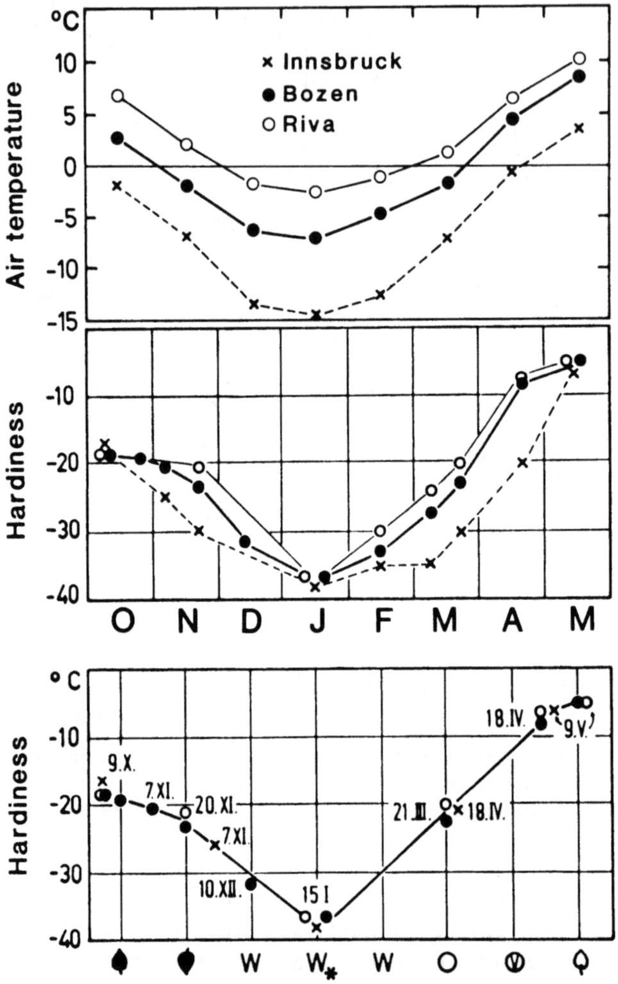

Fig. 6.6. Hardening ability of twig cambium of the Submediterranean tree *Fraxinus ornus* from various provenances, related to the date of sampling and to the phenological phase (*lower figure*). The hardening ability is expressed as frost resistance (LT_{50}) after precooling. *Bozen* is situated near the northern distribution limit of Submediterranean woody plants; *Riva* represents a transition from Submediterranean to Mediterranean climate; *Innsbruck* is situated beyond the natural distribution range of *Fraxinus ornus*. Air temperatures are mean monthly minima. *Symbols: dark leaf* means leaf discolouration, *inverse leaf* abscission; *W* winter periods without continuous frost; *W** main frost period; o bud swelling; ⊗ bud burst. *White leaf* shooting. (From Larcher and Mair 1968)

(Smithberg and Weiser 1968). The timing of cold hardening in the autumn differed among clones: northern or inland clones hardened earlier than those from southern or coastal regions, which sustained winter injury to branch tips in late fall and early winter, although all clones withstood -90 °C by early December in Minnesota. The influence of the geographic origin on the winter survival of *Populus deltoides* seedlings was determined by Mohn and Pauley (1969) in a study of 540 seedlings grown in a uniform plot at St. Paul from seeds collected from the central area of the species range (30°-40° N). The southern sources (30°-35° N) are characterized by high growth rates and late growth cessation. After the first winter, almost all of the seedlings from southern sources were killed to the ground level. They sprouted from the roots the following spring. Four years later, however, all seedlings from latitude 30°-35° N had died, while almost all of the seedlings from latitude 38°-46° N had grown to a height of 3 to 5 m. Frost hardiness of twigs of young trees of *Quercus rubra* grown from seeds collected from 38 different geographic origins and growing on a single site (Weston, Massachusetts) varied widely from October to December (Flint 1972). Plants from

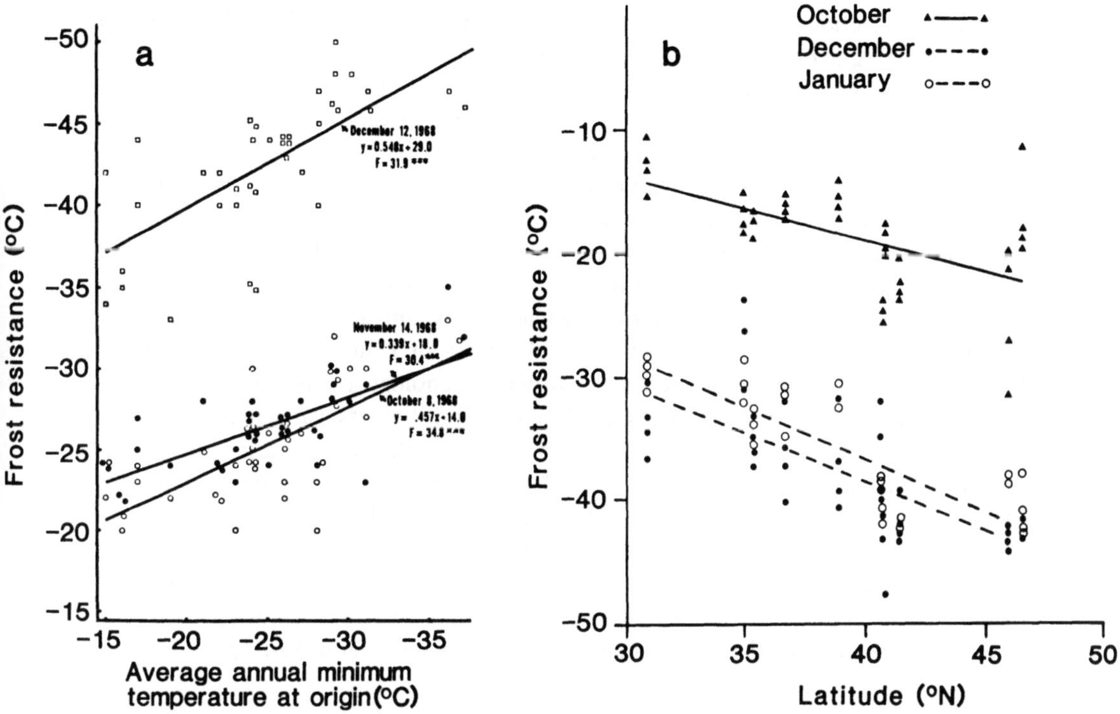

Fig. 6.7. a Regression of lethal temperature (LT_{10}) on estimated average minimum temperature for three sampling dates at the origin of 38 provenances of *Quercus rubra*. (From Flint 1972). **b** Lethal temperatures (LT_{15}) for twig tissues of *Fraxinus americana* as a function of latitude of the provenance for three sampling dates. Samples were cold-acclimated at 6 °C for 2 days prior to the freezing test. (From Alexander et al. 1984)

warmer provenances hardened more slowly than those from cold regions, and were less hardy in early winter. Frost hardiness was strongly related to the estimated average annual minimum temperature of the provenance (Fig. 6.7a). Average annual minimum temperature, extreme minimum temperature and length of the frost-free period were highly correlated with each other and with the latitude of the provenance. All were strongly related to frost resistance in autumn. Stem tissues of *Fraxinus americana* varied in frost hardiness according to the latitude of origin (Fig. 6.7b); northern plants were more resistant than southern plants after cold acclimation in autumn and winter. Such differences disappeared by March. These results suggest that the heterogeneity in frost resistance of *F. americana* of different geographic origins may primarily result from different levels of hardiness rather than from inappropriate timing.

Uncommonly severe cold in the late fall and winter of 1950–1951 in Illinois caused different amounts of damage to southern pines *(P. echinata, P. taeda)* depending upon their geographic and seed sources (Minckler 1951). Experimental field plantings of *P. echinata* and *P. taeda* showed results similar to those in the nursery. *Pinus taeda* from the more southern and southeastern sources (Mississippi, South Carolina) showed by far the greatest freeze damage. *Pinus echinata* from sources in northern Arkansas, Missouri, Kentucky, Ohio and Oklahoma and hybrids of *P. echinata* with parental seed sources in Virginia and North Carolina showed no damage.

Coastal collections of *Pseudotsuga menziesii, Thuja plicata* and *Tsuga heterophylla* survived freezing to only about -20 °C, but those from Idaho (from altitudes of 1200 m) and Colorado were much hardier, resisting to -30° and -50 °C (Sakai and Weiser 1973). Douglas fir segregates into the two varieties *P. menziesii* var. *'menziesii'*, the coastal variety, and var. *'glauca'*, the interior variety. Adaptation to a maritime climate has endowed the coastal variety with growth rates superior to the interior variety. Whereas the interior variety is adapted to the cold winter of the Rocky Mountains and intermountain regions, the coastal variety sustains high mortality due to frost damage (Wright et al. 1971; Stern 1974; Larsen and Ruetz 1980). To assess possibilities for increasing growth potentials of the interior variety, while maintaining adaptation to a relatively cold inland environment, intervarietal hybridizations of Douglas fir were performed by Rehfeldt (1977). Nine traits related to growth, phenology and freezing tolerance in 4-year-old trees were compared. The growth potential of hybrids was generally superior to that of the interior variety, but similar to those of the coastal variety. For traits related to survival (bud burst, bud set, frost damage in nurseries, tree form and freezing tolerance), hybrids were intermediate, but approached levels characteristic of the interior variety. For each sampling date, seedlings of the coastal variety sustained more injury in the freezing test than those of the interior. The degree of injury to hybrids was intermediate between the two (Fig. 6.8). However, the frost resistance of 17 of the 70 hybrid strains exceeded that of the less resistant individuals of the interior variety.

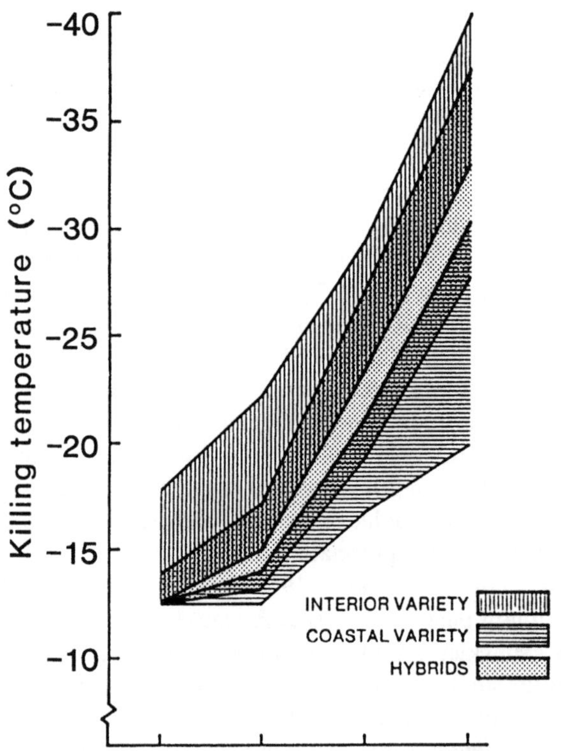

Fig. 6.8. Range in lethal temperatures (LT_{50}) for twigs of *Pseudotsuga menziesii* var. *'menziesii'* (coastal), *P.m.* var. *'glauca'* (interior), and hybrids, for four sampling dates. (From Rehfeldt 1977)

Freezing tests were also conducted to follow cold acclimation in seedlings representing 30 populations of *Pinus contorta* from the northern Rocky Mountains (Rehfeldt 1980). Elevation and geographic region of the seed origin accounted for 78% of the variance in hardiness among populations. In Sweden, in order to find well-adapted and highly productive populations for sylvicultural practice, two extensive series of provenance experiments with *P. sylvestris* were conducted by Eiche (1966), Stefansson and Sinko (1967) and Eiche and Andersson (1974). The results obtained convincingly showed the genecological variation of hardiness, capacity for survival and growth rate of different populations. The interaction between genotype and environment emerges quite obviously. Only the provenances of Norrbotten, a northern district of Sweden, were hardy enough to survive the harsh climate of this region. The authors concluded that only autochthonous provenances would ensure sufficient capacity of acclimation associated with a high growth rate in particularly unfavourable environments.

In *Fraxinus pennsylvanica*, Williams (1984) demonstrated that northern populations from Nebraska, North Dakota and Vermont exhibited little or no significant variance in hardiness, while southern populations were consistently variable. Populations near the northern limit of the species' natural range may be more homogeneous for the loci governing cold hardiness than populations further south. This feature is particularly obvious for tree species which have adapted to high northern climates characterized by a short growing season, a large seasonal variation of photoperiod and severe winter temperatures. In adapting to these special habitats, they may have lost the ability to grow vigorously and to compete at lower latitudes with the local populations. This is an expression of the ecological principle that selection for ability to withstand stress often proceeds at the cost of productivity (Mooney 1972; Grime 1979; Larcher 1987c).

6.1.3 Inheritance of Frost Resistance

The inheritance of frost hardiness was studied first by Nilsson-Ehle (1912) who crossed two winter wheat varieties intermediate in winter hardiness and found transgressive segregation for the character. He concluded that winter hardiness behaved similarly to other quantitative characters controlled by polygenes.

An eighteen-parent diallel cross of barley, which was tested for winter hardiness in six field locations and under controlled conditions, was analyzed by Rohde and Pulham (1960) and reanalyzed by Eunus et al. (1962), who observed that winter hardiness was controlled by genes ranging from completely dominant to completely recessive. Jenkins (1969) tested a five-parental oat diallel for frost hardiness under controlled freezing conditions. In one severe freezing test, frost hardiness was largely determined by recessive genes. Under less severe conditions the data indicated that hardiness was controlled by dominant genes. In a six-parental diallel cross, Sutka (1981) also reported that frost-sensitive varieties of wheat have the largest number of dominant genes, while frost-resistant varieties have the highest proportion of recessive genes. Chromosomal studies have also been employed in the analysis of genetic control of resistance characters. Goujon et al. (1968), using a monosomic set of winter and spring types of wheat, found that the chromosomes 5A, 2D and 5D carried the genes for hardiness, while the chromosomes 7A and 1B were responsible for frost sensitivity. Further results of genome analyses of *Triticum, Secale* and *Triticale* have been reported by Fowler et al. (1977), Dvorak and Fowler (1978) and Limin and Fowler (1982).

The inheritance of cold hardiness has also been studied with various woody plants. Genetic analysis of the resistance properties of fruit trees has been carried out, e.g. by Rudolf and Nienstaedt (1962), Murawski (1968) and Watkins and Spangelo (1970). Important contributions to the breeding of frost-resistant varieties of fruit were made by Schmidt (1942), Zwintscher (1957) and Murawski (1962, 1968). Resistance to winter frost in roses is apparently governed by a very small number of genes, or by coupled genes (Svejda 1979).

Russian apple cultivars and hybrids of *Malus baccata* have been used to develop frost hardy cultivars for the Canadian prairies (Quamme 1978). Crosses of hardy *M. pumila* types yielded progenies with high frost resistance (Stushnoff 1972). Hardiness of pear can be increased by using *Pyrus ussuriensis* in hybridization with *P. communis*. In contrast to the relatively frost-susceptible *Vitis vinifera*, the hybrids derived from *V. riparia* and *V. labrusca* are much hardier. *Vitis riparia* is one of the hardiest species of grape which survived -40 °C at Manitoba without snow cover. *Vitis amurensis* has also been used to improve cold hardiness in breeding programmes. Very hardy *V. amurensis* hybrids which have been produced in China can overwinter in Peking and Shenyang wihout snow or soil cover.

Citrus breeding began about 40 years ago, and new varieties with better cold resistance were selected in Japan. In the unusually cold winter of 1977, 1511 hybrid seedlings from 23 combinations were segregated for cold resistance by exposure to natural freezing conditions (Ikeda et al. 1980). Some of the progenies were more resistant than their parents, and the progenies of Hassaku (H-3) × Clementine (H-2), and Tanikawa buntan (H-6) × Mukakukishu (H-2) were more hardy than those of other combinations (Table 6.10; cf. also Table 3.3). Hybrid seedlings from the combinations of the Satsuma mandarin Imamura unshu (H-2) × Fukuhara orange (H-5), Page (H-5) and Nova (H-3) exhibited segregation for cold resistance when they were exposed to natural freezing conditions (Fig. 6.9); some seedlings were more cold hardy than the female parent. Progress and prospects of breeding citrus cultivars with improved frost hardiness are comprehensively reviewed by Yelenosky (1985).

Inheritance of leaf frost resistance of tea hybrids (controlled pollinated progenies) was analyzed by Toyao (1982) in an artificial freeze test. Parental material used in this study consisted of tea clones belonging to var. *sinensis*, var. *assamica* and selections from hybrids between both varieties. Plants of the first group, which were selected in

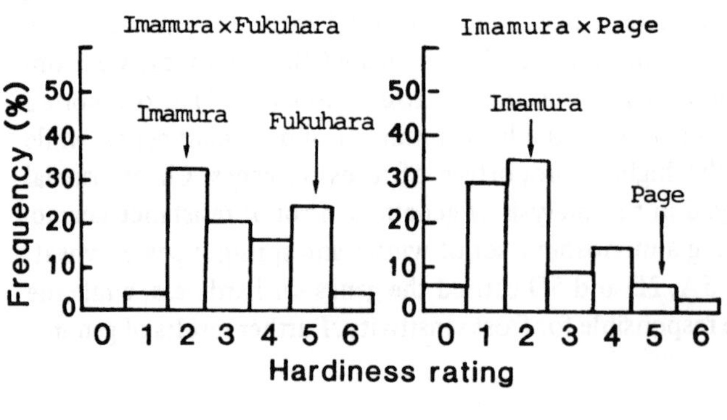

Fig. 6.9. Segregation for frost resistance of 4-year-old seedlings of Satsuma mandarin hybrids to −10.4 °C. *Arrows* indicate resistance of the parents. Hardiness rating: 0 most hardy, 6 least hardy. (From Ikeda 1982)

Table 6.10. Segregation in frost resistance of 10-year-old hybrid seedlings of *Citrus* varieties exposed to −9.1 °C in a severe winter. (From Ikeda et al. 1980)

Parents		Number of plants	Percentage of seedlings					
			Recovery rating[a]					
			0	1	2	3	4	5
Iyo	× Valencia	42			7.1	11.9	31.0	50.0
	Person Brown	103		1.9	5.8	17.5	22.3	52.4
	Fukuhara	25		4.0	4.0	16.0	36.0	40.0
	Sampson	32			12.5	12.5	43.8	31.5
Hassaku	× Valencia	31		6.5	19.4	19.4	29.1	25.8
	Trovita	159	0.6	13.2	18.9	20.1	27.7	19.5
	Fukuhara	38			7.9	29.0	39.5	23.7
	Iyo	50		12.0	22.0	26.0	12.0	28.0
	Clementine	207	2.4	22.2	26.1	20.8	17.9	10.6
Kinukawa	× Trovita	68		1.5	13.2	16.2	17.7	51.5
	Person Brown	69		4.4	10.1	23.2	18.8	43.5
	Hirakishu	14		7.1	7.1	7.1		78.6
Tanikawa buntan	× Valencia	3				66.7		33.3
	Trovita	179		0.6	8.4	21.2	32.4	37.4
	Person Brown	142			5.6	28.2	33.8	32.4
	Iyo	113			13.3	19.5	18.6	48.7
	Mukakukishu	19	5.3	26.3	21.1	26.3	10.5	10.5
Hirakishu	× Trovita	47		4.3	12.8	6.4	19.2	57.5
Naruto	× Sampson	27		11.1	18.5	3.7	18.5	48.2
	Trovita	60			3.3	3.3	18.3	75.0
Okitsu #20	× Trovita	16				18.8	31.3	50.0
Clementine	× Iyo	15			13.3	20.0	20.1	46.7
	Sampson	52		1.9	17.3	11.5	21.1	48.1

[a] 0 = no damage, 5 = dead.

Japan and China, generally showed a high resistance to freezing, those in the second group, which were introduced from India, were susceptible to freezing, and those in the third group had an intermediate hardiness. Regression coefficients of the mean hardiness score of the hybrid progeny on the mid-parent exceeded 0.9 (Fig. 6.10). Thus, the leaf frost resistance of the progenies of tea hybrids in winter is predictable from those of their parents. The heritability of the basal stem hardiness which determines the probability of bark injury of young tea plants in early winter also exceeds $r = 0.9$.

One of the genetic aspects of frost resistance to which much attention has been directed is the influence of polyploidy, probably in the hope that the greater size would favourably influence the resistance of the plants. However, a general trend in this direction could not be recognized. Freezing tests carried out by Oldén (1957) with di- and hexaploid plum cultivars showed that the amplitude of variation is conspicuously broader in the diploid than in the hexaploid varieties. The most resistant, and also the most sensitive, representatives of the plum varieties investigated were diploid. A wide overlap in the range of resistance and a greater average resistance of the diploid plants was registered for tea plants by Simura (1957) and for roses by Svejda (1979). Diploid rye plants in the fully hardened state were significantly more resistant than the cor-

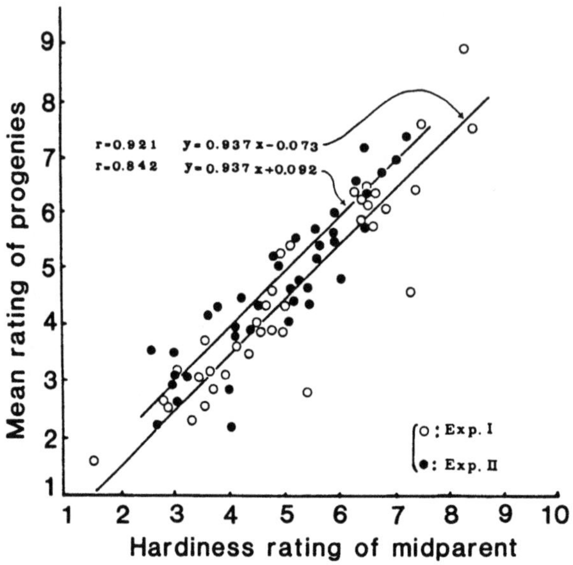

Fig. 6.10. Correlation between leaf frost resistance of progeny and mid-parent of tea hybrids. Hardiness rating: 1 uninjured, 10 dead. (From Toyao 1982)

responding autotetraploid plants (Dvorak and Fowler 1978). The diploid populations of *Festuca pratensis* were also found to be more resistant than tetraploids (A. Larsen 1979; Tyler et al. 1981; Fig. 6.11).

Since experiments on frost resistance have been performed under different conditions, on various genetic materials and in different stages of development, it is difficult to compare and draw general conclusions from the available data. How the chromosomes and genes so far associated with frost hardiness relate to certain physiological and biochemical processes controlling resistance will be an important aspect of further studies.

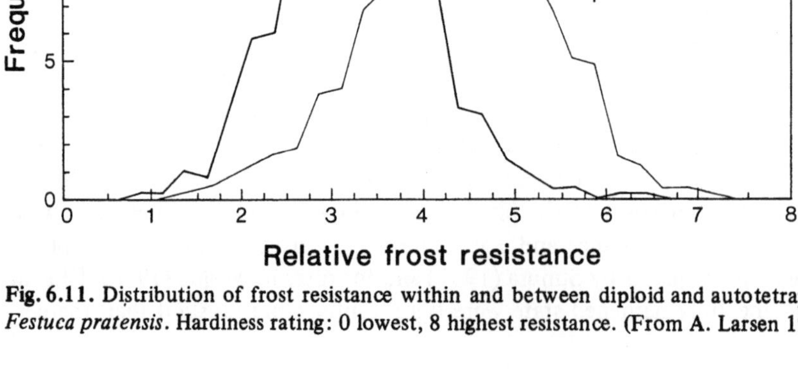

Fig. 6.11. Distribution of frost resistance within and between diploid and autotetraploid plants of *Festuca pratensis*. Hardiness rating: 0 lowest, 8 highest resistance. (From A. Larsen 1979)

6.2 Differences in Frost Resistance of Various Plant Organs and Tissues

Frost hardiness differs greatly from organ to organ in one and the same plant, and even from one tissue to another in the same organ. The differences in organ- and tissue-specific resistance can be quite considerable and may result from different freezing mechanisms. This is strikingly illustrated in the case of many deciduous trees of the temperate zone in winter when the bark parenchyma of the shoot axes is freezing-tolerant, the xylem parenchyma can supercool to low temperatures, whilst in the flower buds the freezing of the floral primordia is delayed and displaced to a lower temperature by translocated ice formation. Thus, in classifying plant species and varieties on the basis of their frost resistance, differences in the response of the individual organs have to be taken into consideration.

Depending upon species, life form, age, season and state of hardening, the differences in resistance between various organs and tissues may range from barely discernable to extremely large. As a rule, however, it can be assumed that the reproductive organs are more sensitive than the vegetative organs and, of the latter, in winter the

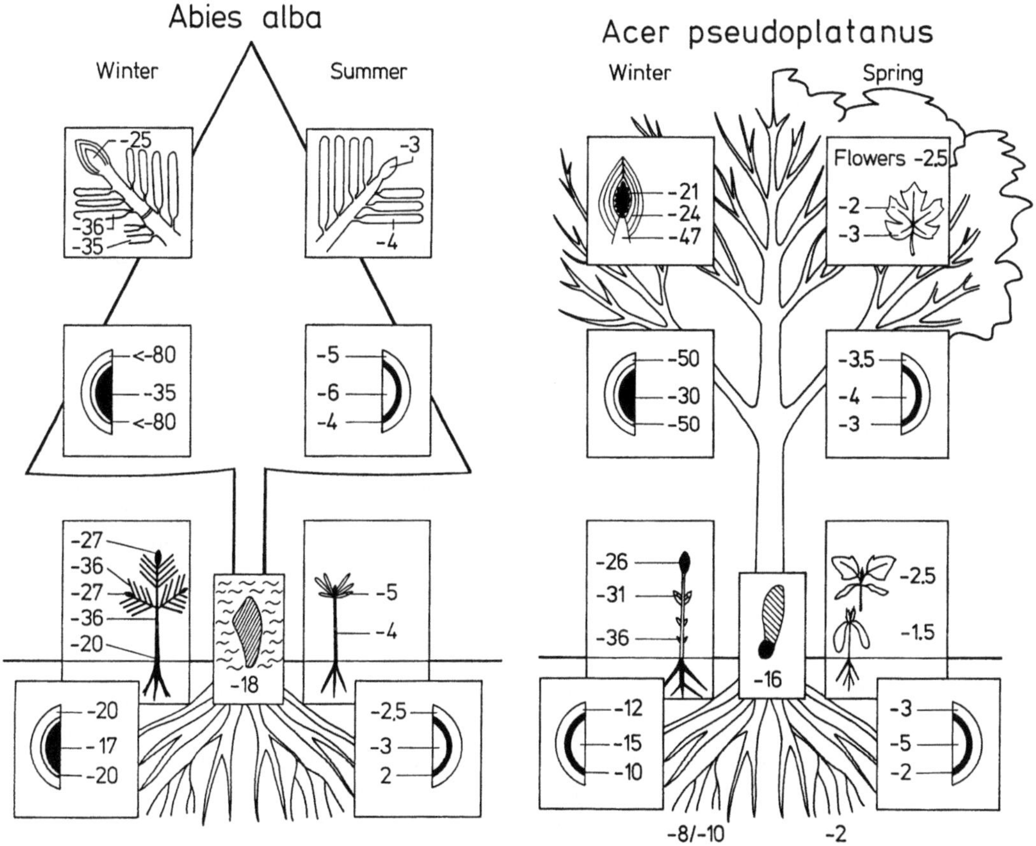

Fig. 6.12. Frost resistance (LT_i at °C) of various organs, tissues and life stages of *Abies alba (left)* and *Acer pseudoplatanus (right)* in winter and during growing season. (After G. Bendetta and J. Harrasser, from Larcher 1985a)

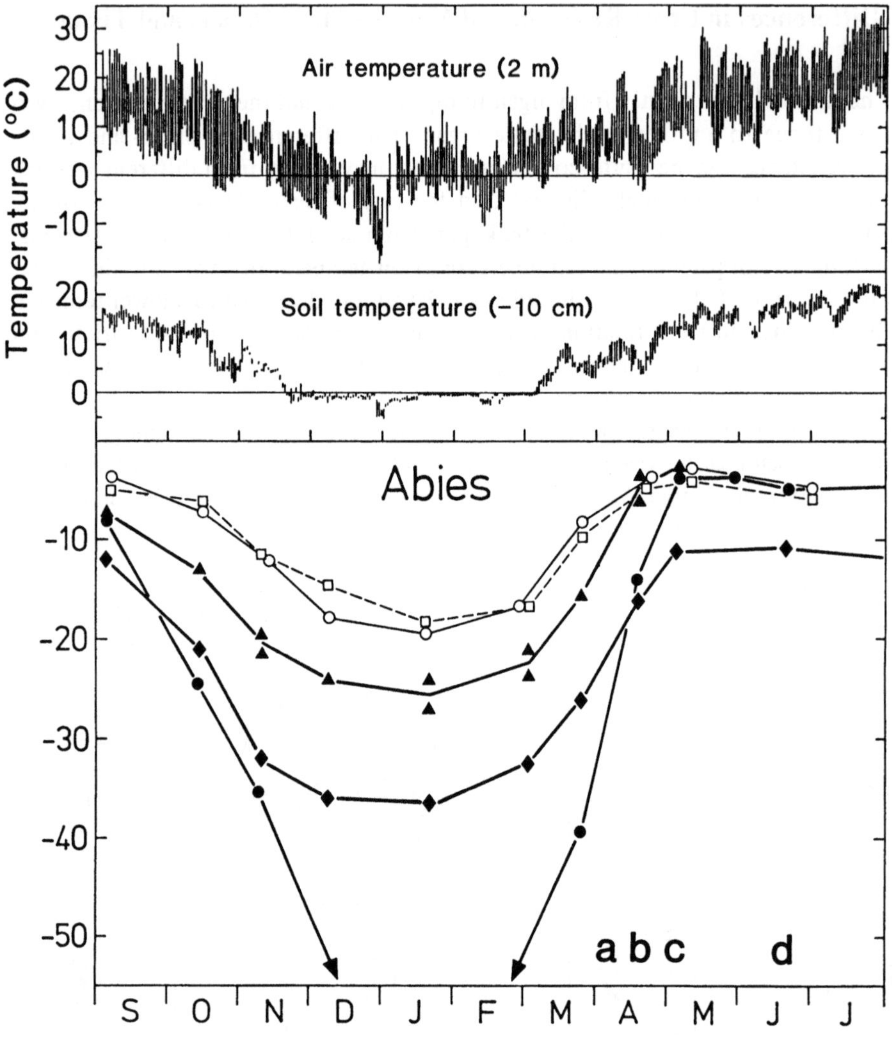

Fig. 6.13. a Annual variation of frost resistance of various tissues of 5-year-old plants of *Abies alba* at environmental temperatures. ○ Root cambium; □ root xylem; ▲ bud meristem; ◆ needles; ● shoot cambium. *a* Bud swelling; *b* bud opening; *c* sprouting; *d* formation of winter buds. (After G. Bendetta, from Larcher 1985a)

roots and underground storage organs, in the growing season the shoot tips and cambium, are especially sensitive (Fig. 6.12, see also Figs. 6.13, 6.14, 6.19, 6.20, 7.3, 7.4 and 7.27 to 7.29). A comparison of the corresponding columns in Tables 6.5, 6.6, 6.14, 7.1, 7.6, 7.8, 7.9 and 7.12 to 7.14 reveals further examples.

The seasonal progress of resistance also may differ from one part of the plant to another. In such cases, the autumn rise in hardiness frequently sets in earlier in the shoot than in the underground parts of the plant and, conversely, dehardening begins earlier in the spring in the roots than in the aerial parts (in fir and maple: Bauer et al.

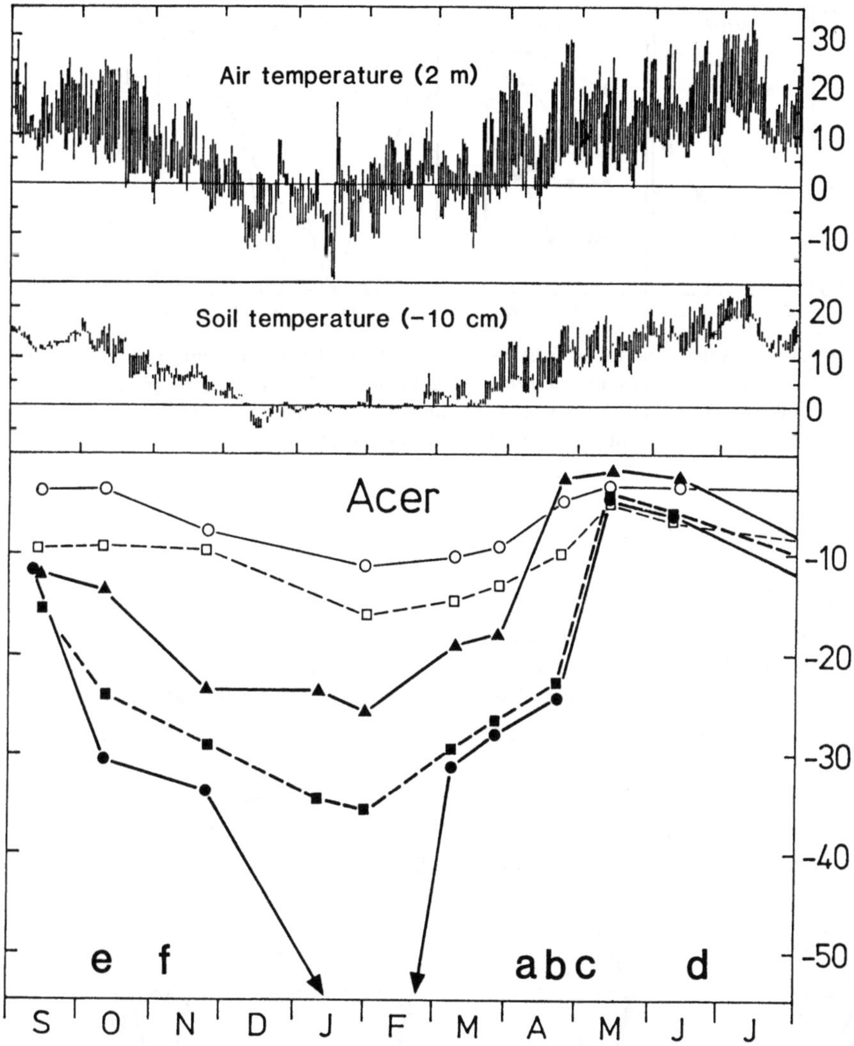

Fig. 6.13. b Annual variation of frost resistance of various tissues of 5-year-old plants of *Acer pseudoplatanus*. Symbols as in Fig. 6.13a; ▪ shoot xylem; *e* yellow leaves; *f* leaf falls. (After J. Harrasser, from Larcher 1985a)

1971; various herbs: Yoshie and Sakai 1981a; spring geophytes: Till 1956, Goryshina 1972; ferns: Sato 1982; see Figs. 6.13 and 6.14). The underground organs are therefore in a state of maximum resistance for a shorter period of time than the aerial parts. In autumn, the steep rise in resistance seen in the terminal buds usually begins slightly later than in the twig axes, in evergreen conifers slightly later than in the needles (Sakai and Okada 1971; Larcher 1985a). In spring, the resistance decreases first in the flower buds (in Douglas fir: Timmis 1977; apple: Pisek 1958, Kohn 1959; plum: Oldén 1956), then in the leaf buds and finally in the stems (Fig. 6.13).

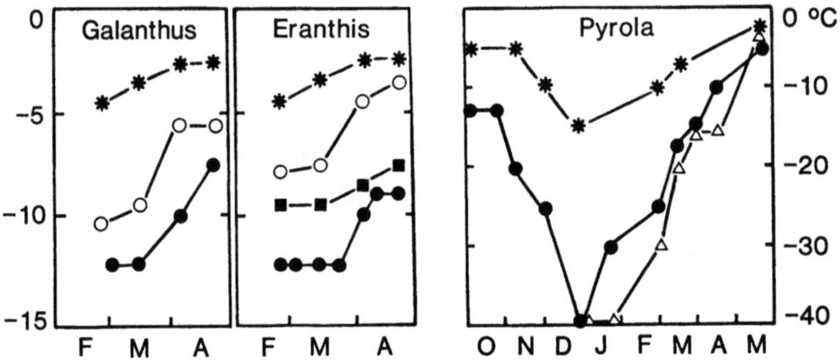

Fig. 6.14. Seasonal changes in frost resistance (LT_i at °C) of different organs of the spring ephemeroids *Galanthus nivalis* and *Eranthis hyemalis*, and of *Pyrola incarnata* growing on the floor of a deciduous forest. △ Buds; ■ above-ground shoot axis; ● leaves; ○ below-ground leaf bases; (*) rhizomes or bulbs. (From Till 1956; Yoshie and Sakai 1981a)

6.2.1 Vegetative Organs

Woody Plants. *Leaf* frost resistance varies according to the degree of differentiation and age, the actively expanding leaves being the most susceptible. Different regions and tissues (e.g. vascular parenchyma) within a single leaf may differ measurably in their resistance (Larcher 1954; Cox and Levitt 1972; Holt and Pellett 1981). The petiole is not infrequently more sensitive to frost than the blade (e.g. in *Cinnamomum*; Larcher 1971).

In the case of *buds*, the position on the twig and the priority in development determines the degree of resistance acquired (Mair 1968; Fig. 6.15). Usually the terminal buds are less frost resistant than the axillary buds and especially the adventitious buds.

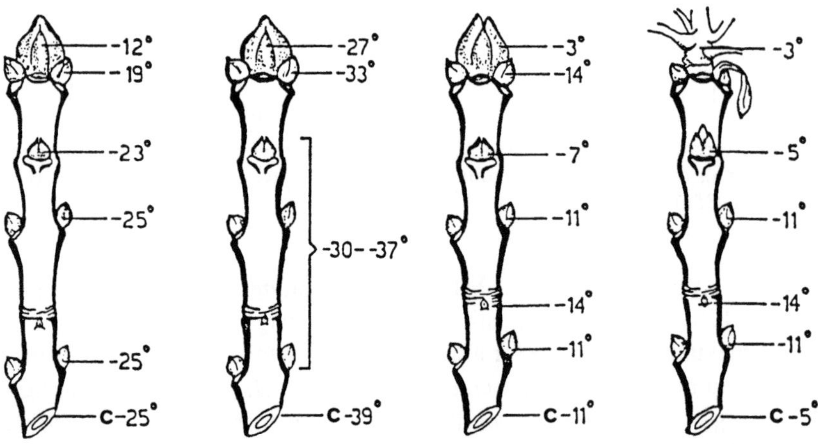

Fig. 6.15a–d. Changes in frost resistance of the buds of *Fraxinus ornus*, depending on winter dormancy and phenological development in the spring. **a** Autumn, onset of hardening; **b** winter, fully hardened; **c** spring, beginning of bud swelling; **d** spring, sprouting. *c* Cambial zone. (From Mair 1968)

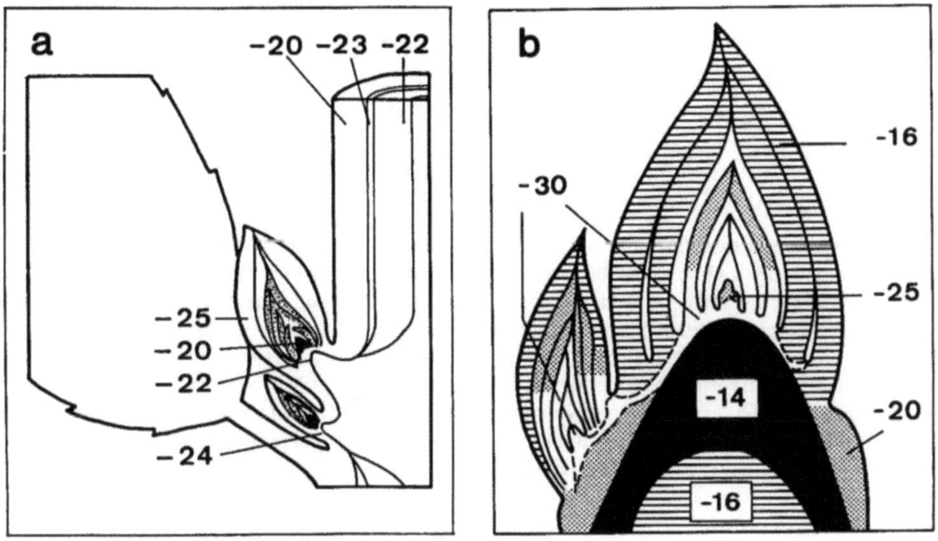

Fig. 6.16 a,b. Frost resistance (°C) of different tissues in winter buds. *Black:* Least resistant part. **a** *Phillyrea latifolia* has buds with greater resistance in the basal region; **b** terminal buds of *Ficus carica* are particularly susceptible at the region of attachment to the twig. (**a** From Larcher 1970; **b** W. Larcher, unpubl.)

In winter buds the scales are almost always the most resistant part. Two types are distinguished by Mair (1967), according to the situation of the most sensitive region, with significance for regeneration processes following injury:

1. Buds with greater resistance in the basal region: the apical dome and the leaf primordia are the most sensitive tissues, the rib meristem with the cambial attachment of the shoot axis is especially resistant. The most resistant are the meristem clusters at the base of older leaf primordia, from which a new apical meristem regenerates following partial damage to the bud. This type is found, e.g. in *Pinus, Abies, Acer* and *Phillyrea* (Fig. 6.16a).
2. Buds with greater resistance in the apical region: these are most sensitive at the bud basis and at the cambial connection between bud and shoot axis. As a consequence, after partial injury the bud will decay. Examples of this type are buds of *Fraxinus* and *Ficus* (Fig. 6.16b).

Woody stems are usually more resistant to frost than leaves and buds. Immature twig tips are, in winter, 2-3 K and sometimes even 10 K more sensitive than mature stems (Pisek 1958; Larcher 1970). This can be observed especially well in young plants and in plants whose growth has not been completed in time, due to either a cool, wet summer, or, as may happen at higher altitudes, if the growing season has been too short. In woody plants of regions with cold winters, in the frost-hardened state, the most resistant tissue is the cambium, and the least resistant is the xylem (Fig. 6.17, see also Fig. 6.12). Whereas the phloem resembles the cambium in its behaviour, the outer cortex may differ. The pith is nearly always far more sensitive to frost than the other tissues of the shoot axis. In trees and shrubs from regions with mild winters the bark may remain

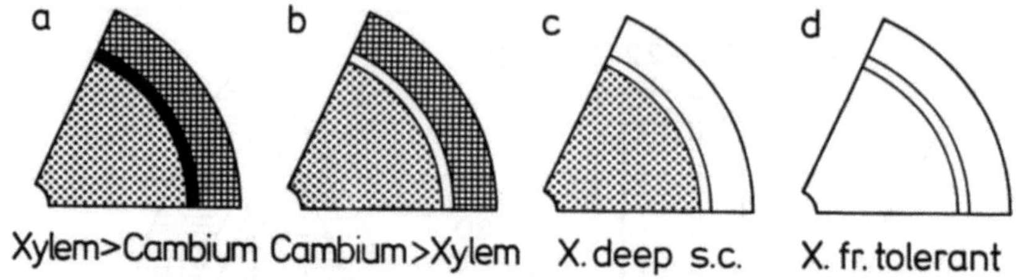

Fig. 6.17a–d. Patterns of frost resistance of woody stems. *Dark:* Most susceptible tissue; *white:* most resistant tissue. **a** All stem tissues are freezing-sensitive, xylem parenchyma supercool persistently; injuries occur between –7° and –15 °C; examples: *Ceratonia, Laurus, Nerium.* **b** Cambium capable of developing limited freezing tolerance when inactive, xylem and cortex remain freezing-sensitive even in winter; injuries occur between –15° and –20 °C; examples: *Citrus* species, *Quercus ilex.* **c** Cambium and cortex freezing-tolerant during dormancy, xylem survives by deep supercooling (down to –50 °C); examples: many deciduous trees of the temperate zone. **d** All stem tissues freezing-tolerant during dormancy; examples: extremely hardy conifers and certain deciduous woody plants of high latitudes. (From Larcher 1982)

more tender than the woody tissues (Larcher 1970, 1971). *Citrus* species and other woody plants of tropical and subtropical origin show only very small differences in resistance between cambium, wood and bark, whereby the wood parenchyma is the most likely to attain a somewhat higher degree of resistance due to its persistent supercooling properties. During growth the cambium is the most sensitive tissue in woody plants from all climatic regions.

The resistance of the various tissues of woody stems adjusts to the climatic conditions to different degrees and at different speeds (Fig. 6.18). If the cold season begins

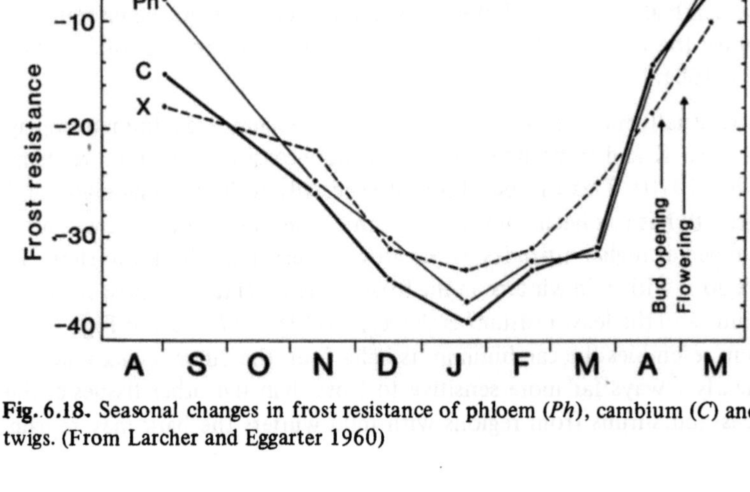

Fig. 6.18. Seasonal changes in frost resistance of phloem (*Ph*), cambium (*C*) and xylem (*X*) of pear twigs. (From Larcher and Eggarter 1960)

early or ends late, either due to local conditions or an exceptional year, cambium and phloem may remain at a higher level of resistance longer than the xylem. If climatic conditions are favourable, however, they may relinquish their state of maximum hardiness earlier than the xylem. The latter gives the impression of being less responsive to climatic conditions than cambium and cortex.

Underground parts of the plant are at all seasons more susceptible to frost than the shoot organs. Root tips freeze at $-1°$ to -3 °C. In winter the permanent roots, dependent upon species and state of activity, already freeze between $-5°$ and -20 °C, and only rarely at lower temperatures. Threshold values for the root hardiness of some woody plant species are shown in Table 6.11.

The conspicuous sensitivity of the roots has repeatedly provoked the question as to whether this is a property specific to the organ itself or whether it is due to the lesser

Table 6.11. Frost resistance (°C) of roots of woody plants. (From Havis 1976)

Species	LT_{50}
Magnolia x soulangeana	− 5.0
Magnolia stellata	− 5.0
Cornus florida	− 6.7
Daphne cneorum	− 6.7
Ilex crenata	− 6.7
Ilex opaca	− 6.7
Pyracantha coccinea	− 7.8
Cryptomeria japonica	− 8.9
Cotoneaster horizontalis	− 9.4
Viburnum carlesii	− 9.4
Cytisus x praecox	− 9.4
Buxus sempervirens	− 9.4
Ilex glabra	− 9.4
Euonymus fortunei	− 9.4
Hedera helix	− 9.4
Pachysandra terminalis	− 9.4
Vinca minor	− 9.4
Pieris japonica	− 9.4
Acer palmatum cv. *Atropurpureum*	−10.0
Cotoneaster adpressa praecox	−12.2
Taxus x media cv. *Nigra*	−12.2
Rhododendron cv. *Gibraltar*	−12.2
Rhododendron cv. *Hinodegiri*	−12.2
Pieris japonica	−12.2
Leucothoe fontanesiana	−15.0
Pieris floribunda	−15.0
Euonymus fortunei cv. *Colorata*	−15.0
Juniperus horizontalis	−17.8
Rhododendron carolinianum	−17.8
Rhododendron catawbiense	−17.8
Rhododendron P.J.M hybrids	−23.3
Potentilla fruticosa	−23.3
Picea glauca	−23.3
Picea omorika	−23.3

degree of cold experienced below ground. By exposing the roots before the onset of winter Chandler (1954) and Tumanov and Khvalin (1967) brought about a rise in resistance to nearly that of the shoot system. By storing at −7 °C for several weeks Pellet (1971) was able to double the root resistance of 2-year seedlings of *Lonicera tatarica, Cotoneaster horizontalis* and *Euonymus europaeus*. The interesting point of this experiment is that the hardening treatment in these species resulted in an equal rise in the resistance of the shoots, so that in the state of maximum hardening the resistance gradient between root and shoot remained the same. In *Ligustrum obtusifolium*, on the other hand, only the shoot could be hardened, the level of resistance of the roots remained unchanged. Havis (1976) was unable to provoke hardening of the roots of *Magnolia, Ilex, Cornus* and *Cotoneaster* in the dark, although this was later achieved by Johnson and Havis (1977) in *Potentilla fruticosa* and *Picea glauca* under the influence of short day and cold. These observations lead to the conclusion that the low freezing resistance of roots is indeed a specific characteristic; in addition, the different timing of shoot and root growth and the lesser impact of frost action in the soil are also involved.

Herbaceous Plants. On the *shoot* of herbaceous dicotyledons in the hardened state the aboveground stems, the renewal buds, the leaf primordia and the young leaves are the most resistant, whereas the old, and especially senescent, leaves are the most sensitive (Fig. 6.19). Therefore, towards the end of winter, extensive freezing damage is frequently seen in green, overwintering herbs, often only the youngest leaves surviving. In the shoot axes, especially of rosette plants, resistance decreases from top to bottom and from outside to inside, as well as in the vincinity of the vascular bundles. The shoot apex of resting plants is more resistant to freezing than the medullary parenchyma of the crown (Marini and Boyce 1977). Adventitious rosettes are usually more resistant than the main shoot (see Fig. 7.29).

In graminoid plants the intercalary growth zone at the basis of the leaf sheaths and the crown are the shoot tissues most vulnerable to winter kill (Fig. 6.20). Ice spreading

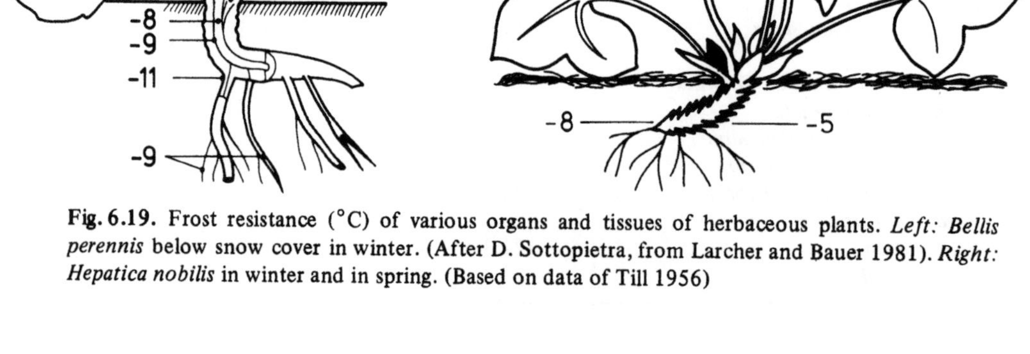

Fig. 6.19. Frost resistance (°C) of various organs and tissues of herbaceous plants. *Left: Bellis perennis* below snow cover in winter. (After D. Sottopietra, from Larcher and Bauer 1981). *Right: Hepatica nobilis* in winter and in spring. (Based on data of Till 1956)

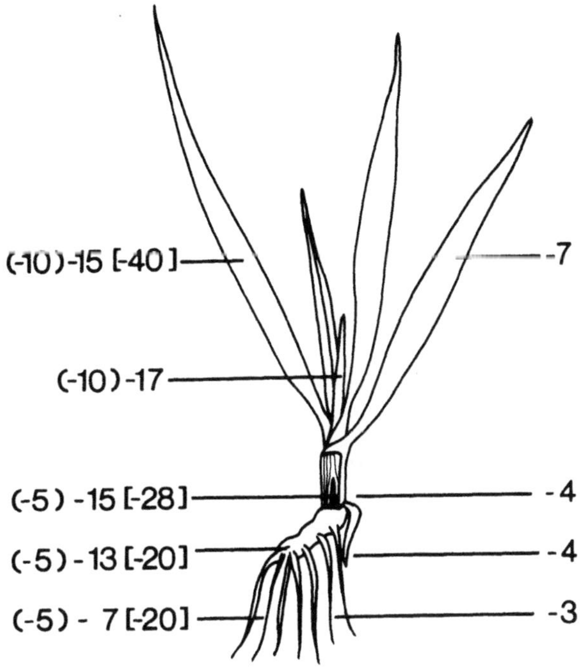

Fig. 6.20. Frost resistance (°C) of various organs of *Poa pratensis* in winter. High values in square brackets indicate resistance of the hardiest varieties, low values in parenthesis apply to mild winter conditions. (After data of Noshiro and Sakai 1979; Gusta et al. 1980, from Larcher 1985a)

from the vessels of the vascular transition zone at the basal portion of the crown can split the tissue (Olien 1964). The apical meristem is the most resistant tissue of cold-acclimated tillers (Krasavtsev and Khvalin 1982). The loss of a great part of the leaves and roots is of little consequence for survival if meristems remain intact (Olien and Marchetti 1976; Tanino and McKersie 1985).

In herbs, too, the *roots* are more sensitive than the shoot, and species with taproots acquire less resistance than those with adventitions and filamentous roots (Noshiro and Sakai 1979). An exception is seen in plants in localities where the frost penetrates deeper into the ground as, for example, certain rock plants, whose taproots survive winter temperatures of -30 °C and below (cf. Fig. 7.28).

Bulbs, corms and rhizomes seldom survive more than -20 °C (Table 6.12). Corms of *Anemone* and *Spiloxene* which tolerate severe dehydration do not survive freezing to -5 °C in a hydrated state (Sakai 1960b). *Eranthis cilicica*, which is also very tolerant to dehydration, survived -10 °C in mid-winter. In early October bulbs of *Gagea lutea* developed roots and elongated their shoot tip to 0.5 cm. At this stage of development they did not survive freezing to -5 °C, but by mid-winter the resistance increased to -10 °C. The hardiness of the underground organs of geophytes is distinctly correlated with the degree of stress from winter cold (Sakai and Yoshie 1984; see also Table 4.1). Species of *Iris* in the understorey of Siberian larch forests, and *Fritillaria camtschatcensis* and *Lilium dauricum* from E. Siberia, NE. China, inland Hokkaido and Alaska survive -20° to -25 °C. In these areas they overwinter in continuously frozen soil for 2 to 4 months. In highland regions of Asia Minor, which has a continental climate with dry summers and very cold winters (-20° to -24 °C), bulbous plants, such as species of *Tulipa, Iris, Muscari, Ornithogalum, Scilla* and *Galanthus* are considerably hardier than species from the western Mediterranean coastal areas. Bulbs and corms of *Leuco-*

Table 6.12. Frost resistance (LT_i at °C) of bulbs and corms in winter. (From Lundquist and Pellett 1976; Sakai and Yoshie 1984)

Species	Vegetative tissues	Flower primordia	Original distribution area
Lilium dauricum	–23	–17	Hokkaido, NE China
Fritillaria camtschatcensis	–20	–15	Hokkaido, E. Siberia
Allium oreophilum	–20	–17	W. Turkestan
Ixolirion tataricum	–18		Asia Minor, Caspis
Lilium canadense	–17	–17	Canada (Quebec)
Camassia quamash	–16		Canada (BC, Alberta)
Muscari armeniacum	–15	–13	Armenia, W. Iran
Muscari botryoides	–13	–13	C. and S. Eur., Transcaucasia
Camassia cusickii	–13	–13	Oregon
Fritillaria imperialis	–13		NW Himalaya, Afghanistan, Iran
Ornithogalum umbellatum	–12	–12	Caucasus, Asia Minor, N. Africa, S. Europe
Scilla sibirica	–12	–12	E. Europe, SW Asia
Tulipa gesneriana	–12	–12	C. Asia, Turkey
Allium moly	–12	– 8	S. Europe
Puschkinia scilloides	–11		Caucasus, Asia Minor
Eranthis hyemalis	–11		S. Europe
Eranthis cilicica	–10	–10	Turkey, Asia Minor
Scilla mischtschenkoana	–10	–10	NW Iran
Chionodoxa luciliae	–10	– 8	Asia Minor
Gagea lutea	–10	– 7	Europe, Caucasus, Himalaya, E. Asia
Tulipa fosterana	–10		C. Asia
Tulipa kaufmanniana	– 9	– 5	Turkestan
Galanthus nivalis	– 9		Europe
Crocus speciosus	– 9		S. Russia, Iran
Hyacinthus orientalis	– 9	– 5	E. Mediterraneis
Lilium speciosum var. rubrum	– 8	– 7	Japan (cult.)
Crocus chrysanthus	– 8	– 5	E. Mediterraneis
Galanthus elwesii	– 8	– 5	SW Asia
Narcissus poeticus	– 7	– 7	S. Europe
Leucojum vernum	– 7		C. and S. Europe
Triteleia bridgesii	– 7	– 5	Oregon, California
Colchicum autumnale	– 7	– 5	Europe, N. Africa
Anemone blanda	– 3	– 3	SE Europe, SW Asia
Ranunculus asiaticus	– 3		S. Europe, SW Asia
Sparaxis tricolor	– 3	– 3	S. Africa
Dipidax triquetra	– 0.5	– 3	S. Africa
Freesia refracta var. leichtlinii	– 0.5	– 3	S. Africa
Ixia leucantha	– 0.5	– 3	S. Africa
Leucocoryne ixioides	– 0.5	– 3	S. Africa
Tritonia crocata	– 0.5	– 0.5	S. Africa
Babiana stricta	– 0.5	– 0.5	S. Africa
Lachenalia pendula	0[a]	0[a]	S. Africa
Watsonia beatricis	0[a]	0[a]	S. Africa

[a] Killed at –0.5 °C.

coryne, Lachenalia, Babiana, Dipidax, Freesia, Spiloxene, Gladiolus, Watsonia and *Tritonia* from South Africa, where mild winter climate prevails, are already killed at temperatures below -1 °C.

In hardy species survival depends primarily on the resistance of the base plate, which is, together with the roots, the most susceptible part of the bulb. Stems and leaves developing from bulbs with a damaged base plate and dead roots are still able to expand, although the shoot elongation is markedly reduced. Some bulbs of *Allium, Muscari, Tulipa* and corms of *Crocus* have very resistant lateral buds which produce numerous small daughter bulbs as the mother bulb/corm decays. This regenerative capacity may permit such species to survive in areas where they are occasionally subjected to temperatures which kill basal plate tissues. Flower primordia in bulbs are less frost resistant than vegetative buds, scales and stems (Lundquist and Pellett 1976; Sakai and Yoshie 1984). Flowers developing from partially injured bulbs often abort or are deformed (Van der Valk 1970). In tender bulbs and corms, all tissues are killed at essentially the same temperature.

6.2.2 Reproductive Organs

It is usually the reproductive organs of a plant that are most sensitive to frost, with the exception of the *gametophytes* of ferns, which are as resistant as the sporophytes (see Sect. 7.3.2).

The *blossoms* of most herbaceous and woody plants of the temperate zone freeze between -1° and -3 °C, those of plants flowering in winter or very early spring between -5° and -15 °C (Table 6.13). In flowers, the style and the seed primordia are usually

Table 6.13. Frost resistance (LT_i at °C) of flowers in relation to flowering time. (From Larcher and Bauer 1981; after Till 1956; Pisek 1958; Timmis 1977; Proebsting and Mills 1978; Reader 1979; Sakai and Hakoda 1979; D. Sottopietra, unpubl. data; cf. also Table 7.10)

Plant species	Flowering time	First injury below °C
Viola wittrockiana	Winter	−15 to −18
Senecio vulgaris	Winter	−15
Anemone hepatica	Early spring	−11
Camellia vernalis	Winter to spring	− 5 to −7
Corylus avellana ♂	February	−16
Corylus avellana ♀	April	− 4
Acer pseudoplatanus	April	− 2 to −3
Quercus robur	April/May	− 2 to −3
Fraxinus excelsior	April/May	− 2
Rosacean fruit trees	April/May	− 2 to −3
Pseudotsuga menziesii ♂	May	− 3
Pseudotsuga menziesii ♀	May	− 2 to −3
Chamaedaphne calyculata	May	− 9
Andromeda glaucophylla	May/June	− 5
Kalmia polifolia	June	− 4
Ledum groenlandicum	June/July	− 3
Vaccinium macrocarpon	July	− 1

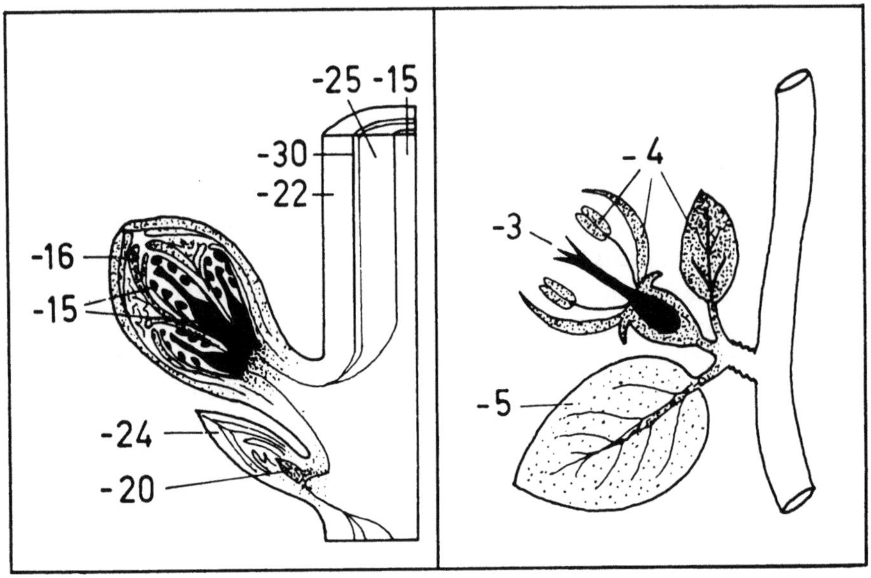

Fig. 6.21. Frost resistance (TL_{50} at °C) of the various tissues in a floral bud of *Prunus dulcis* in winter (*left*) and of a flowering shoot of *Malus sylvestris* in spring (*right*). *Black:* most sensitive parts. (From Larcher 1983a)

the first to undergo damage, and the perianth is frequently the last to be injured (Fig. 6.21). The epithelium of the style canal is particularly sensitive, as well as the inner epidermis of the cavity and of the vascular bundles in the ovary (Marro and Deveronico 1979). The anthers often remain largely undamaged even if the pistil is severely frozen. Male flowers of *Pseudotsuga menziesii* are 1.5 K more resistant than the females (Timmis 1977). In an air-dry state, pollen is completely frost resistant, and even the well-hydrated, freshly released pollen can tolerate freezing to low temperatures if cooling takes place slowly (e.g. pollen of *Cedrus deodara* to -20° to -25 °C; Otsuka 1971). This is important for anemochorous plants, since their pollen is borne up to the higher air layers where negative temperatures prevail even in summer.

The *flower primordia* in winter buds are at least 2-5 K, sometimes as much as 10-20 K more sensitive than vegetative buds of the same plant, so that after experiencing temperatures below -30° to -40 °C only few plant species are capable of flowering in the following year (Table 6.14). In this case the primordia of pistil and anthers are less resistant than the enveloping tissues (Larcher 1954, 1970; Till 1956; Bittenbender and Howell 1976; Fig. 6.21, see also Fig. 3.8).

6.3 Ontogenetic Variation in Frost Resistance

The various developmental stages differ with respect to sensitivity to cold and hardening capacity. Whereas dormant stages are characterized by high, often absolute, freezing tolerance, germinating seeds and seedlings are tender in proportion to the intensity

Table 6.14. Potential frost resistance (LT_i at °C) of flower buds as compared with vegetative buds. (From Sakai 1978b,c, 1982c)

Plant species	Floral primordia	Shoot primordia
Aesculus hippocastanum	–25	–40
Cercis chinensis	–20	–25
C. canadensis	–25	–30
Cornus florida	–20	–30
Cotinus coggygria	–30	–30
Deutzia crenata	–20 ... –25	–50
Forsythia intermedia	–20	–40
F. koreana	–20 ... –25	–70
Hamamelis japonica	–30	–40
H. mollis	–25 ... –30	–40
Kerria japonica	–20	–25
Prunus padus	–40	–70 >
P. sargentii	–25	–30 (–70)
P. sargentii var. *jamasakura*	–20	–25
Spiraea thunbergii	–25 ... –30	–30
Viburnum opulus	–70	–70 >
V. tomentosum var. *plicatum*	–30	–60
Weigelia hybrids	–20	–25
Wisteria floribunda	–20	–20 ... –25

of growth. Adult individuals, in many cases even juvenile stages, attain the highest capacity for hardening, and senescent plants again become more susceptible. For ecological purposes it is important to understand the response of seedlings and early life stages as compared with that of adult individuals, since seedling establishment is the most curcial phase in the life cycle. There are the most susceptible stages in ontogeny which determine the persistence of a species in a given habitat (Thienemann's law). In addition, seedlings and juvenile individuals close to the ground are exposed to a microenvironment that differs considerably from that of fully grown plants (cf. Sect. 1.4).

6.3.1 Variation of Frost Resistance During the Cell Cycle

In cultured cells, the highest survival is obtained for cells in the lag phase or early exponential phase (Fig. 6.22). The capacity of the surviving cells to embark upon cell division is significantly affected by the stage in the batch culture growth cycle at which they are frozen (Withers and Street 1977). *Acer pseudoplatanus* cells exhibited the highest rate of survival if they were in the G_1 phase of the cell cycle, shortly after a mitosis, at the time of frost treatment, and when the ratio of cytoplasma volume to vacuole volume was high (Withers 1978). Cells in the premitotic G_2 phase, mitotic cells and ageing cells were considerably more sensitive. In synchronous cultures of *Saccharomyces cerevisiae* the survival rate was highest following deep freezing at the beginning of the phase of DNA synthesis, i.e. also in the vicinity of the G_1 phase (Cottrell 1981). Yeast cells in the exponential phase of growth are highly vacuolized and

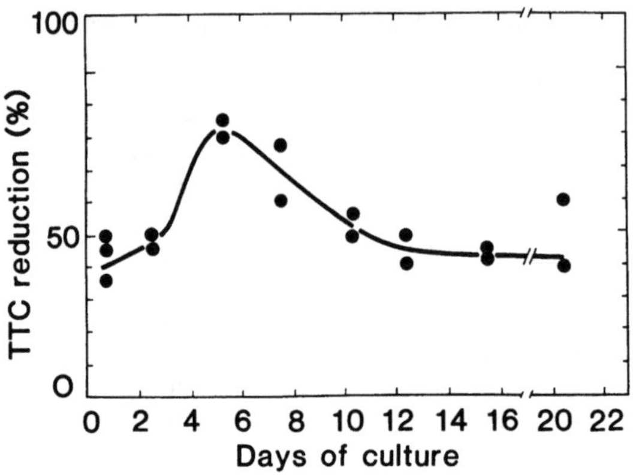

Fig. 6.22. Frost survival of suspension-cultured cells of *Acer pseudoplatanus* depending on culture stage. Survival is expressed as reduction of triphenyltetrazolium chloride (TTC) in cells frozen slowly to -30 °C in the presence of cryoprotectants (24% DMSO, 10% glucose) as compared with that of unfrozen cells with the same additives. The 5th or 6th days of cell culture correspond to the later lag phase or the early cell division phase. (From Sugawara and Sakai 1974)

freezing-sensitive: at -2° to -5 °C ice forms in the vacuoles and the cells die (Moor 1964). Cells of the resting phase, however, have very small vacuoles and survive at temperatures to -15 °C, after which ice forms in the cytoplasm. Cells rich in glycogen are tolerant to freezing and are only damaged below -25 °C. In suspension cultures small cells filled with a highly dense cytoplasm resulting from protein neosynthesis are more resistant than cells in the advanced phases of the cell cycle (Sutton-Jones and Street 1968). Nag and Street (1973) also observed that a reduction in survival during freezing and thawing is primarily due to death of the large, more highly vacuolated, free cells which do not give rise to colonies even in unfrozen suspension.

6.3.2 Frost Resistance of Seeds

Dormant seeds are very resistant to frost. Seeds with a low water content can, as a rule, survive cooling to very low temperatures (see Sect. 4.5). Seeds with a water content of more than about 20% of their fresh weight (e.g. acorns with large, starch-filled cotyledons; Table 6.15) or imbibed, but not yet germinating seeds can be cooled to about -10° to -20 °C.

Table 6.15. Frost resistance (LT_i at °C) of overwintering acorns and chestnut seeds (samples collected below litter were frozen at test temperatures for 12 to 16 h after slow cooling)

Species	LT_i	Reference
Quercus ilex[a]	– 6	Larcher and Mair (1969)
Qu. phillyraeoides[a]	– 8	Sakai (unpubl.)
Qu. myrsinaefolia[a]	–10	Sakai (unpubl.)
Qu. acutissima	–10	Sakai (unpubl.)
Qu. dentata[b]	–10	Sakai (unpubl.)
Qu. grosseserrata[b]	–12	Sakai (unpubl.)
Qu. rubra	–15	Sakai (unpubl.)
Castanea sativa	–11	Bendetta (unpubl.)

[a] Evergreen species.
[b] With developed radicle (ca. 5 cm).

6.3.3 Frost Susceptibility During Germination and at the Early Seedling Stage

At the beginning of germination the originally high resistance of the dormant state is quickly lost. Processes that result in the breaking of dormancy, e.g. cold stratification of apple seeds, do not immediately bring about a lowering of resistance (Kacperska-Palacz et al. 1980); only after germination has set in does resistance drop in proportion to the elongation of the radicle (Kemmer and Thiele 1955; Levitt 1956; Larcher 1969a). During emergence the plants exhibit maximum sensitivity at the period of rapid elongation (Fig. 6.23). The most sensitive phase in dicotyledons is during unfolding of the cotyledons and primary leaves, and in graminoids at the beginning of coleoptile elongation. When the primary leaf has half emerged from the coleoptile, winter wheat plants are again capable of hardening (Schwarzbach 1972). However, full ability to harden is not achieved until the tillering stage (J.E. Andrews et al. 1960; Roberts and Grant 1968; Fuller and Eagles 1978; A. Larsen 1978b). Although this is a generally observed trend, it does not preclude the possibility that plants germinating in winter and spring may also survive frost even at a very early stage in development. During sudden outbreaks of bad weather in summer, seedlings of alpine plants can readily gain frost resistance (see Sect. 7.5.1). Thus, the previously held view (Larcher 1973) that growing and metabolically active plants were unable to become freezing-tolerant can no longer be maintained.

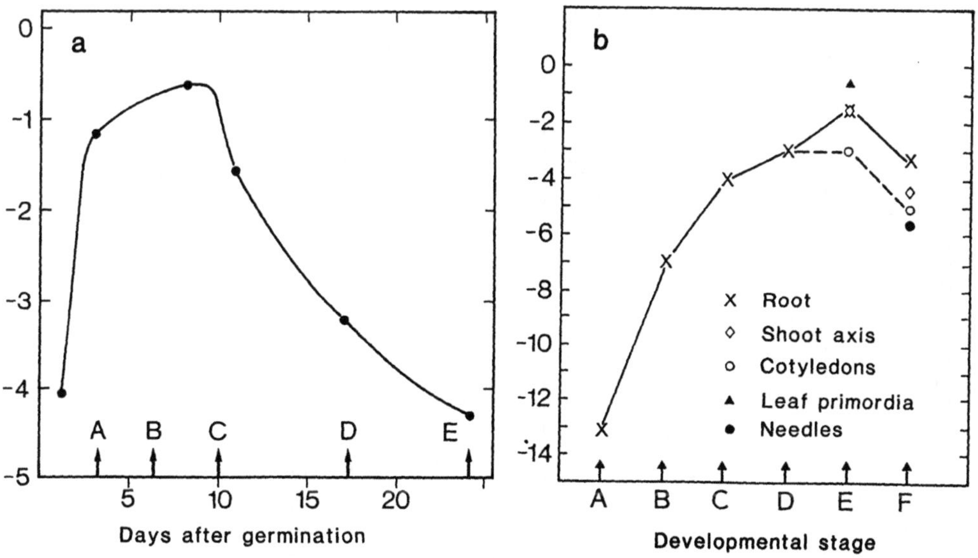

Fig. 6.23a,b. Frost resistance (LT_{50}) during germination and early development of seedlings. **a** *Beta vulgaris: A* root length 1 cm; *B* stem 1 cm; *C* cotyledons in light; *D* two true leaves. (From Cary 1975). **b** *Abies alba: A* seeds during stratification; *B* onset of germination; *C* radicle 4 mm; *D* radicle 8 mm; *E* cotyledons developed; *F* seedling with lignified xylem in July. (After G. Bendetta, from Larcher 1985a)

6.3.4 Frost Resistance During Establishment and Ageing

Some woody plants already attain complete resistance in the *juvenile stage*, so that overwintering 1-year-old saplings are nearly or quite as resistant as adult plants. Examples are provided by most species of *Pinus* and *Picea* of winter cold regions, as well as many species of *Abies* (Sakai and Okada 1971), apple (Karnatz 1956) and probably also many other tree species of the temperate zone. Alberdi and Rios (1983) found that species of Proteaceae whose young plants are exposed to severe radiation frost, acquire very early their full ability to harden, or may even be more resistant to frost in the juvenile stage than adults (Fig. 6.24). On the other hand, juvenile plants can be less resistant than the adults (Fig. 6.25), e.g. *Quercus ilex* (Larcher 1969a), certain conifers (*Abies balsamea, A. nordmanniana, A. veitchii*; Sakai and Okada 1971) and subtropical palms (Larcher and Winter 1981). In *Agave deserti* and *Ferocactus acanthodes* Nobel (1984) found that 5-year-old, cold-acclimated plants had a frost resistance about 3 K lower than that of the adult plants. In *Quercus ilex* all parts of the young plants are more sensitive. In conifers there is only a marked difference in resistance between young and adult plants in the shoot axis, and above all in its cambium. A lower resistance in tree saplings might be attributable to inadequate quiescence, since in the juvenile stage growth ceases later in autumn than in adult trees, and sprouting occurs earlier in spring.

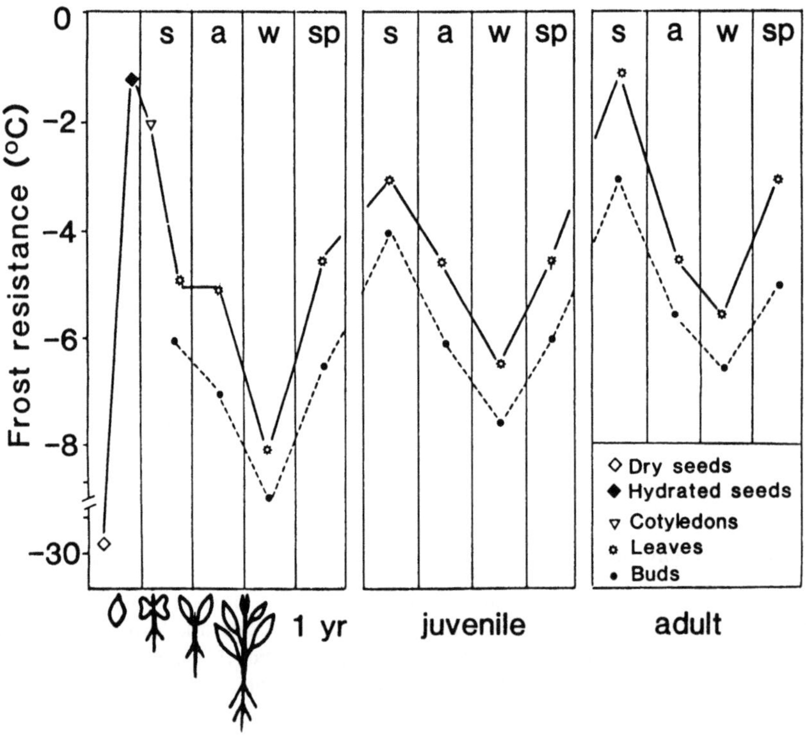

Fig. 6.24. Frost resistance (LT_{50}) of different ontogenetic stages of *Embothrium coccineum*, a tropophytic woody plant whose seedlings establish in habitats exposed to frost. Seasons: *s* summer, *a* autumn, *w* winter, *sp* spring. (From Alberdi and Rios 1983)

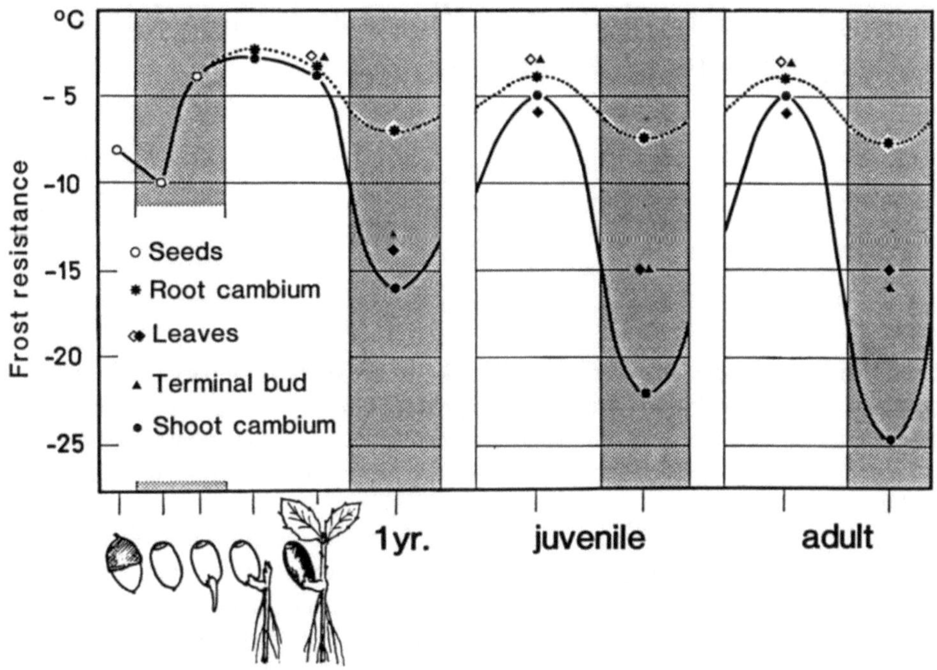

Fig. 6.25. Increasing frost resistance (LT_{50}) during development and aging of *Quercus ilex*, an evergreen oak whose seedlings grow protected by tree canopy. *Shaded areas:* winter. (From Larcher 1969a)

Hardening capacity apparently decreases with *ageing* as is clearly recognizable after a severe winter by the greater extent of freezing injury in older trees. A thorough analysis of the underlying causes of the decrease in resistance associated with senescence has not yet been made. Of interest in this context are not only events at the cellular and subcellular levels, but processes affecting the plant as a whole, particularly the effect of phytohormones. The results of freezing experiments employing shoots from the basal and apical zones of apple trees (Lapins 1961) and *Carya illinoënsis* (Sparks and Payne 1977) have shown that the tissues of the basal zone, laid down during the juvenile stage, were more able to harden than the tissues of the terminal parts that developed later in ontogeny.

7. Regional Distribution of Plants and Their Adaptive Responses to Low Temperatures

Distribution boundaries of plant groups and biomes are set by a complexity of ecological (climatic, edaphic and biotic) and historical (phytohistorical and anthropogenic) factors which interact with the demands made on its environment by the species involved, and its capacity to withstand stress (Boysen-Jensen 1949; Good 1974; Whittaker 1975; Walter 1984).

Whether or not a plant species is able to establish itself in a particular locality depends upon (1) its *survival capacity*, i.e. its ability to survive hazardous situations and adverse seasons; (2) its *production capacity*, which depends upon sufficient dry matter production to ensure growth and successful competition and (3) its *reproductive capacity*, i.e. its ability to multiply. In addition, the life-cycle timing has to be tuned to local seasonality. Although a particular environmental factor may be decisive in limiting the occurrence of a species along certain parts of the distribution boundaries, it is almost impossible to unravel the complexity of delimiting factors for larger stretches. Even if we deliberately confine ourselves to climatic factors, a very wide variety of parameters still has to be considered (for reviews, see Hintikka 1963; Tuhkanen 1980). Even for the most conspicuous of the biome boundaries, the subarctic and the subalpine *timberlines*, which are assumed to be set primarily by the direct and indirect effects of low temperature, we have no simple, generally applicable explanation (Ellenberg 1966; Holtmaier 1973; Walter 1976; Turner and Tranquillini 1985; for arctic forest lines: Tikhomirov 1970; J.A. Larsen 1974; Black and Bliss 1980; Tukhanen 1984; Oechel and Lawrence 1985; for alpine forest lines: Daubenmire 1974; Wardle 1974, 1985; Tranquillini 1979a; W.K. Smith 1985a). The explanation of biome limits is also made more difficult by the facts that they are inhomogeneous, i.e. they are formed by different species with differing response patterns, that they are dynamic, i.e. unstable on a long-term scale, and that they are seldom determined by one, definable environmental factor.

The influence of cold stress becomes obvious when a comparison is made between the distribution range of a species or a category of plants, characterized by a certain response pattern (e.g. chilling-sensitive plants, freezing-sensitive and freezing-tolerant plants; see Sect. 4.1.3) and the area over which their resistance to cold is potentially adequate, and over which the plant could survive if the only limiting factor were in fact its resistance to cold. At points where the actual limits of distribution coincide with those of the potential range with respect to frost resistance, it is evident that cold stress can represent the limiting factor. The possibility of limits being set by cold was already considered by Grisebach (1838), De Candolle (1855) and Shreve (1911, 1914).

The main categories with respect to a resistance mechanism can roughly be related to low temperature thresholds (cf. Fig. 1.1). Table 7.1 gives an impression of the upper and lower thresholds of the potential cold resistance for plant groups constituting the principal terrestrial biomes. Species lists giving data for frost resistance are available for European forest trees and herbs (Till 1956), for N. American and E. Asiatic forest trees (Sakai and Okada 1971; Sakai and Weiser 1973; Sakai 1978b,c), for the predominantly evergreen tree flora of the Pacific region (Sakai 1971b, 1978a; Sakai and Wardle 1978; Bannister 1984b; Alberdi et al. 1985); for Mediterranean sclerophylls (Larcher 1970), for tropical rain forest plants (Biebl 1964) and woody plants at the upper tree line of a tropical high mountain (Table 7.2), for subtropical grassland (Rowley et al. 1975; Ludlow 1980), for temperate grassland (Noshiro and Sakai 1979), for heath plants (Larcher and Bauer 1981), for succulent vegetation in N. American deserts (Nobel 1982, 1985; Nobel and Smith 1983), for high mountain plants (Ulmer 1937; Tyurina 1957; Sakai and Otsuka 1970; Beck et al. 1982; Goldstein et al. 1985a) and for arctic tundra plants (Biebl 1968; Riedmüller-Schölm 1974; Yoshie and Sakai 1981b).

Distribution boundaries related to low temperature stress are most likely to occur where a group of plants has not succeeded in switching to the more efficient category of a resistance mechanism. This can limit the zonal distribution of whole families of plants. For example, apparently not a single palm species is able to efficiently develop freezing tolerance and the family is consequently mainly confined to the tropics and subtropics: in no case can it spread to regions with temperatures below -10° to $-15\ ^\circ C$ (Sect. 7.1.2). Of the Lauraceae, as far as we know, all evergreen members are freezing-sensitive and confined to regions with mild winters (Larcher 1980c). Temperature minima are also probably responsible for the northern limits of distribution of certain bamboo species (Numata 1979). Those trees of the temperate zone that become only partially freezing-tolerant, due to deep supercooling of their wood, are limited to regions in which the mean annual minimum temperature does not drop below $-40\ ^\circ C$ (see Fig. 4.11).

We shall now consider – in the form of case studies – some examples of low temperature resistance in connection with plant distribution, with particular reference to our own investigations.

7.1 Woody Plants Within the Tropics

Very little quantitative data is available, so far, on the sensitivity of tropical woody plants to low temperatures and frost. An experimental screening of the leaves of trees, shrubs and lianas of a rain forest in Puerto Rico, and of mangroves and tropical ornamental plants were carried out by Biebl (1964). The susceptibility to frost of trees of tropical islands in E. Asia was studied by Sakai (1978a,d), that of provenances of *Jacaranda mimosifolia* by Chauvin (1984) and of the genus *Ficus* by Hummel and Johnson (1985). The results of studies on mangroves and palms are described in more detail below.

The available data indicate that tropical woody plants are killed by frost at temperatures of -1° to $-4\ ^\circ C$; they are permanently freezing-sensitive. Some of them are already severely injured by exposure to low temperatures above freezing, i.e. they are

Table 7.1. Frost resistance (°C) of terrestrial vascular plants. [Compiled from data of numerous publications as cited in Larcher and Bauer (1981), Larcher (1985a) and in Tables and Figures of Chaps. 4, 6, 7. Data for bamboos: Sakai (1976), Ishikawa (1984)]

Climate and life forms	Leaves	Buds[a]	Stem	Below-ground organs
Tropical climate				
Rain forest trees	X to −3	X to −5	X to −5	
Lianas and epiphytes	X to −2			
Forest understorey herbs	X to −3 (−5)			
Savanna grasses	X to −4 (−8)	−4 to −6 (−10)		
Mountain tree line species	−3 to −8 (−10)	−3 to −10	−3 to −10	
High mountain plants	−7 to −15 (−20)			
Humid subtropical and warm-temperate climates				
Subtropical woody plants	−3 to −10	−3 to −12	−5 to −15	
Laurophyllous woody evergreen plants	−8 to −20	−8 to −20	−10 to −22	−5 to −12
Deciduous trees and shrubs		−10 to −15	−10 to −25	
Southern hemisphere conifers	−10 to −23	−13 to −23	−10 to −23	
Warm-temperate ferns	−7 to −17			−3 to −12
Semi-arid to arid climates with mild winters				
Sclerophyllous woody plants	−5 to −15	−8 to −18	−8 to −22	Down to −10
Drought deciduous trees		−8 to −15	(−15 to −20)	
Succulents	−4 to −20			
Geophytes				0 to −10
Desert winter annuals	−6 to −10			
Temperate climates				
Conifers	−25 to −70	−25 to −50	−30 to −70	−15 to −25
Deciduous trees and shrubs		−25 to −35	−30 to −50	−15 to −25
Forest understorey herbs	−10 to −20	−10 to −20 (−40)		−7 to −15
Cool-temperate ferns	−10 to −40			−5 to −40
Cool-temperate bamboos	−15 to −25	−15 to −20	−20 to −30	−5 to −15
Heathland shrubs	−15 to −25	−15 to −30	−15 to −30	−10 to −20
Meadow grasses and forbs	−10 to −25 (−30)	−10 to −30		−5 to −20
Halophytes	−10 to −20		−20 to −22	−10 to −20
Geophytes	−5 to −12		−20 to −22	−10 to −20
Hydrophytes	−5 to −10	−5 to −10	−5 to −10	−3 to −7

Arctic and alpine climates				
Boreal conifers	Down to −70 (LN_2)	Down to −70 (LN_2)	−50 to −70 (LN_2)	−20 to −35
Deciduous trees and shrubs		Down to −70 (LN_2)	−40 to −70 (LN_2)	
Dwarf shrubs	−30 to −50 (−70)	−30 to −40	−30 to −60 (LN_2)	−20 to −35
Rosette and cushion plants	−20 to −50 (LN_2)	−30 to −50 (LN_2)		−20 to −60 (LN_2)

[a] Shoot buds; flower buds are by 5–10 (30) K less resistant (cf. Table 6.14); ( ) exceptional values; X chilling-sensitive species damaged at +5° to 0 °C; (LN_2) survive immersion in liquid nitrogen.

Table 7.2. Frost resistance (LT_i at °C) of timberline trees and shrubs on Mt. Wilhelm, Papua New Guinea. The samples were collected at end of August and tested after hardening at 0 °C for 10 days. (From Sakai, unpubl.)

Plant species	Family	Altitude (m)	Leaves	Buds	Cortex	Xylem	Phenology[c]
Dacrycarpus compactus	Podocarpaceae	3600	−4	−4	−4	− 6	A
Dacrycarpus compactus	Podocarpaceae	3400	−3	−3	−3	− 5	A
Podocarpus brassii	Podocarpaceae	3500	−4	−6	−6	−10	A
Papuacedrus papuana	Cupressaceae	2700	−5	−3	−5	− 5	
Phyllocladus hypophyllus	Podocarpaceae	2700	−3	−3	−3	− 3	
Drimys piperita	Winteraceae	3700	−4	−4	−4	− 6	A
Rhododendron womersleyii	Ericaceae	3500	−4	−6	−6	−10	A, B
Rhododendron gaultherifolium	Ericaceae	3500	−3	−3	−3	−10	A, B
Rhododendron atropurpureum	Ericaceae	3400	−4	−5	−5	−10	A, B
Gaultheria mundula	Ericaceae	3400	(−6)[a]	(−6)[a]	(−6)[a]	(− 6)[a]	A, C
Diplycosia morobensis	Ericaceae	3400	−5	−5	−5	−10	A
Vaccinium keysseri	Ericaceae	3400	(−5)[b]	(−5)[b]	(−5)[b]	− 5	A
Vaccinium cruentum	Ericaceae	3400	(−5)[b]	(−5)[b]	(−5)[b]	− 5	A
Styphelia suaveolens	Epacridaceae	3400	−5	−5	−5	(−10)[a]	A, C
Pittosporum pullifolium	Pittosporaceae	3400	−4	−4	−4	− 4	A
Olearia cabrerae	Asteraceae	3400	−3	−3	−4	−10	A
Coprosma divergens	Rubiaceae	3400	(−5)[b]	(−5)[b]	(−5)[b]	− 5	A

[a] Uninjured.
[b] Killed at the indicated temperature.
[c] Phenology: A = terminal bud formation; B = growth cessation; C = flowering.

Table 7.3. Chilling and freezing injury to the leaves of tropical plants after 24 h exposure to the indicated temperatures. (From Biebl 1964)

Plant species	Low temperature treatment[a] (°C)			
	−4/−2	−1.5/0	+0.5/+1	+2/+4
Trees and shrubs				
Guarea guara	•	•	•	•
Duchatrea sintenisii	•	•	•	x
Cecropia peltata	•	•	•	o
Cordia borinquensis	•	•	•	o
Psychotria berteriana	•	•	x	o
Croton poecilanthus	•	o	o	o
Calycogonium squamulosum	•	o	o	o
Ficus nitida	•	o	o	o
Mangifera indica	•	o	o	o
Allamanda cathartica	•	o	o	o
Syngonium auritum	•	o	o	o
Codiaeum variegatum	•	o	o	o
Cestrum macrophyllum	•	o	o	o
Bougainvillea glabra	•	o	o	o
Hibiscus rosa-sinensis	x	o	o	o
Lianas				
Macgravia sintenisii	•	•	•	•
Marcgravia rectiflora	•	•	•	o
Schlegelia portoricensis	•	x	x	o
Exogonium repandum	•	x	o	o
Herbaceous plants				
Pilea obtusata	•	•	•	x
Ruellia coccinea	•	•	•	x
Psychotria uliginosa	•	•	•	o
Peperomia hernandifolia	•	•	x	o
Alpinia antillarum	•	o	o	o
Begonia decandra	•	o	o	o
Ichnanthus pallens	•	o	o	o
Pilea krugii	•	o	o	o
Guzmania berteroniana	•	o	o	o
Acalypha wilkesiana	•	o	o	o
Heliconia psiticorum	•	o	o	o

[a] o uninjured, x partially injured, • killed.

additionally chilling-sensitive (Table 7.3). Unfortunately, most studies concerned only the leaves or, if other plant organs were considered, then only with respect to their frost sensitivity and not their chilling susceptibility. Differences in chilling susceptibility between organs and tissues in a tropical plantation species, *Coffea arabica*, have been revealed by Bodner and Larcher (1987) and are presented in Fig. 7.1. It still remains to be proved whether or not and to what extent developmental phases characterized by particularly high chilling susceptibility, such as germination and fruit ripening, are of ecological significance, i.e. whether the same degree of cooling at either of these stages

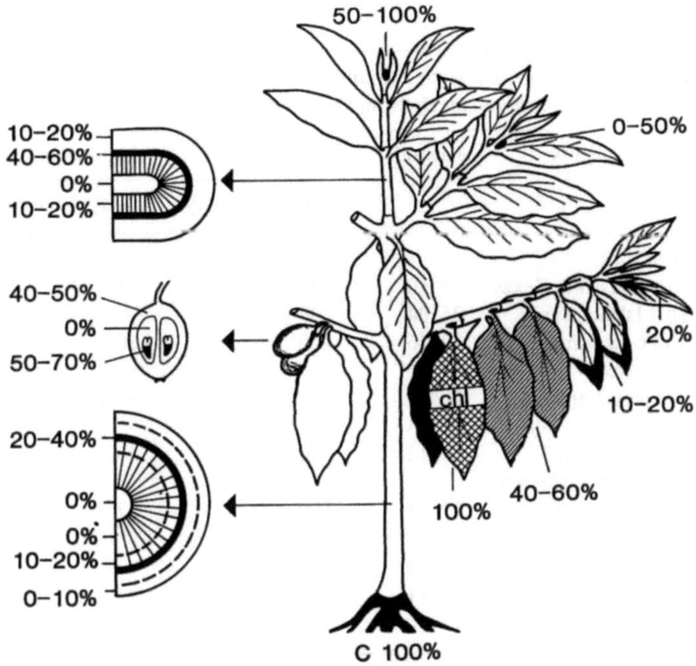

Fig. 7.1. Distribution of injuries (percent damage) after chilling at +1 °C for 36 h of 3-years-old plants of *Coffea arabica*. The most chilling-susceptible organs and tissues are the roots, even of the developing embryo in unripe fruits, and the cambial zone, including phloem in the differentiating stage; *C* cambial zone of lignified roots. Chlorotic (*chl*) and senescent leaves are considerably more susceptible than mature and young leaves. Axillary buds were damaged if ready for sprouting, resting buds remained uninjured under experimental conditions. (According to Bodner and Larcher 1987 and unpublished data of A. Comploj)

does the same amount of damage to plants growing in the field and what consequences this would have on propagation (Larcher and Bauer 1981).

7.1.1 Mangroves

Mangroves extend from tropical habitats to 32° N and to 38° S, i.e. they reach extratropical latitudes on the Satsuma peninsula (Hosokawa et al. 1977) at the Gulf of Mexico, and the Gulf of Akaba (Chapman 1970), in spite of their pronounced chilling sensitivity (Biebl 1964). The latitudinal differentiation among populations of three mangroves with respect to chilling response was investigated by McMillan (1965) and Markley et al. (1982). Of the mangroves tested, *Avicennia germinans* was the least sensitive to chilling and was altogether the species with the widest range of variation (Table 7.4). The populations of *Laguncularia racemosa* were intermediate in adaptive differentiation. *Rhizophora mangle*, the mangrove that occupies the coastal habitat furthest from land, exhibited the narrowest range of variation. Within each species the more northern populations had a greater tolerance to chilling temperatures than those of more equatorial origin.

Table 7.4. Survival of mangrove plants from various latitudes after exposure to +2° to +4 °C. (From Markley et al. 1982)

(a) Percentage of surviving seeds and unrooted seedlings of *Avicennia germinans, Laguncularia racemosa* and *Rhizophora mangle* from various provenances

Collection locality	Latitude (° N)	*Avicennia*		*Laguncularia*		*Rhizophora*	
		3 days	6 days	3 days	6 days	3 days	6 days
Western sites							
Harbor Island, Texas	27°50′	100	96	–	–	–	–
Port Island, Texas	26°04′	100	88	–	–	–	–
La Pesca, Tamps., Mexico	23°46′	100	55	60	60	100	100
Tuxpan, Veracruz, Mexico	20°57′	48	22	66	66	–	–
Anton Lizardo, Veracruz, Mexico	19°12′	66	0	40	40	100	60
Belize City, Belize	17°31′	100	11	7	0	50	50
Eastern sites							
Biscayne Bay, Florida	25°22′	100	22	–	–	100	100
Key Largo, Florida	25°11′	83	38	73	93	100	90
Port Royal, Jamaica	17°50′	38	0	20	13	–	–

(b) Mean percentage of leaf injury and survival after chilling treatment of 5.5-year-old plants of *Avicennia germinans* which were cultivated under constant laboratory conditions

Provenance	Leaf injury after		Survival after	
	24 h at 2° to 4 °C	48 h at 2° to 4 °C	24 h at 2° to 4 °C	48 h at 2° to 4 °C
Western sites				
Harbor Island, Texas	0	0	100	100
La Pesca, Tamps., Mexico	4	11	100	100
Tuxpan, Veracruz, Mexico	17	86	100	80
Anton Lizardo, Veracruz, Mexico	54	100	75	20
Belize City, Belize	87	100	75	40
Eastern sites				
Biscayne Bay, Florida	17	79	100	100
Key Largo, Florida	10	81	100	60
Port Royal, Jamaica	97	100	75	20

After chilling pretreatment of *Avicennia* from habitats at 27°50′N and 17°29′N those from the more northerly latitude showed a higher degree of unsaturation of glycolipids and phospholipids than those from the lower latitude. This evidence and the fact that plants kept 5 years under laboratory conditions maintained their characteristic chilling sensitivity, suggest that the diversification resulting from adaptive selection is based on inheritance. The studies on mangroves thus reveal that the first step in the acquisition of tolerance to low temperatures is via the development of less chilling-sensitive ecotypes. *Conocarpus erecta*, a shrub that still occurs within the mangrove belt, but also withstands drier conditions, is already tolerant to low temperatures above zero (Biebl 1964).

7.1.2 Palms

Palms have never succeeded in establishing themselves in regions with severe frost. Arecaceae, therefore, are considered as a plant family typical for the tropical zone, as defined by the Tropics of Cancer and Capricorn (Good 1974). Only few palm genera extend beyond the tropics into warm-temperate regions (Moore 1973; Fig. 7.2). Among the coryphoid palms, *Chamaerops humilis* advances as far as 40° N in Spain and Sardinia thus delimiting the extreme northward deviation of the distribution area of palms. Species of *Sabal* and *Washingtonia* in N. America and of *Trachycarpus* in E. Asia advance up to 35° N. Among the feather palms, *Phoenix canariensis* is native to the Canary Islands and Madeira (30° N), naturalized *Phoenix dactylifera* is found in Crete (35° N) and *Phoenix sylvestris* at 35° N in the Punjab (Meusel 1965). In the Southern Hemisphere, *Jubaea chilensis* delimits at 33° S the distribution area of palms in South America, *Phoenix reclinata* at 32° S in S. Africa, and arecoid palms at 40° S in New Zealand. Altitudinal limits are 2400 m a.s.l. for *Trachycarpus* species in the Himalayas (Beccari 1933) and 3000-4000 m a.s.l. for *Ceroxylon* species in the tropical Andes (McCurrach 1960; K. Mägdefrau, pers. commun.).

In order to compare thermal distribution limits with the family's potential to survive low temperatures, the specific frost resistance of various organs and developmental

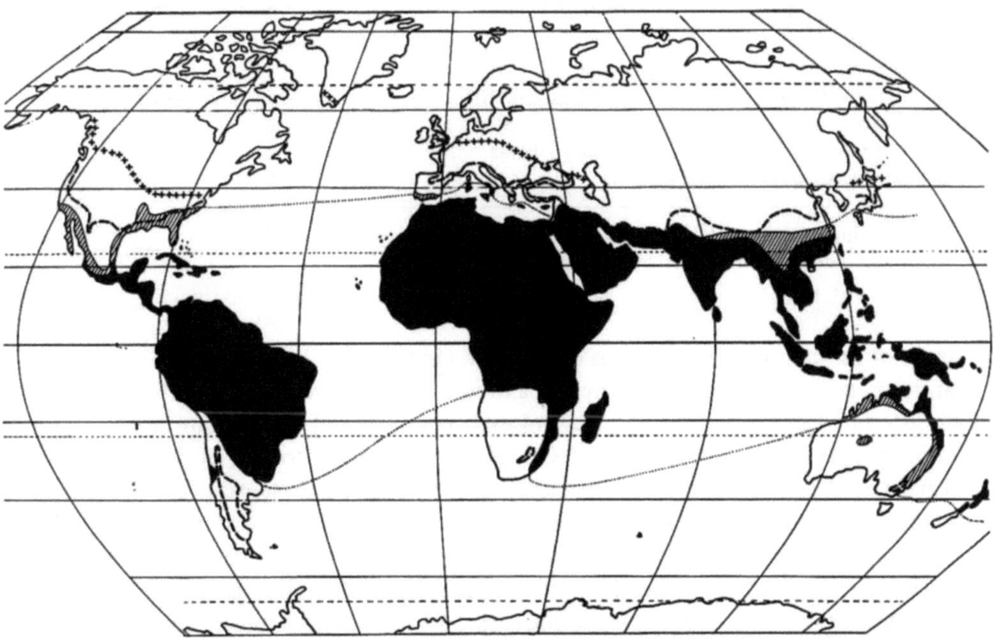

Fig. 7.2. Geographic distribution of palms. *Black:* fan and feather palms; *hatched area:* only fan palms; (+++) northern boundary of Tertiary palm fossils. *Broken line:* −15 °C annual minimum isotherm. (From distribution maps of Irmscher 1922; Moore 1973; isotherm after Hoffmann 1960)

Table 7.5. Freezing susceptibility of palm leaves. The threshold freezing temperatures indicate incipient frost injury if no previous disturbance at low positive temperatures occurred. At *mean freezing temperature* large parts of the leaves are killed. (From Larcher and Winter 1981)

Palm species[a]	Distribution[b]	T_f (°C)	Mean freezing temperature (°C)
Subtropical fan palms			
Washingtonia filifera (s)	S. California, Arizona (33° N)	– 4	– 6
Washingtonia filifera (a)		– 8	–10
Livistona australis (a)	Australia (40° S)	– 8	– 9
Livistona chinensis (a)	S. China (30° N)	– 9	–10.5
Rhapis excelsa (a)	S. China	– 8	–10.5
Trithrinax acanthocoma (a)	S. Brazil (30° S)	– 9	–10.5
Chamaerops humilis (a)	S. Mediterranean (40° N)	– 9	–11.5
Serenoa repens (s)	S. Carolina, Florida (32° N)	–10	–12.5
Sabal minor (a)	N. Carolina, Texas (33° N)	–10.5	–13.5
Trachycarpus fortunei (s)	China, Japan (32° N)	– 9	–12
Trachycarpus fortunei (a)		–11	–14
Feather palms			
Elaeis guineensis	W. Africa	– 3	– 3.5
Cocos nucifera (j)	trop. Pacific islands	– 3	– 3.8
Chrysalidocarpus lutescens (j)	Madagascar	– 3	– 4
Caryota urens (a)	Indomalaya	– 4	– 5.5
Chamaedorea costaricana (a)	C. America	– 4.5	– 6
Euterpe edulis (s)	Brazil	– 5	– 6.5
Howea forsterana (a)	S. Australia (40° S)	– 5.5	– 6
Phytelephas macrocarpa (a)	trop. S. America	– 6	– 6.5
Jubaea chilensis (s)	Chile (33° S)	– 7	– 7.5
Jubaea chilensis (a)		– 7	– 9
Phoenix roebelenii (a)	Assam	– 7	– 8
Phoenix reclinata (a)	S. Africa (32° S)	– 7	–10.5
Phoenix dactylifera (a)	N. Africa (35° N)	– 8	– 9.5
Phoenix canariensis (s)	Canary Islands	– 6	– 7.5
Phoenix canariensis (a)		– 9	–10.5

[a] (s) = seeding, (j) = juvenile, (a) = adult plants.
[b] Latitudes indicate extreme distribution limits.

stages of selected palm species was determined (Larcher and Winter 1981). The temperatures below which frost injuries appear and which cause severe freezing of palm leaves are listed in Table 7.5. Leaf damage and even complete defoliation can be, but is not necessarily, lethal for palms. Recovery of *Trachycarpus fortunei* after severe frost damage of the foliage has been repeatedly observed. In all these cases the shoot apex was still living due to better protection and probably also higher frost resistance. From Fig. 7.3 it becomes evident that the shoot apex indeed attains higher resistance than the leaves, provided preceding cold weather has forced the plant to arrest its growth processes. Palm roots, especially the root tips, are particularly susceptible to freezing. Even the crown roots above the soil surface, where during a radiative frost the lowest temperatures normally occur, are more sensitive than most of the shoot tissues. As a consequence, palms are severely affected by ground frost.

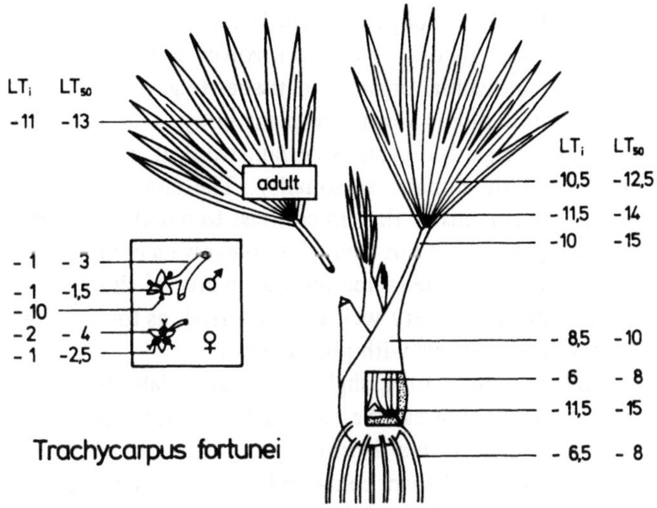

Fig. 7.3. Frost resistance (°C) of juvenile plants and of leaves and flowers of adult *Trachycarpus fortunei*. (From Larcher and Winter 1981)

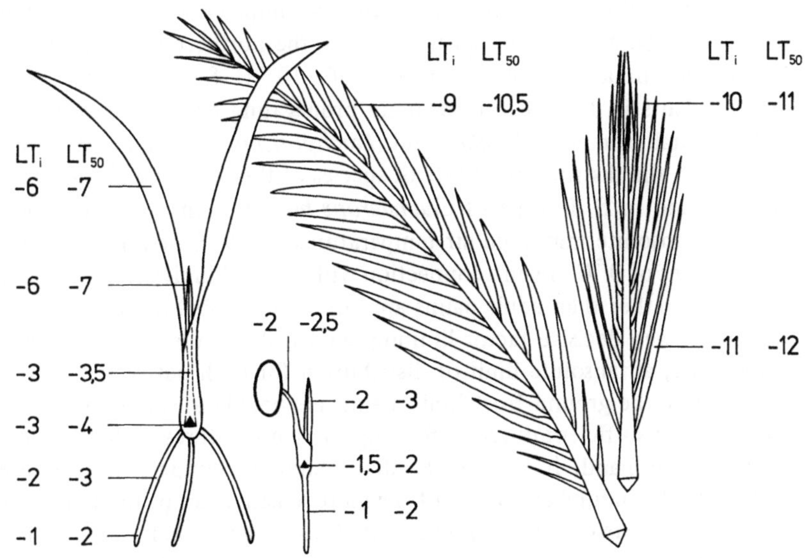

Fig. 7.4. Frost susceptibility (°C) of emerging and 1-year-old seedlings and of leaves of adult plants of *Phoenix canariensis*. (From Larcher and Winter 1981)

Seeds are very resistant to frost. If air-dry seeds of *Trachycarpus fortunei* and of *Washingtonia filifera* were cooled to −70 °C for 12 h, most maintained their ability to germinate. Hydrated seeds are considerably less resistant: 50% watersoaked seeds of *Trachycarpus* were killed after cooling to −20 °C, of *Washingtonia* to −10 °C. After cooling to −16 °C, 20% of hydrated seeds of *Washingtonia* remained alive. As a rule, germinating seeds and young seedlings are the weakest developmental stages (Fig. 7.4).

A peculiar feature of palms is the lack of large seasonal changes in frost susceptibility, except in those tissues which are involved in extension growth. In palms the state of activity of individual organs and tissues (e.g. of the shoot apex) apparently does not affect the rest of the vegetative plant body. This kind of autonomy of the individual parts of the organism typifies certain tropical trees, in which development is not synchronized with climatic seasonality. The analogous behaviour of *subtropical* palms leads to the conclusion that, in contrast to woody angiosperms that penetrate to higher latitudes, palms have not been able to evolve a complete timing of their annual pattern of growth to match the increasing seasonality of climate. Nevertheless, certain processes in development are very well synchronized, as for example the process of flowering in *Trachycarpus fortunei* with photoperiod.

If plants exhibit only slight seasonal variability of their frost susceptibility, little effect on hardening can be expected from lowering temperatures. Leaves of juvenile plants of *Trachycarpus fortunei* were only 1–2 K less frost resistant after 10 days' exposure to 20 °C than cold-conditioned plants. Combined drought and cold treatment resulted in a shift of the injurious temperature level from about −10 °C to about −12.5 °C. Thus, the greatest adaption amplitude of juvenile *Trachycarpus* does not exceed 3–4 K. The seasonal and adaptive effect on frost resistance of mature leaves of *Washingtonia filifera* and *Chamaerops humilis* was only 1 K or less.

If the natural distribution area of palms is compared with frost resistance levels and isotherms, good agreement appears between the northern boundary for fan palms and the average annual minimum isotherm of −10 °C. Since the frost susceptibility of juvenile plants is decisive for naturally propagating plants, the temperature limit for seedling establishment of *Sabal* and *Trachycarpus* should be at about −12 °C. In N. America the distribution limits of species of the *Livistona* alliance (including *Washingtonia*) and in S. Europe of *Chamaerops humilis* seem to correspond with the isotherm for absolute minimum air temperatures of −5 °C and average annual minimum temperatures of 0 °C. Again a connection with survival limits of seedlings (and probably also roots) rather than adult palms appears. Very probably, therefore, frost is an important or even decisive factor limiting the distribution of the palm family as a whole. In any case, an absolute limit of distribution for both native and cultivated palms is undoubtedly the ground frost limit. Palms are unable to survive in regions where frost periods are of sufficient duration for negative temperatures to develop in the soil itself. On the other hand, it is unlikely that minimum temperatures represent the main obstacle to the natural expansion of the more sensitive fan palms and the feather palms. *Phoenix canariensis*, which survives −10 °C as an adult and −5 °C as a juvenile plant, can be cultivated, but does not naturally distribute at elevations much higher than 600 m a.s.l. on the Canary Islands. At this altitudinal level the lowest temperatures are on an average +6 °C and never lower than 0 °C. If the distribution areas (as presented by Moore 1973) of the other feather palms are compared with isotherms below 0 °C (e.g. Hoffmann 1960) no clear correlation can be recognized.

7.2 Broad-Leaved Evergreen Trees and Shrubs

Broad-leaved evergreen woody plants are typical of regions with a favourable temperature all year round; they occur mainly under three climatic regimes: (1) in the wet tropics; (2) in the warm-temperate zone on islands and in coastal regions exposed to oceanic influence (Fig. 7.5a,b,d) and at the humid side of low latitude mountain ranges; (3) in regions with Mediterranean-type climate with dry summer and a wet season from late autumn to spring (Fig. 7.5c).

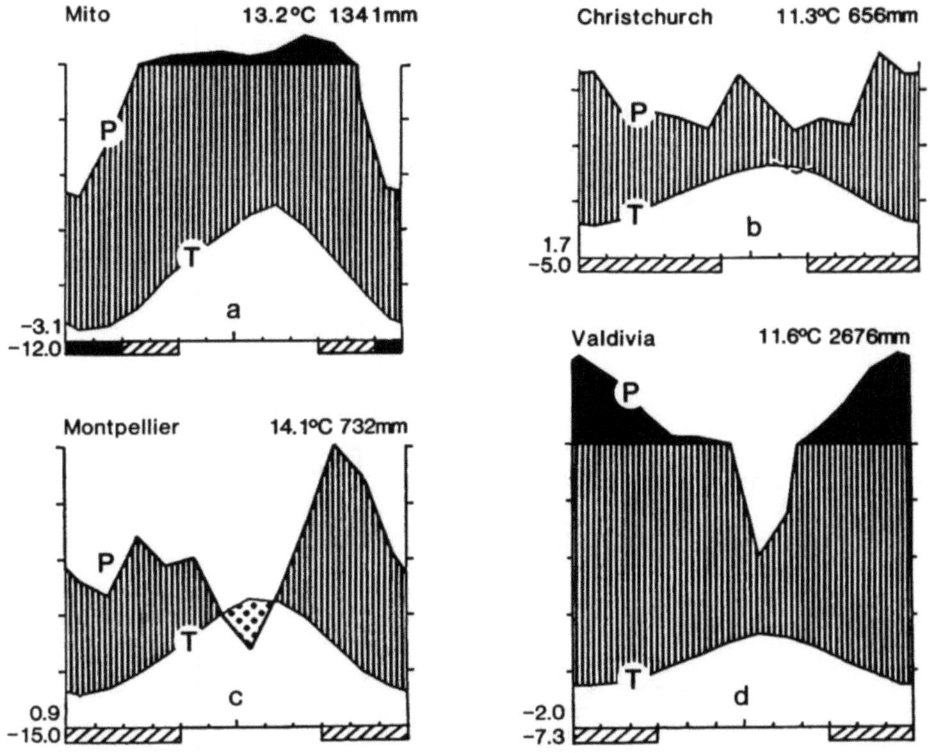

Fig. 7.5a–d. Climate diagrams for: **a** Mito, Japan, representative of the northern limit of humid, warm-temperate climate in E. Asia; **b** Christchurch, New Zealand, Southern Hemisphere type of highly oceanic climate; **c** Montpellier, France, near the northern limit of Mediterranean climate; **d** Valdivia, Chile, oceanic climate near the transition to Southern Hemisphere Mediterranean-type climate. *T* mean air temperature (°C) and *P* mean precipitation (mm) annual means are given at top for each locality. *Abscissa:* for Northern Hemisphere stations, months from January to December, for Southern Hemisphere, from July to June; *numbers upper left:* mean daily minimum of the coldest month; *numbers lower left:* absolute temperature minimum. *Black bar:* months with mean daily minimum below 0 °C; *hatched bar:* months with absolute minimum below 0 °C. *Ordinate:* one subdivision represents 10 °C or 20 mm precipitation. *Vertical shading:* humid season; *black:* perhumid; *stippled area:* arid season. (From Walter and Lieth 1967)

7.2.1 Temperate Rain Forest Species

In the northern temperate zone, broad-leaved evergreen rain forests are confined to oceanic winter-mild areas as seen in coastal regions of E. Asia and N. America. Evergreen forests also extend from the south side of the E. Himalaya mountains and the uplands of NW. Yunnan eastwards in a broad belt across southern China to Taiwan and Japan. In the warm-temperate zone, between 1000 m and 2500 m altitude at the

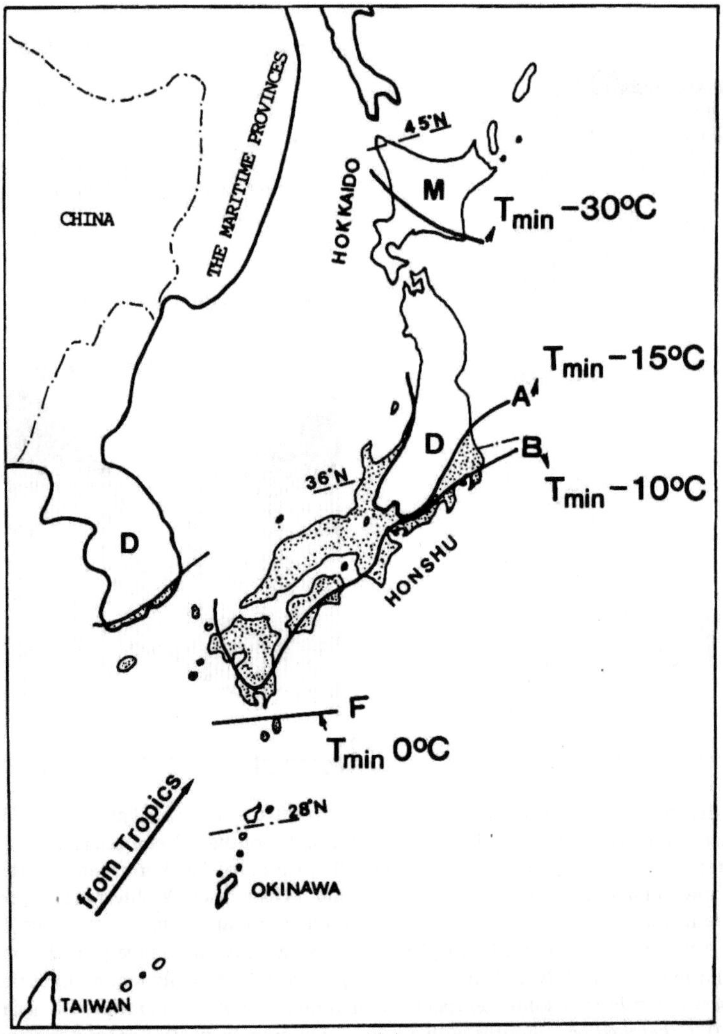

Fig. 7.6. Tree vegetation and frost lines in Japan. *M* mixed deciduous broad-leaved and subboreal coniferous forest in Hokkaido; *D* temperate deciduous forest; *A* northern borderline of the natural distribution of most warm-temperate evergreen tree species; *B* northern limit for evergreen trees extending from tropical and subtropical ES. Asia; *F* northern limit of the frost-free region of Japan, Yakushima and Tanegashima. (From Sakai 1978a)

E. Himalayas and the Yunnan uplands, the predominant plant communities are vigorously growing broad-leaved and coniferous forests rich in epiphytes. The climate of this region is characterized by a narrow annual range of temperature (see Fig. 7.26), high rainfall in summer and rather high air humidity at moderate temperatures in winter, which are similar to the southern oceanic climates (Sakai et al. 1981).

In Japan, different climatic regimes, from severe cold to subtropical, range from north to south along the archipelago. There is ample precipitation for plants. Temperature is thus the dominant factor determining the vegetation zones (Fig. 7.6). With increasing winter cold, the evergreen species except for plants on the forest floor, decrease considerably in number and in diversity as deciduous trees increase to become dominant at mid- or northern latitudes, especially at inland mountainous areas. Most of the evergreen trees near the northern limits of their natural range are hardy between -15° and -18 °C (Table 7.6). Their natural ranges concide with the mean air temperatures of 1° to 2 °C in the coldest month, which is comparable to extreme minimum temperatures between -15° and -18 °C. Some shrubs, such as *Camellia japonica, Daphniphyllum macropodum, Euonymus japonicum* and *Pieris japonicum*, which extend far north along the seacoast, were found to be the hardiest of the evergreen woody plants in Japan, resisting between -20° and -23 °C. Their potential frost resistance roughly coincides with the minimum air temperature of their natural ranges (Sakai 1978a).

The temperature rain forests of the Southern Hemisphere even in latitudes far south, are characterized by the nearly total absence of deciduous hardwoods (Wardle 1963; Sakai and Wardle 1978). Broad-leaved evergreen forests and warm-temperate coniferous forests prevail under southern temperate climates which are characterized by a very narrow seasonal range of temperature, high precipitation well distributed throughout the year and mild winters. New Zealand, as a whole, experiences an oceanic climate with high rainfall and a relatively narrow annual range of temperature (Sakai et al. 1981). The temperature range is larger in the interior (Craigieburn Range) and east (Christchurch) than in the west (Hokitika) of the South Island. This is reflected in the distribution of low-altitude tree species which, in conformity with their specific freezing resistance (Table 7.7) are more numerous along the west coast than the east, with few reaching the interior highlands (Sakai and Wardle 1978). Among the subalpine coniferous species, the hardiest, *Dacrydium bidwillii*, ranges even on the floor of an intermontane basin. *Nothofagus solandri*, which forms pure stands extending to the upper tree limit on adjacent slopes, does not reach the lowest parts of this basin, where minimum temperatures fall below the freezing resistance of *N. solandri* seedlings (Wardle and Campbell 1976a). As the climates became warm during early, post-glacial times, there was a sequence of pollen dominance from *Dacrydium* (probably *D. bidwillii*) through *Phyllocladus* (probably *P. asplenifolius* var. *alpinus*) to *Podocarpus* and *Dacrycarpus*, and this sequence fits well with the order of decreasing freezing resistance (Wardle and Campbell 1976a; Sakai and Wardle 1978).

On the Australian continent the annual temperature ranges at Canberra and in the Snowy Mountains are considerably greater than those experienced in New Zealand. This, in turn, agrees with the greater freezing resistance measured in *Eucalyptus pauciflora* as compared with the subalpine evergreen *Nothofagus* of New Zealand (Table 7.7). *Eucalyptus pauciflora* extends to the upper tree limit in the Snowy Mountains at about

Table 7.6. Freezing resistance (LT_0 at °C) of broad-leaved evergreen trees in Japan. (From Sakai 1978a, 1982c)

(a) Species approaching the northern limits of the warm-temperate zone (cf. Fig. 7.6a:A)

Plant species	Leaf	Bud	Cortex	Xylem
Cinnamomum camphora	−10	− 8	−10	−12
Cinnamomum japonicum	−10	−10	−10	−12
Neolitsea sericea	− 8 to −10	− 8	−10	−15 to −17
Machilus thunbergii	−12	−10	−10	−12
Pasania edulis	−10	−12	−15	−15 to −17
Castanopsis cuspidata var. *sieboldii*	−13	−13	−13	−15
Ligustrum japonicum	−15	−15	−15	−17
Osmanthus ilicifolius	−15	−17	−17	−20
Ilex integra	−15	−15	−15	−18
Cleyera japonica	−15	−15	−15	−15 to −17
Ternstroemia japonica	−17	−17	−17	−20
Quercus gilva	−10	−10	−13	−13
Quercus salicina	−15	−15	−15	−15
Quercus phillyraeoides	−15	−15	−15	−20
Quercus myrsinaefolia	−15	−15	−15	−15 to −17
Quercus glauca	−15	−15	−15	−18
Quercus acuta	−15	−17	−17	−20
Quercus sessiliflora	−15	−15	−15	−18
Eurya japonica	−15	−15	−17	−17
Illicium religiosum	−20	−20	−18	−20
Camellia japonica	−15 to −20	−15 to −18	−20	−20
Daphniphyllum macropodum	−20	−20	−20	−
Pieris japonica	−22	−22	−25	−

(b) Species from subtropical regions extending to the southwestern warm seacoast of Japan (cf. Fig. 7.6:B)

Plant species	Leaf	Bud	Twig	Distribution
Ficus superba var. *japonica*	(− 3)[a]	− 3	− 3	SW. Japan, SE. Asia
Schefflera arboricola	(− 3)[a]	− 3	− 3	S. Kyushu, SE. Asia
Syzygium buxifolium	− 3	− 5	− 5	S. Kyushu, Okinawa, SE. Asia
Turpinia ternata	− 5	− 5	− 7	Shikoku, Kyushu, Ryukyu, Taiwan
Actinodaphne lancifolia	− 7	− 7	− 7	SW. Japan, Taiwan
Distylium racemosum	−10	−10	−10	S. Japan, Taiwan
Diospyros morrisiana	− 5	− 5	− 5	S. Japan, Taiwan, S. China
Elaeocarpus sylvestris	− 7	− 7	− 7	S. Japan, Taiwan, S. China
Helicia cochinchinensis	− 8	− 7	− 7	Ryukyu, Taiwan, SE. Asia
Litsea japonica	−10	− 8	− 8	Ryukyu
Myrica rubra	−10	−10	− 8	Ryukyu, Taiwan, SE. Asia

[a] Injured at −3 °C.

1900 m a.s.l., where it becomes stunted (Fig. 7.7). Snow that may persist for 5 months, compared with 3 months in New Zealand, gives undergrowth species such as *Podocarpus lawrencii* more constant protection than elsewhere. On valley floors and frost hollows, where trees are absent due to ponding of cold air, an inverted tree line may have developed (Geiger 1957; Wardle 1974; Moore and Williams 1976).

Table 7.7. Frost resistance (LT_0 at °C) of trees of the southern temperate zone. (From Sakai et al. 1981)

Plant species	Leaf	Bud	Cortex	Xylem	Collection site
Conifers					
Dacrydium bidwillii	−23	−23	−23	−23	New Zealand, 670 m a.s.l. (frost basin)
Dacrydium cupressinum	− 8	− 8	−10	−10	New Zealand, Westland 90 m a.s.l.
Phyllocladus asplenifolius	−10	−13	−13	−13	New Zealand, Westland 30 m a.s.l.
Phyllocladus asplenifolius var. *alpinus*	−22	−20	−23	−23	New Zealand, 1200 m a.s.l.
Podocarpus nivalis	−22	−20	−22	−22	New Zealand, 910 m a.s.l.
Podocarpus ferrugineus	−10	−17	−10	−10	New Zealand, Westland 30 m a.s.l.
Athrotaxis cupressoides	−20	−20	−20	−20	Tasmania, 1000 m a.s.l.
Angiosperms					
Eucalyptus pauciflora	−15	–	−18	−20	Snowy Mountains, 1760 m a.s.l.
Eucalyptus pauciflora	−16	−20	−20	−20	Snowy Mountains, 1900 m a.s.l.
Eucalyptus gunnii	−15	–	−18	−18	Tasmania, 1000 m a.s.l.
Eucalyptus cinerea	− 8	− 8	−12	−12	Canberra
Nothofagus fusca	− 8	−10	−10	−17	New Zealand, 230 m a.s.l.
Nothofagus solandri var. *cliffortioides*	−13	−15	−15	−15	New Zealand, 1370 m a.s.l.
Nothofagus gunnii (deciduous)	–	−17	−17	−17	Tasmania, 1000 m a.s.l.
Nothofagus antarctica (deciduous)	–	−22	−22	−22	Chile (planted in Christchurch)

Most Australasian conifers are comparable in freezing resistance only with the least hardy Japanese conifers, such as *Podocarpus macrophyllus* (potential frost resistance −13 °C), that are native to warm temperate seacoast and subtropical districts. Figure 7.8 shows a histogram of frost resistance of conifers of the Southern Hemisphere and Japan. Even the four hardiest southern conifers (cf. Table 7.7) show only the same order of hardiness as species native to warm-temperate parts of Japan, such as *Cryptomeria japonica* and *Abies firma*. *Nothofagus* likewise compares with the Japanese warm-temperate evergreen Fagaceae, but is much less hardy than the temperate deciduous species.

The low level of frost resistance in Southern Hemisphere broad-leaved trees is paralleled by their evergreen habit and tendency to lack specialized resting buds. Unprotected buds are especially characteristic of *Eucalyptus*, including subalpine species. Presumably, the evergreen habit and the absence of dormant buds are related to the small and variable photoperiodic responses in the genus (Paton 1978). In the hardiest *Eucalyptus* species at least, the ability to respond readily to hardening temperatures appears partially to compensate for the absence of the short-day induction of winter dormancy. Compared with *Eucalyptus, Nothofagus* shows evidence of closer adaptation to seasonally cold climates. *Nothofagus solandri* buds exhibit true dormancy that is broken by chilling (Wardle and Campbell 1976b).

Fig. 7.7. Stunted *Eucalyptus pauciflora* at the upper tree line on the Snowy Mountains at 1900 m a.s.l. (Photo: A. Sakai)

Chilean climatic stations show a small range in annual temperature (Fig. 7.26) which is consistent with the presence of broad-leaved evergreen trees and shrubs. In a comparative study Steubing et al. (1983) and Alberdi et al. (1985) found five laurel woodland species of Proteaceae resistant to -7° to -9 °C (leaves), -14° to -18 °C (buds) and -10° to -19 °C (stems), and the phylogenetically primitive low shrub *Drimys winteri* var. *andina* resistant to -10 °C (leaves) and -13 °C (buds, cortex, cambial zone and xylem).

Figure 7.9 and Table 7.8 give an idea of the range of variation of the potential frost resistance of evergreen leaves of trees and shrubs of warm-temperate regions of the Northern and Southern Hemisphere. It can be seen that very hardy evergreen woody plants whose leaves withstand freezing below -30 °C were not evolved in the warm-temperate zone.

An exception is provided by the genus *Rhododendron* (Sakai et al. 1986); its approximately 600 species, in sizes from dwarf shrubs to trees of 15 m or more, extend

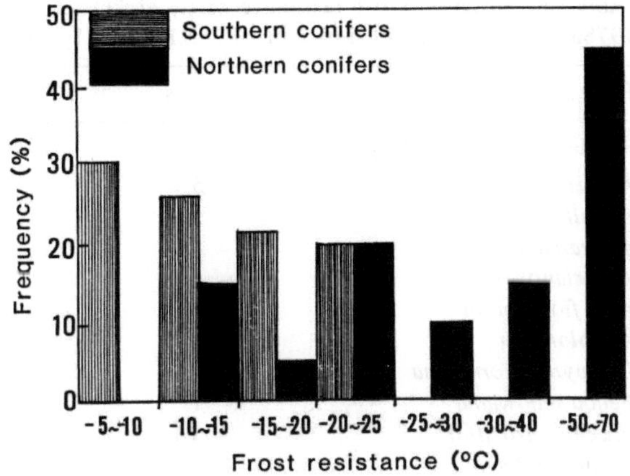

Fig. 7.8. Frequency diagram for potential leaf frost resistance of conifers of Japan and the Southern Hemisphere. Frequency: Number of species of the various resistance classes expressed as percentage of a total of 25 (Japan) or 30 (S. Hemisphere) species (A. Sakai, unpubl.)

over a large distribution area which includes environmental conditions as different as subtropic, warm-temperate, cool-temperate, alpine and subarctic climates. The genus *Rhododendron* originates from a limited area in SW. China which covers NW. Yunnan, SE. Tibet and W. Szechwan between 22° and 30° N. Most of the *Rhododendron* species are moderately hardy to between -20° and -25 °C (Table 7.9), including those growing near the timberline at the E. Himalaya (Sakai et al. 1981b) and from high altitudes of NW. Yunnan, where moderate temperatures and high humidity prevail during winter. Very hardy rhododendrons which, except for the dwarf forms, survive below -30 °C are *R. brachycarpum, R. metternichii* and its varieties native to mountainous areas in Japan, and *R. catawbiense, R. maximum* and *R. carolinianum* in the Appalachian Mts. (Table 7.9). *R. brachycarpum*, which occurs in subalpine regions in Japan, is the hardiest among the *Rhododendron* species tested, resisting freezing to -60 °C in the leaves and

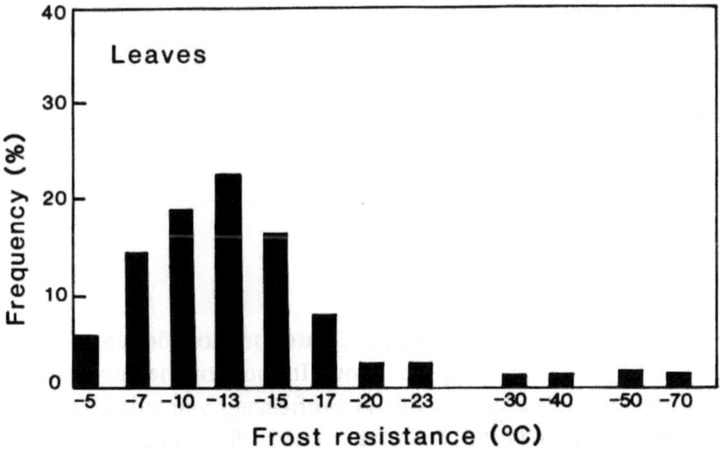

Fig. 7.9. Frequency diagram for leaf frost resistance of broad-leaved evergreen trees and shrubs, based on a total of 90 species. Only leaves of *Kalmia latifolia* and of several *Rhododendron* species survive freezing below -30 °C. (From Sakai 1980)

Table 7.8. Potential frost resistance (LT_0 at °C) of broad-leaved evergreen trees. [From Sakai 1978a,d, 1982c), extended by data of Larcher (1971); see also Tables 7.6, 7.7 and 7.9)]

Plant species	Leaves	Buds	Twigs
Species of subtropical regions			
Acacia baileyana	− 5	− 5	−10
A. cultriformis	− 5	− 5	− 7
A. dealbata	− 5	− 7	− 7
A. melanoxylon	− 8	−10	−12
Acer floridanum	− 8		
A. oblongum	− 8		
Diospyros morrisiana	−10		
Feijoa sellowiana	− 7	− 6	− 7
Fraxinus griffithii	− 5		
Nerium indicum	− 8	− 8	− 8
Prunus spinulosa	− 7		
P. zippeliana	−10	−10	− 8
Pyrus koehnei	− 8		
Rosa wichuraiana	−13		
Species of regions with warm-temperate and maritime climates			
Aucuba japonica	−15	−20	−20
Banksia marginata	−10		
Eucalyptus cinerea	− 8	− 8	−10
E. globulus	− 5	− 7	− 7 to −10
E. gunnii	− 8 to −15	− 8 to −15	−10 to −15
E. pauciflora	−15	−15 to −20	−17 to −20
E. viminalis	− 8	− 8	−10
Euonymus japonicus	−22	−22	−22
Gardenia jasminoides	−10	−10	−10
Ilex aquifolium	−15	−18	−23
I. crenata	−22	−22	−22
I. opaca	−20	−17	−25
Jasminum odoratissimum	−10	−10	−10
Magnolia grandiflora	−20	−17	−20
Myrica californica	−15		
Nandina domestica	−12	−12	−15
Osmanthus fragrans	−13		−13
Photinia glabra	−15	−15	−15
Ph. serrulata	−17	−18	−17
Viburnum odoratissimum	−10	−10	−10

−30 °C in the flower buds. Even the hardiest rhododendrons are confined to habitats protected by a forest canopy or snow. In most of the very hardy rhododendrons the leaves curl and move downwards at subfreezing temperatures (Fig. 7.10). In a quantitative study, Nilsen (1985) found that leaf curling of *Rhododendron maximum* started at 0 °C and was fully developed at about −6 °C, whereas leaf angle depression was related to temperature and water potential. The question of whether and how leaf curling might favour winter survival of *Rhododendron* requires further investigations.

Table 7.9. Frost resistance (LT_0 at °C) of the genus *Rhododendron*. (From Sakai and Malla 1981; Sakai et al. 1986)

Species	Series, subseries	Leaves	Flower bud	Shoot bud	Cortex	Xylem	Distribution
Cool-temperate zone							
**R. camtschaticum*	Camtschaticum	D	–34	–60	–60	–40	NE. Asia
**R. parvifolium*	Lapponicum	–60	–32	–60	–60	–50	E. Asia
R. brachycarpum	Ponticum	–60	–30	–60	–60	–35	N. Japan
R. catawbiense	Ponticum	–60	–30	–60	–60	–40	E. North America
R. maximum	Ponticum	–60	–27	–60	–60	–35	NE. North America
R. dauricum	Dauricum	–50	–30	–60	–60	–40	NE. Asia
R. metternichii var. *pentamerum*	Ponticum	–50	–24	–50	–60	–30	N. Japan
R. mucronulatum	Dauricum	D	–30	–50	–50	–40	E. Asia
R. carolinianum	Carolinianum	–40	–27	–50	–50	–35	SE. North America
R. ponticum	Ponticum	–35	–23	–30	–30	–30	S. Europe, Caucasus, Asia Minor
R. viscosum	Azalea, Luteum	D	–30	–50	–60	–40	E. North America
R. arborescens	Azalea, Luteum	D	–30	–40	–60	–40	E. North America
R. canadense	Azalea, Canadense	D	–30	(–40)[z]	(–40)[z]	(–40)[z]	E. North America
R. japonicum	Azalea, Luteum	D	–25	–40	–50	–40	NE. Japan
R. schlippenbachii	Azalea, Schlippenbachii	D	–28	–35	–35	–30	C. and N. China
R. albrechtii	Azalea, Canadense	D	–	–35	–35	–30	Japan
High elevations at intermediate latitudes							
**R. aureum*	Ponticum	–60	–27	–60	–60	–50	C. Japan (2500–3000 m), NE. Asia
**R. impeditum*	Lapponicum	–50	–25	–50	–50	–	Yunnan (3300–3900 m)
**R. dasypetalum*	Lapponicum	–50	–25	–50	–50	–35	Yunnan (3200–4000 m)
**R. anthopogon*	Anthopogon	–50	–23	–50	–50	–40	Nepal (3000–4800 m)
**R. setosum*	Lapponicum	–50	–23	–50	–50	–50	Nepal, Tibet (3300–4800 m)
**R. fastigiatum*	Lapponicum	–40	–24	–40	–40	–30	Yunnan (3100–4100 m)
**R. scintillans*	Lapponicum	–40	–20	–40	–40	–35	Yunnan (3300–4200 m)
**R. russatum*	Lapponicum	–40	–20	–40	–40	–30	Yunnan (3300–4000 m)

Table 7.9 (continued)

Species	Series, subseries	Leaves	Flower bud	Shoot bud	Cortex	Xylem	Distribution
**R. lepidotum*	Lepidotum	−40	−18	−30	−40	−30	Nepal, Yunnan (2500–4500 m)
**R. intricatum*	Lapponicum	−35	−23	−40	−50	−35	Yunnan (3600–4500 m)
**R. fimbriatum*	Lapponicum	−30	−25	−40	−40	−40	Szechwan (3600 m)
**R. ferrugineum*	Ferrugineum	−30	−25	−25	−35	−30	European Alps (1500–2200 m)
**R. keiskei*	Triflorum	−25	−23	−30	−30	−25	Japan (600–1800 m)
**R. forestii* var. *repens*	Neriflorum	−20	−20	−20	−20	−23	Yunnan Tibet (3000–4300 m)
R. smirnowii	Ponticum	−30	−23	−40	−40	−30	Caucasus (1500–2000 m)
R. wallichii	Campanulatum	−28	−20	−30	−30	−25	Nepal, Sikkim (3300–4000 m)
R. campylocarpum	Tomsonii	−25	−20	−30	−25	−25	Nepal, Sikkim (3300–4000 m)
R. hodgsonii	Falconeri	−25	–	−30	−25	−25	Nepal, Bhutan (3300–3900 m)
R. campanulatum	Campanulatum	−25	−20	−30	−30	−25	Nepal, Bhutan (2800–2900 m)
R. griffithianum	Griffithianum	−23	−20	−23	−23	−23	Nepal, Sikkim (2100–2800 m)
R. wardii	Thomsonii	−20	−20	−27	−27	−23	Yunnan, Tibet (3000–4000 m)
R. cinnabarinum	Cinnabarinum	−18	−18	−18	−20	−20	Nepal, Sikkim (3000–3600 m)
R. barbatum	Barbatum	−18	−18	−18	−18	−18	Nepal (3000–3400 m)
Warm-temperate and subtropical zones							
R. metternichii	Ponticum	−50	−25	−50	−50	−35	E. Japan
R. metternichii var. *hondoensis*	Ponticum	−50	−25	−60	−60	−35	C. Japan
R. makinoi	Ponticum	−40	−23	−50	−50	−27	C. Japan
R. yakushimanum	Ponticum	−40	−23	−50	−50	−27	Yakushima (1000–1800 m)
R. kiusianum	Azalea, Obtusum	−30	−23	−30	−30	−25	S. Japan (1300–1700 m)
R. yedoense var. *poukhanense*	Azalea, Obtusum	−30	−25	−30	−30	−30	Korea
R. kaempferi	Azalea, Obtusum	−25	−23	−25	−30	−30	S. Japan
R. micranthum	Micranthum	−25	−23	−30	−30	−25	Szechwan, Kansu (2000–2500 m)
R. morii	Barbatum	−23	−23	−23	−23	−23	Taiwan
R. occidentale	Azalea, Luteum	D	−20	−25	−25	−23	W. North America
R. simsii	Azalea, Obtusum	D	−18	−25	−25	−25	Yunnan (1000–2600 m)

R. ripense	Azalea, Obtusum	−20	−15	−20	−20	−20	S. Japan
R. macrophyllum	Ponticum	−20	−20	−25	−25	−20	W. North America
R. discolor	Fortunei	−20	−18	−20	−20	−20	Szechwan (1300–2000 m)
R. chapmanii	Calorianum	−18	−20	−20	−20	−20	Florida
R. yunnanense	Triflorum	−18	−15	−20	−25	−25	Yunnan, Burma (2200–3000 m)
R. fortunei	Fortunei	−18	−15	−25	−23	−23	E. China
R. augustinii	Triflorum	−15	−15	−20	−20	−20	Yunnan (1500–2700 m)
R. griersonianum	Griersonianum	−15	−17	−17	−17	−17	Yunnan, Burma (2100–2800 m)
R. racemosum	Scabrifolium	−15	−17	−18	−18	−18	Yunnan (1700–3700 m)
R. scabrum	Azalea	−15	−13	−15	−15	−15	Ryukyu
R. tashiroi	Azalea, Obtusum	−15	−13	−15	−15	−15	S. Japan, Ryukyu
R. eriocarpum	Azalea, Obtusum	−15	−13	−15	−15	−15	S. Japan, SE China
R. lutescens	Triflorum	−13	−13	−18	−18	−18	Yunnan, Szechwan (600–3000 m)
R. arboreum	Arboreum						
var. *cinnamomeum*		−13	−13	−13	−13	−13	Nepal (1500–2800 m)
var. *campbelliae*		−10	−10	−10	−10	−10	Nepal (1550–2500 m)
Tropical zone							
R. kawakami		−10	–	−10	−10	−10	Taiwan
R. atropurpureum		− 5	–	− 5	− 5	−10	Papua New Guinea (3600 m)
R. womersleyi		− 4	–	− 6	− 6	−10	Papua New Guinea (3600 m)
R. gaultherifolium		− 3	–	− 3	− 3	−10	Papua New Guinea (3500 m)
R. javanicum		X	–	X	X	X	Malaysia (800–2400 m)
R. ravum		X	–	X	X	X	Malaysia (1600–3500 m)

Symbols: (*) Dwarf shrubs; D = deciduous; X = not surviving −4 °C for 3 h; [z] = uninjured at the indicated temperature.

Fig. 7.10. Downward moving and curling of leaves of *Rhododendron brachycarpum* at an air temperature of $-10\ ^\circ C$. (Photo: A. Sakai)

7.2.2 Mediterranean Sclerophylls

Mediterranean sclerophyllous trees and shrubs are considered to represent a life form which, by evolution and natural screening, has adapted to climates with summer drought and wet, mild winters (for detailed information on Mediterranean-type vegetation and climate, see Di Castri and Mooney 1973). Obviously drought stress proves to be the climatic factor essentially responsible for the restriction of productivity and survival of woody evergreen plants in Mediterranean-type regions. In addition, low temperature effects in marginal districts, particularly in the northern and eastern parts of the Mediterranean basin and near the altitudinal limit of the maquis vegetation, should not be neglected (Larcher 1981c).

The Mediterranean flora originates from a laurophyllous forest around the Tethys (Pignatti 1978). The Tertiary Mediterranean Basin belonged to the tropical zone, but since the climate was not constant, the intermittent arid phases stimulated evolutionary selection and drought adaptation. The thermal crisis of the Pleistocene with cool periods, probably 5° to 10 °C colder than at present, forced selection for low temperature resistance. An impression of the range of frost resistance of the laurophyllous ancestors can be obtained from relict species of the Tertiary Mediterranean flora which survived on frost-free Atlantic islands (Table 7.10). The actual sclerophyllous vegetation in the Mediterranean Basin, the Quercetea ilicis associations, can be considered as the remnants of the laurophyllous forests that survived the radical climatic changes during the Pleistocene.

Table 7.10. Frost resistance (°C) of leaves of relict species of the Tertiary Mediterranean laurophyllous flora; the samples were tested in February on the Canary island Tenerife. (From Larcher 1980c)

Plant species	Family	LT_0	LT_{50}
Persea indica	Lauraceae	−6	− 7
Heberdenia excelsa	Myrsinaceae	−7	− 8
Apollonias barbujana	Lauraceae	−7	−10
Ocotea foetens	Lauraceae	−7.5	−10
Myrica faya	Myricaceae	−7.5	−10
Picconia excelsa	Oleaceae	−8	−10
Viburnum tinus var. *rigidum*	Caprifoliaceae	−8	−10
Laurus azorica	Lauraceae	−9.5	−11.5

The recent vegetation of the Mediterranean regions is endangered by frost during severe winters, which are to be expected every 8–10 years (Fig. 7.11). In such winters, minimum temperatures of −10° to −12 °C are observed near the northern and eastern limits of the region and −6 °C in the western and central parts. Such temperatures cause striking damage to evergreen trees and shrubs (for reports, see Gola 1929; Oppenheimer 1949; Larcher 1954, 1964; Lavagne and Muotte 1971), clearly demonstrating that the resistance of Mediterranean sclerophylls is inadequate for the lowest temperatures which may recur in their natural habitats.

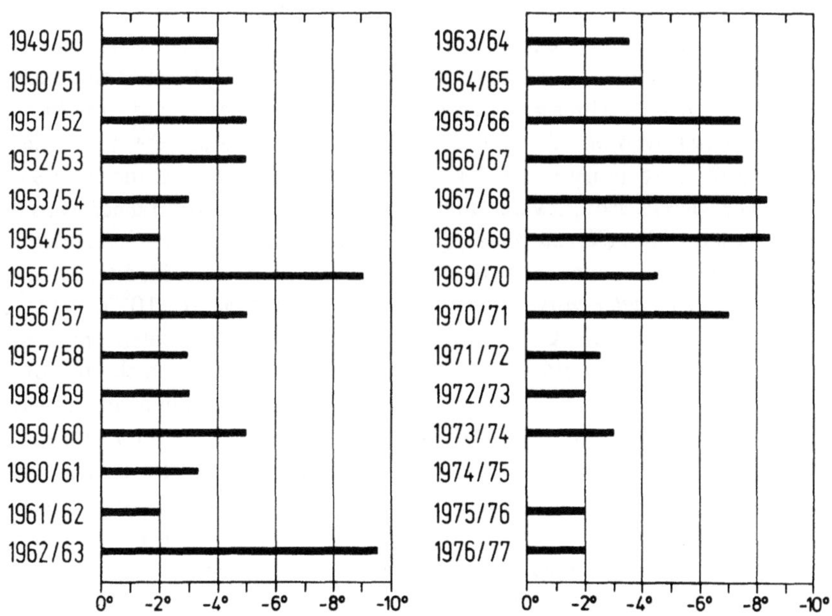

Fig. 7.11. Lowest air temperatures recorded at Arco (Lake Garda region in Italy) in 28 winters. Arco is situated at the northern borderline of the natural distribution of Mediterranean sclerophyllous trees in Europe. (From Larcher 1978)

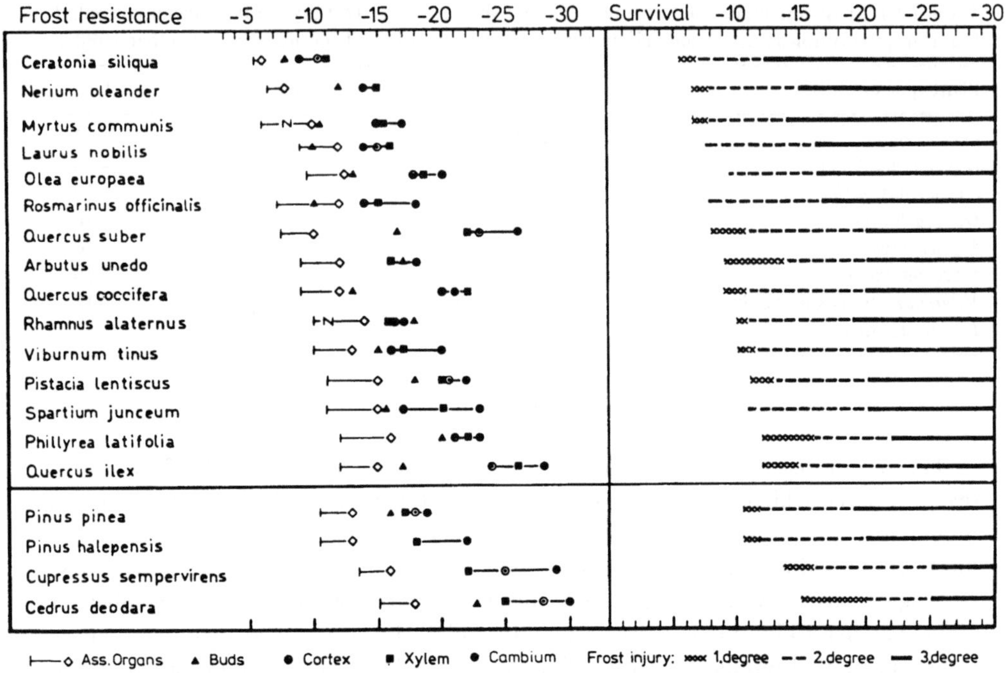

Fig. 7.12. Frost resistance (LT_{50} at °C) and survival capacity of Mediterranean sclerophyllous woody plants. Degree of injury: *1* defoliation; *2* dieback and poor regrowth; *3* severe to total damage of plants. (Rearranged from Larcher 1970)

A selection of data from experimental determinations of frost resistance of Mediterranean sclerophylls and conifers (Larcher 1954, 1964, 1970; Larcher and Mair 1969) is presented in Fig. 7.12. Three groups can be distinguished with regard to frost resistance: *Ceratonia siliqua, Nerium oleander*, and *Myrtus communis* are the most sensitive species with 50% frost injury to the leaves at -6° to -8 °C and to the shoots at -9° to -15 °C. The second group, which includes most of the Mediterranean woody plants, such as *Laurus nobilis, Olea europaea, Quercus coccifera, Quercus suber, Arbutus unedo, Rhamnus alaternus, Viburnum tinus, Pistacia lentiscus,* and the Mediterranean conifers *Pinus pinea* and *P. halepensis*, shows damage to the foliage at -10° to -14 °C and to the stems at -15° to -20 °C. The third group consists of particularly resistant species (*Quercus ilex, Phillyrea latifolia* and *Cupressus sempervirens*) which are not seriously damaged until -15° to -25 °C. These groups reflect the latitudinal and altitudinal gradients in the floristic composition of the maquis vegetation in the European Mediterranean region. Within this series of species a transition can also be traced from the freezing-sensitive type, which relies upon avoidance mechanisms (e.g. leaves of *Ceratonia siliqua, Nerium oleander, Myrtus communis, Laurus nobilis* and *Olea europea*), to the type whose leaves already possess a demonstrable, albeit modest, degree of freezing tolerance (e.g. *Arbutus unedo, Quercus ilex* and the conifers; Larcher 1970). Shoot axes of the less resistant species belonging to the association of Oleo-Ceratonion, which forms the climax vegetation in the mildest and altitudinally lowest positions of the Mediterranean Basin, survive -10° to -20 °C, those of the more resistant representatives of the Quercion ilicis survive −20° to −25 °C, whereas the roots are severely injured at −5° to −10 °C.

The resistance of adult individuals is not necessarily indicative of the survival capacity of a species in a certain area. Observations have confirmed that trees of the Mediterranean region recover from loss of foliage and even from severe bud damage (e.g. Morettini 1961). Only if frost persists for some weeks and penetrates the thicker tree trunks and the xylopodia of shrubs is it lethal for adult stands. Indeed, several Mediterranean sclerophylls can be cultivated far north of their proper climatic zone, provided the lowest temperatures are not below the specific resistance limit. It is more likely that the distribution area of species is to be determined by less resistant propagative organs and developmental stages, such as overwintering flower buds and immature fruits (e.g. *Laurus nobilis, Olea europaea* and *Arbutus unedo*) or germinating seeds and growing seedlings. The data listed in Table 7.11 suggest that none of the Mediterranean sclerophylls would be able to produce flowers and fruits after a winter frost with temperatures below −7° to −15 °C. Stages particularly susceptible to frost are germinating seeds, seedlings and young plants (Fig. 7.13). Mediterranean species germinate preferentially during autumn and winter, which are the wet seasons, favourable with respect to water availability, but not necessarily as to temperature conditions. Thus, the developmental stages most susceptible to frost are also those most endangered (Iversen 1944; Larcher 1973). As a consequence, a frost of −5° to −10 °C will destroy all the germinated seeds of that year; if this occurs in successive winters, the propagation of the species becomes severely impaired.

The attempts to explain the northern and altitudinal limit of Mediterranean evergreens on the basis of climate (e.g. De Philippis 1937) have employed both the absolute winter temperature minima from extremely severe winters and the mean winter minima. The *absolute* temperature minima, especially in the case of the most northerly occurring Mediterranean sclerophyllous species, *Quercus ilex*, are seldom useful. This is readily seen by comparing distribution maps with climate maps showing absolute temperature minima: the northern limit for *Quercus ilex* in southern France, for example,

Table 7.11. Frost resistance of reproductive organs of Mediterranean sclerophyllous trees during winter. (From Larcher 1981b, updated)

Species	LT_{50} (°C)[a]
Olea europaea	− 5 (S)
Laurus nobilis	− 7 (F)
Laurus nobilis	− 7 (S)
Rhamnus alaternus	− 8 (F)
Viburnum tinus	− 8 (S)
Viburnum tinus	−10 (F)
Quercus coccifera	−12 (B)
Quercus suber	−14 (B)
Quercus ilex	−15 (B)
Phillyrea latifolia	−15 (B)

[a] B = Flower primordia; F = flower buds ready for opening, and flowers; S = ripening seeds and fruits.

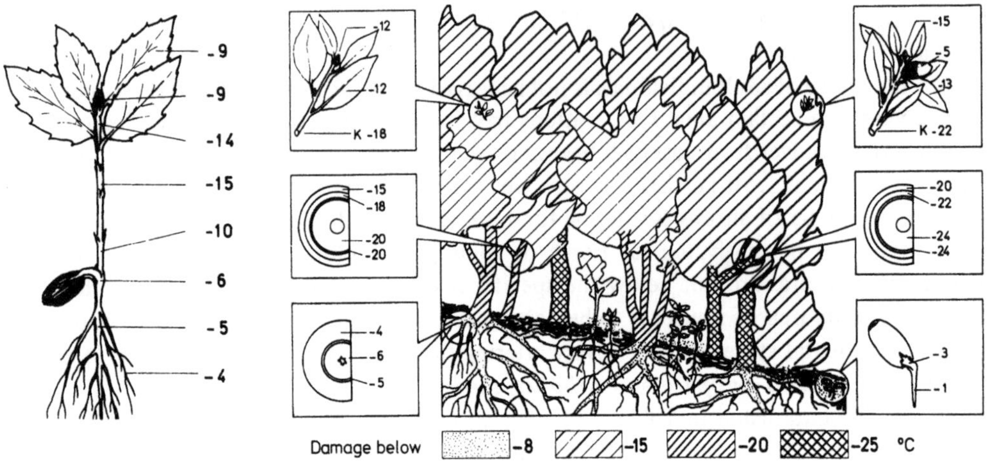

Fig. 7.13. Frost resistance (LT_0 at °C) of the various age groups in a population of *Quercus ilex*. (From Larcher and Mair 1969)

accompanies in places the isotherm for absolute minima down to -20 °C, on the Apennine and Balkan peninsula the isotherm for -15 °C, whilst in the Atlantic regions of western France there is no clear connection at all. These discrepancies result from the fact that, although the natural vegetation is severely damaged by the extremely cold winters that occur a few times each century, it is not, in fact, totally exterminated. Distribution is not limited by catastrophic drops in temperature, but rather by the regular recurrence of an average stress. This is why *mean* temperature minima are more useful in comparing resistance data than absolute minimum values (Emberger 1932; Iversen 1944; Brisse and Grandjouan 1974; Tuhkanen 1980). Thus, it appears that the northernmost occurrence of natural stands of *Quercus ilex* on Lake Garda in Italy is attributable to the absence of ground frost and of air temperatures below -6 °C in normal winters. Mature trees, on the other hand, thrive in locations where minima of -12 °C are recorded each winter (Fig. 7.14).

7.3 Transition from an Evergreen to a Deciduous Flora and Its Consequences for Frost Survival

The evolution of the deciduous habit provides an example of adaptation to large-scale changes in time and space. Deciduous hardwoods first appeared in the Northern Hemisphere during the Cretaceous era (Axelrod 1966). The progressive cooling of the climate during the Tertiary brought with it a change in vegetation from the predominantly evergreen paleotropical flora to the mainly summer-green arctotertiary flora (Mägdefrau

Fig. 7.14. Small-scale distribution of *Quercus ilex* communities (*black area*) in the mild Lake Garda region, and occurrence of adult trees planted for ornamental purposes outside their natural distribution range (*star symbols*). *Numerals* indicate mean annual minimum temperatures and absolute lowest recorded temperatures (*italics*). For explanation, see text. (From Larcher and Mair 1969)

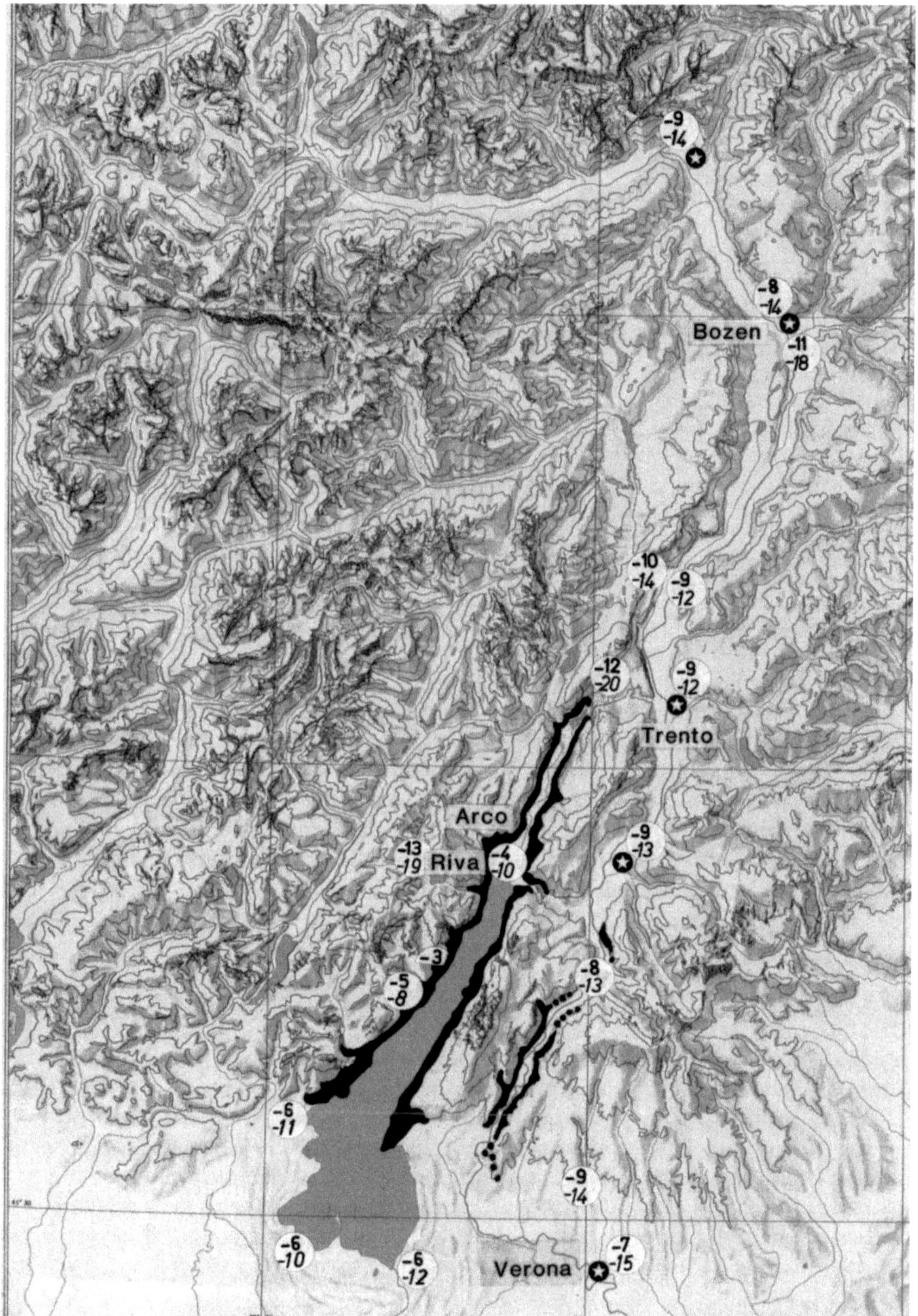
-9
-14
-8
-14
Bozen
-11
-18
-10
-14
-9
-12
-12
-20
-9
-12
Trento
Arco
-13
-19
Riva
-4
-10
-9
-13
-3
-5
-8
-8
-13
-6
-11
-9
-14
-6
-10
-6
-12
Verona
-7
-15

Table 7.12. Frost resistance (LT_0 at °C) of tree species of the Arcto-Tertiary flora which were extinguished in Japan in late Tertiary. (From Sakai 1971b)

Plant species	Leaves	Buds	Twig	Distribution of the species or allies at present
Conifers				
Taiwania cryptomeroides	−15	–	−15	Taiwan (1800–2000 m), S. China
Sequoia sempervirens	−15	−15	−15	W. North America
Keteleeria davidiana	−15	−20	−20	N. and S. Taiwan (500–900 m), SW. China (1600–2600 m)
Glyptostrobus pensilis	−18	−20	−20	S. China
Cunninghamia lanceolata	−13	−13	−13	C. China (1000 m), S. China (1000–2000 m)
Cunninghamia konishii	−25	−23	−25	C. Taiwan (1300–1800 m)
Pseudolarix amabilis	–	−25	−20	SE. China (1000 m)
Metasequoia glyptostroboides	–	−30	−30	C. China (750–1500 m)
Taxodium distichum	–	−30	−30	SE. North America
Broad-leaved trees				
Liquidambar formosana	–	−17	−17	C. Taiwan, S. China
Liquidambar styraciflua	–	−25	−25	SE. North America
Nyssa sylvatica	–	−25	−25	SE. North America
Liriodendron tulipifera	–	−25	−25	North America
Ginkgo biloba [a]	–	−30	−30	C. and S. China

[a] Mesozoic relict.

1968; Mai 1981). During Pleistocene many genera, which were incapable of adapting to lower and more variable temperatures and to seasonal and reduced precipitation, became extinguished at higher latitudes and altitudes (Table 7.12); they shifted southwards and were substituted by geoelements with better fitness for the new conditions. A transition from the evergreen to the deciduous habit is also discernable in connection with recent changes in latitude and altitude. The consequences of this tendency for woody plants and ferns will be discussed below.

7.3.1 Woody Plants

Deciduous trees and shrubs of moist-tropical to warm-temperate derivation are able to live (1) in more continental climates, (2) in cool climates at higher latitudes with marked seasonal photoperiodicity and (3) in mountain climates.

Many deciduous woody plants that are usually regarded as "temperate" belong to genera that also have evergreen species in subtropical and mild climates. The evergreen members of genera, such as *Acer, Diospyrus, Fraxinus, Pyrus, Prunus* and *Rosa*, only resist frost of −5° to −15 °C (cf. Table 7.8). Thus, the distribution of these evergreen species is confined to a mild winter climate and to the subtropics. Several evergreen species belonging to *Euonymus, Ilex, Magnolia*, and *Quercus* have evolved leaves which tolerate freezing at −15° to −20 °C. In general, the transition to seasonal leaf shedding is accompanied by an increase in the potential frost resistance of the vegetative buds and the stem tissues (Fig. 7.15, Table 7.13; for further data, see Sakai 1978a).

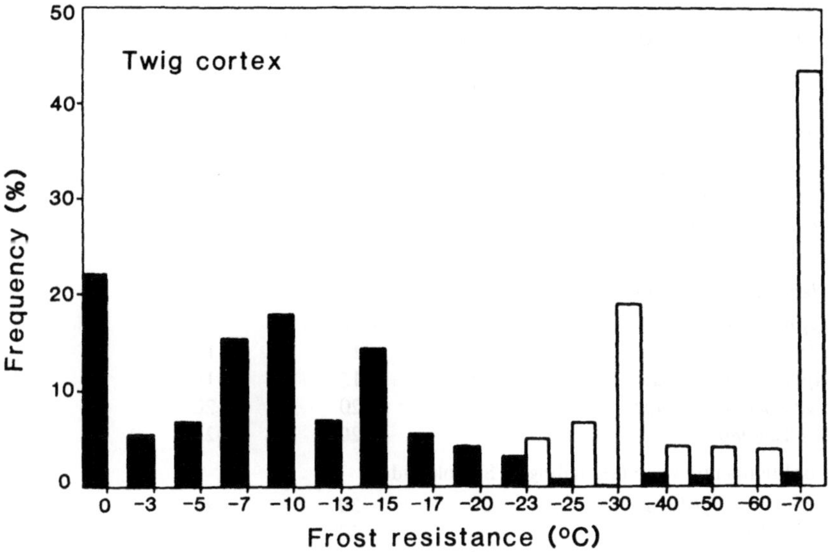

Fig. 7.15. Frequency diagram for frost resistance of twig cortex of evergreen (*black bars*) and cool-temperate deciduous (*white bars*) broad-leaved trees growing in Japan. A total number of 73 evergreen and 73 deciduous species was evaluated. (From Sakai 1980)

Table 7.13. Frost resistance (LT_i at °C) of deciduous species of the genus *Quercus* as compared with evergreen oaks (cf. Table 7.6 and Fig. 7.12). (From Till 1956; Parker 1962; Larcher and Mair 1969; Sakai 1971b, 1972, 1978a; Sakai and Weiser 1973)

Species	Distribution	Buds	Cortex	Xylem
Qu. lyrata	Mississippi region	−20	−20	−20
Qu. serrata	Japan	−25	−30	−27
Qu. aliena	Japan	−30	−30	−30
Qu. pubescens	S. Europe	−24	−40	−33
Qu. robur	C. Europe	−27		
Qu. acutissima	Japan	−27	−40	−27
Qu. crispula	Sakhalin	−35 to −50	−70	−35
Qu. mongolica	Japan	−30	−30	−30
Qu. mongolica	Vladivostok	−50 to −70	−70	−35
Qu. rubra	NE. America		Below −70	−40 to −45
Qu. macrocarpa	S. Canada	Below −60	Below −60	−46
Evergreen oaks	Warm-temperate regions	−10 to −17	−10 to −17	−13 to −18
	Mediterranean region	−13 to −16	−20 to −24	−22 to −26

Most of the deciduous trees which extend from the tropics of Asia to their northern growth limit on the subtropical Ryukyu Islands do not survive freezing at −5 °C even when cold-treated. Deciduous trees in Japan can be classified into warm-temperate, cool-temperate and boreal hardwoods (Figs. 7.6 and 7.16; Table 7.14). Warm-temperate deciduous genera which range from the subtropics to the warm-temperate zone, such

Table 7.14. Potential frost resistance (LT_0 at °C) of deciduous trees in Japan and E. Asia (see also Fig. 7.6). (From Sakai 1978a)

Plant species	Bud	Cortex	Xylem
Subtropic to warm-temperate zone			
Hibiscus mutabilis	− 5	− 5	− 5
Glochidion obovatum	− 7	− 7	− 7
Melia azedarach	− 8	− 8	− 8
Rhus succedanea	−12	−10	−10
Warm-temperate zone			
Sapindus mukurosii	−10	−10	−10
Firmiana simplex	−15	−15	−15
Aphananthe aspera	−15	−15	−15
Lagerstroemia indica		−20	−20
Albizzia mollis		−20	−20
Cool-temperate zone (extending to S. Hokkaido)			
Acer crataegifolium	−30	−30	−25
Alnus firma	−30	−30	−25
Carpinus tschonoskii	−25	−25	−25
Fagus crenata	−25	−25	−25
Quercus serrata	−25	−30	−27
Zelkova serrata	−30	−30	−27
Boreal zone (inland Hokkaido, Manchuria, Sakhalin)			
Alnus hirsuta	−70	−70	−40
Cornus controversa	−70	−70	−40
Fraxinus mandshurica	−70	−70	−40
Kalopanax septemlobus	−70	−70	−35
Sorbus sargenti	−70	−70	−35
Phellodendron amurense	−70	−70	−35
Tilia japonica	−70	−70	−35
Quercus grosseserrata	−70	−70	−35
Betula ermani	−70	−70	−70
Salix sachalinensis	−70	−70	−70
Populus maximowiczii	−70	−70	−70
Viburnum sargenti	−70	−70	−70
Maackia amurensis	−70	−70	−70

as *Melia, Sapindus, Firmiana, Aphananthe, Albizzia* and *Lagerstroemia*, have the same order of frost resistance as warm-temperate evergreen trees. The cool-temperate deciduous species, which have the northern limits of their natural ranges in N. Honshu and S. Hokkaido, are hardy to −30 °C. In these trees, little or no difference in hardiness exists between cortex, xylem and vegetative buds. In most of the extremely frost-resistant deciduous trees, xylem is the least hardy tissue: in many boreal tree species which extend to inland Hokkaido, Sakhalin, Manchuria and the Maritime Province of the USSR, the xylem sustains injury between −30° and −50 °C (Fig. 7.17). The xylem of *Salix sachalinensis, Populus maximowiczii, Betula ermanii, Maackia amurensis* and *Viburnum sargenti* tolerates freezing at −70 °C. The two families, Betulaceae and Salicaceae, have evolved the hardiest deciduous tree species: these even survive in interior Alaska, N. Canada and Siberia where temperatures fall below −50° to −60 °C.

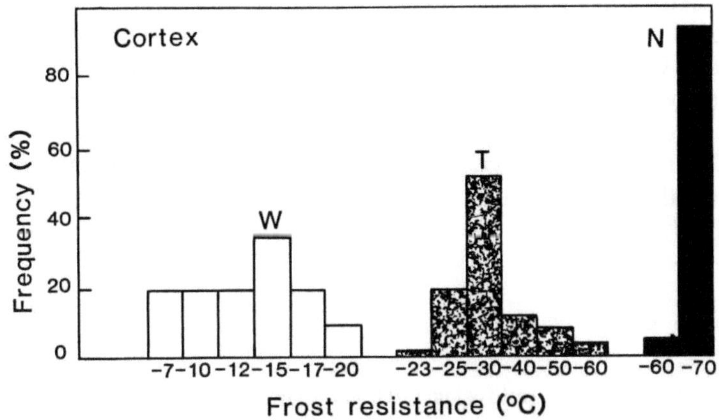

Fig. 7.16. Frequency diagram for frost resistance of twig cortex of warm-temperate (*W*), cool-temperate (*T*) and boreal (*N*) deciduous hardwoods. (From Sakai 1980)

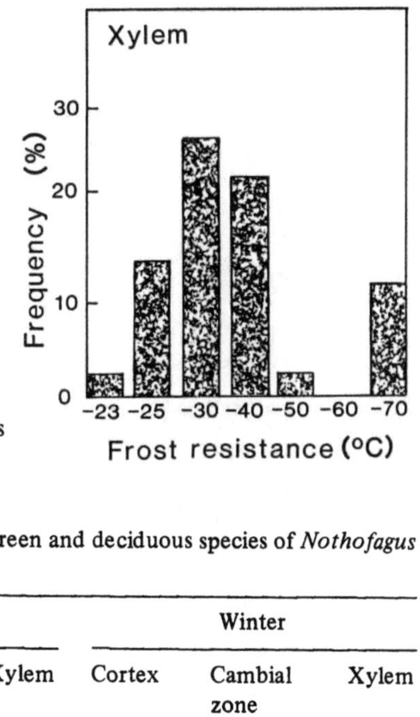

Fig. 7.17. Frequency diagram for xylem frost resistance of cool-temperate and boreal deciduous broad-leaved trees in Japan, based on a total of 55 species. (From Sakai 1980)

Table 7.15. Seasonal frost resistance (LT_{50} at °C) of evergreen and deciduous species of *Nothofagus* in Chile. (From Alberdi et al. 1985)

Species	Altitude (m)	Summer			Winter		
		Cortex	Cambial zone	Xylem	Cortex	Cambial zone	Xylem
Evergreen							
N. dombeyi	500	−5.5	−5.5	−7.5	− 8.0	− 8.0	− 8.0
	1000	−7.5	−7.5	−8.5	−16.0	−16.0	−16.0
N. betuloides	1000	−4.0	−4.0	−7.0	−12.0	−13.0	−13.0
Deciduous							
N. antarctica	120	−3.5	−4.5	−6.0	−16.0	−16.0	−19.0
	220	−3.5	−3.5	−7.0	− 9.0	− 9.5	−12.0
	700	−4.0	−4.5	−6.5	−16.5	−17.0	−19.0
	1080	−2.0	−2.0	−2.0	−17.0	−21.0	−22.0
N. pumilio	1040	−3.0	−2.0	−4.0	−16.0	−18.0	−18.0

Altitudinal gradients of frost resistance in connection with the transition from evergreen to seasonal foliation within the genus *Nothofagus* (Table 7.15) were demonstrated in S. Chile by Alberdi et al. (1985); the subalpine forest zone with abundant snowfall (snow cover 7–8 months per year) and lowest temperatures at −7.5 °C is dominated by the deciduous species *N. pumilio* and *N. antarctica*; both become krummholz above the tree line. *Nothofagus antarctica* is the hardiest southern beech (cf. Table 7.7) and it is the most tolerant to extreme environments, including frost hollows (Veblen et al. 1977).

7.3.2 Ferns

In the Japanese archipelago evergreen fern species decrease towards higher latitudes, while summer-green ferns increase (Fig. 7.18). One of the reasons for this transition towards higher latitudes may be the shorter growing season and more severe frost. Since plant establishment depends on the climatic conditions of sites in which propagules are dispersed, the life history adaptation of fern to cold climates, the leaf development, the sporulation period and the growth pattern of gametophytes and sporophytes were studied by Sato and Sakai (1981a,b) and Sato (1982, 1983) with special reference to frost resistance at each ontogenetic stage. Thorough, year-round investigations of the temperature and desiccation resistance of the gametophytes and

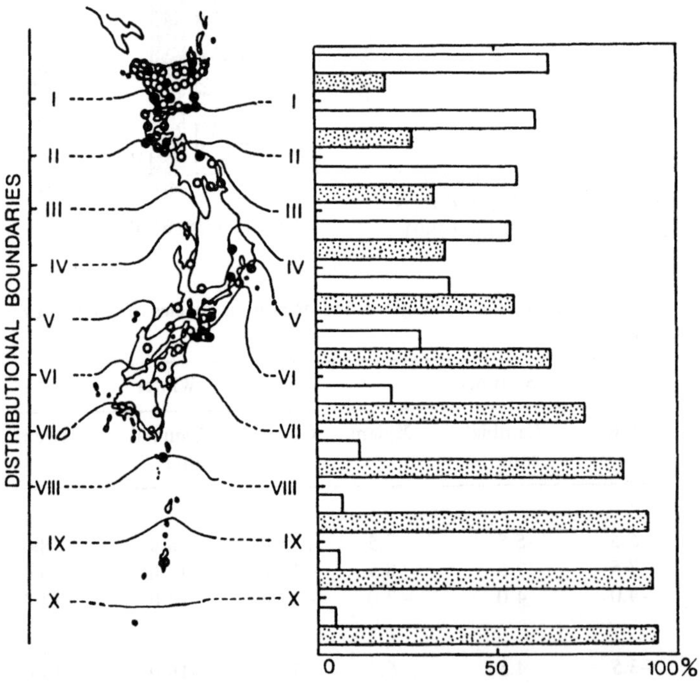

Fig. 7.18. Distribution zones (*left*) and relative abundance of summer-green (*white bars*) and evergreen (*stippled bars*) fern species in Japan. Abundance: Number of summer-green or evergreen species expressed as percentage of total number of fern species of each zone. (From Sato 1982)

sporophytes of European Polypodiaceae had earlier been carried out by Kappen (1964, 1965).

According to the seasonality of the leaves and their life span, four different basic types can be distinguished: evergreen, winter-green, summer-green and semi-evergreen. Most ferns common in Hokaido are summer-green, with leaves developing in late May to June and decaying during October. There are also great differences in the sporulation period. *Drypteris monticola*, a fern of cool-temperate regions, has a sporulation period of 3 to 5 months from late July to September around boundary I in Fig. 7.18, and from June to October around boundary IV. Species distributed in the warm-temperate region of Japan, such as *Dryopteris lacera*, sporulate for 5 to 7 months in southern parts (VI and VII) and for 2 to 3 months in the northern parts of their distribution range. Species restricted to the southern part of Japan, such as *Pteris vittata*, have a long sporulation period covering almost the whole year. In Hokkaido, spores dispersed from June to September germinate before winter begins, forming vegetative prothallia;

Table 7.16. Winter frost resistance (LT_0 at °C) of ferns in Japan. (From Sato and Sakai 1981a; Sato 1982, 1983; selected examples)

Species	Life form[a]	Prothallia[b]	Leaves	Rhizome
Cool-temperate region (Hokkaido)				
Athyrium yokoscense	○	(−40)*	–	(−40)*
Dryopteris austriaca	○	(−40)*	–	−30
Dryopteris laeta	○	(−40)*	–	−25
Dryopteris saxifraga	○		−40	−40
Asplenium incisum	●	(−40)*	−25	−20
Dryopteris amurensis	◑	−40	−12.5	−17
Dryopteris crassirhizoma	◑	(−40)*	−22.5	−15
Dryopteris sabaei	◑	−40	−20	−15
Dryopteris monticola	○	−40	–	−15
Polystichum retroso-palaeaceum	◑	−40	−15	−10
Osmunda japonica	○		–	−10
Matteucia orientalis	○		–	− 7
Coniogramma intermedia	○		–	− 7
Athyrium vidalii	○	−30	–	− 5
Lunathyrium pterorachis	○	−30	–	− 5
Stegnogramma pozoi	○		–	− 5
Warm-temperate region (S. Honshu)				
Struthiopteris nipponica	●		−17	−12
Crypsinus hastatus	●	−30	−15	− 5
Dryopteris lacera	●	−20	− 7	− 7
Coniogramma intermedia	◑●		−10	− 5
Stenogramma pozoi	◑●	−20	− 7	− 5
Cyrtonium falcatum	●	−10	−10	− 5
Pteris vittata	●	− 7	−10	− 5
Woodwardia orientalis	●		−10	− 5
Gleichenia japonica	●		−10	0
Angiopteris lygodiifolia	●	0	0	0

[a] ● evergreen, ◑ semi-evergreen, ○ summer-green.

[b] (*) Not injured at the indicated temperature.

the gametophytes mature in the following summer. Thus, the gametophytes as well as the sporophytes are exposed to severe winter conditions. In some ferns, such as *Dryopteris saxifraga* and *Matteuccia orientalis*, spores are dispersed in November and remain dormant until spring or even summer.

Young fern gametophytes become more frost-resistant than sporophytes in the cold-adapted state (Table 7.16). Gametophytes with mature gametangia are usually less resistant than young gametophytes. The hardiness of the gametophytes increases from October to late November, and reaches maximal values in December (Kappen 1965; Sato and Sakai 1981a). Gametophytes from warm-temperate regions are resistant to only between -10° and -20 °C in mid-winter.

The sporophytes of most of the ferns of N. Japan were tolerant to freezing to -40 °C, the average winter hardiness of leaves and rhizomes of southern species was -7° to -13 °C and -4° to -11 °C respectively (Fig. 7.19). Seasonal changes in the leaf resistance are shown in Fig. 7.20. Of the 12 *evergreen* species tested, 6 tolerated -40 °C, and 6 species tolerated -10° to -30 °C; the leaves of *semi-evergreen* ferns were damaged below temperatures of -15° to -25 °C; of the *summer-green* ferns, the leaves of 9 species were damaged below -5 °C, those of 3 species below -3 °C and the leaves of 18 did not survive frost of only -3 °C. The leaves of young sporophytes of the summer-

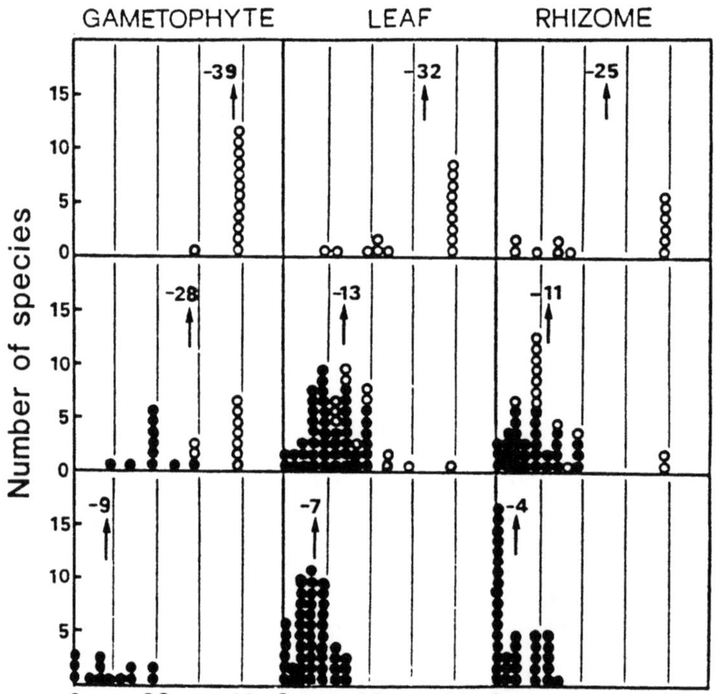

Fig. 7.19. Frequency diagram for potential frost resistance of fern gametophytes and sporophytes depending on their distribution in Japan. *Arrows* indicate the average value of frost resistance. Collection sites in Hokkaido (○); in Honshu (●). (From Sato 1983)

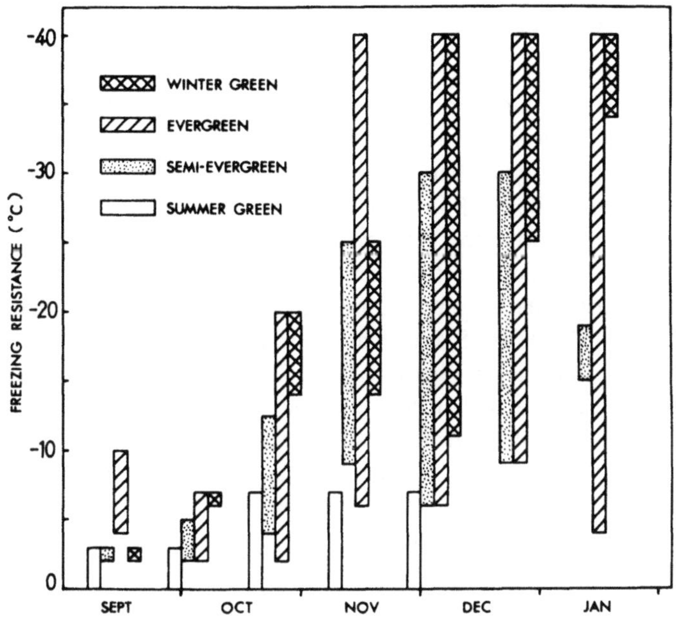

Fig. 7.20. Ranges of leaf frost resistance of ferns of Hokkaido with different seasonal life span of the leaves. (From Sato 1982)

green ferns remain green in winter; they survive -10° and -20 °C, in contrast to the leaves of mature sporophytes which are killed at -5 °C in November. This corresponds to findings of Kappen (1964) for *Cystopteris fragilis*. The hardiness of rhizomes increases from October to December and reaches maximal values from -4° to -40 °C in early December (Sato 1983). Interestingly, ferns of New Zealand, with its highly oceanic climate, show little annual fluctuation in frost resistance and develop less winter hardiness than their counterparts from similar latitudes in the Northern Hemisphere (Bannister 1984b).

Adaptation to cold climates of ferns appears to be governed by seasonality timing of each stage of the life history and different winter survival capacities of gametophytes and sporophytes. Differences in frost resistance in the two generations of the life history of ferns might explain the ecological phenomenon reported by Farrar (1967, 1978) and Wagner and Sharp (1963), who found sporophyteless species of *Vittaria* and some other genera in the Appalachian region. Having evolved vegetative propagation of the gametophytes by gemmae, the gametophytes of cool-temperate ferns growing on the forest floor must be able to survive and persist in their habitat independently of sporophytes, provided they develop sufficiently high frost and desiccation resistance in winter.

Probably, winter frost can contribute to the evolution of a summer-green life mode. *Coniogramma intermedia* and *Stegnogramma pozoi* are evergreen in warm-temperate regions and become summer-green in Hokkaido (Sato and Sakai 1980b). This suggests that one of the adaptations during the spread of species to northern latitudes may be a change of leaf phenology.

7.4 Conifers

A survey of the history of conifers reveals that old, widely distributed families, especially Taxodiaceae, have definitely lost extensive parts of their maximum area; in Cretaceous and early Tertiary times, several genera were widespread, reaching as far north as the arctics (Florin 1963). The general climatic trend toward cooling and increasing continentality during the Tertiary forced them to retreat southwards (Chaney 1940, 1947; Tanai 1967) and most of them were eliminated by the end of the Pliocene or during the Pleistocene. At present, most genera of Taxodiaceae inhabit intermediate latitudes, below 43° N or 46° S, and all are confined as relicts to restricted regions in eastern Asia, North America or Tasmania (Fig. 7.21). Pinaceae, which include four of the largest existing genera of northern conifers, such as *Pinus, Picea, Abies* and *Larix*, gradually became dominant members during the later Miocene, and, in contrast to Taxodiaceae, continued to expand through the Pliocene (Florin 1963; Tanai 1972). They soon returned to areas from which they had retreated after the maximum glaciation in the Pleistocene, and now extend as far as to the subarctic tree line where severe, continental climates have prevented the establishment of other, less resistant species (Fig. 7.21). This became possible because genera with extreme winter hardiness and a short growing period were evolved. In interior Alaska, most trees cease their growth as early as late July, at a 20 h day length; the growing season is as short as 40 to 50 days.

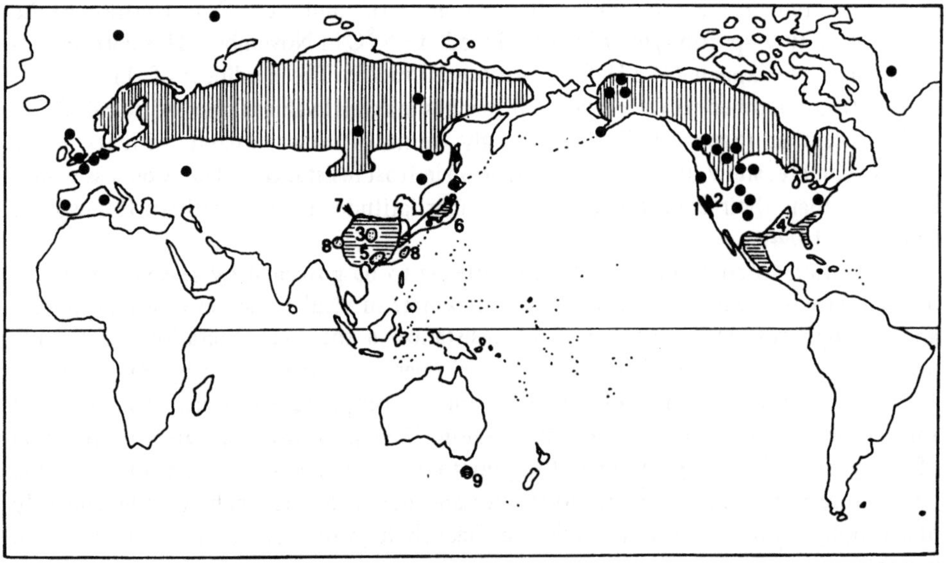

Fig. 7.21. Geographic distribution of boreal conifers of the genera *Abies, Picea, Pinus* and *Larix* (*vertically hatched*) and of relict species of Taxodiaceae (*horizontally hatched*). 1 *Sequoia*; 2 *Sequoiadendron*; 3 *Metasequoia*; 4 *Taxodium*; 5 *Glyptostrobus*; 6 *Cryptomeria*; 7 *Cunninghamia*; 8 *Taiwania*; 9 *Arthotaxis*. *Dots:* locations at which fossils of Taxodiaceae were found. (From Florin 1963, partially revised)

7.4.1 The Freezing Resistance of Various Conifer Families and Its Connection with Their Distribution Range

In extensive studies on frost resistance of conifers from different climatic zones in the Northern and Southern Hemispheres, A. Sakai and his co-workers found large-scale differences in freezing resistance among families and genera (Fig. 7.22). It is apparent that extremely hardy conifers, in which the bud primordia survive freezing to -70 °C, have only evolved from northern genera of Pinaceae, i.e. *Pinus, Picea, Larix* and *Abies. Thuja occidentalis* survives freezing to -70 °C or below, but is very sensitive to winter desiccation (White and Weiser 1964; Sakai 1970b). The large genus *Juniperus* which colonizes

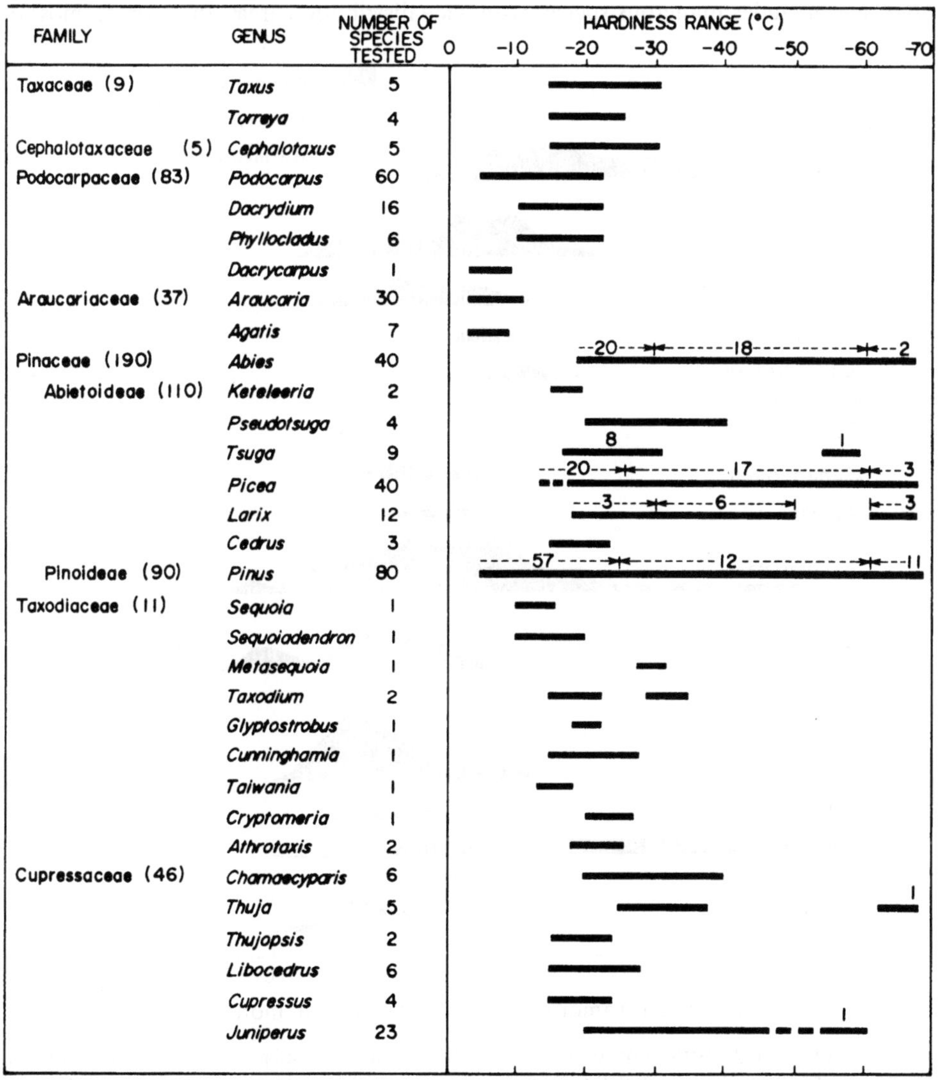

Fig. 7.22. Range of frost resistance of winter buds of different conifer genera. *Numerals on the bars* indicate the number of species falling within the indicated range. (From Sakai 1983)

vast, continuous regions at low to high altitudes in N. America and Eurasia, has also evolved very hardy species, especially in dwarf form, in both high altitudes and latitudes. In general, the genera which have only differentiated into few species have not developed very hardy species and they are confined, at present, to a narrow, discontinuous area with mild winters as in the case of Taxodiaceae (Sakai 1971b). In contrast, most of the genera (except southern conifers), which include numerous, widely ranging species, have evolved very hardy species. The same trend can be observed in the subsections of the genus *Pinus* (Fig. 7.23). Very hardy pines from the subsection Cembrae of the subgenus *Strobus* and from the subsections Sylvestres and Contortae of the subgenus *Pinus* expanded into winter-cold regions, where a severe continental climate has excluded less hardy pines. The subsection Sylvestres is the largest group of the genus *Pinus* and includes numerous species with wide distribution areas. In contrast, most of

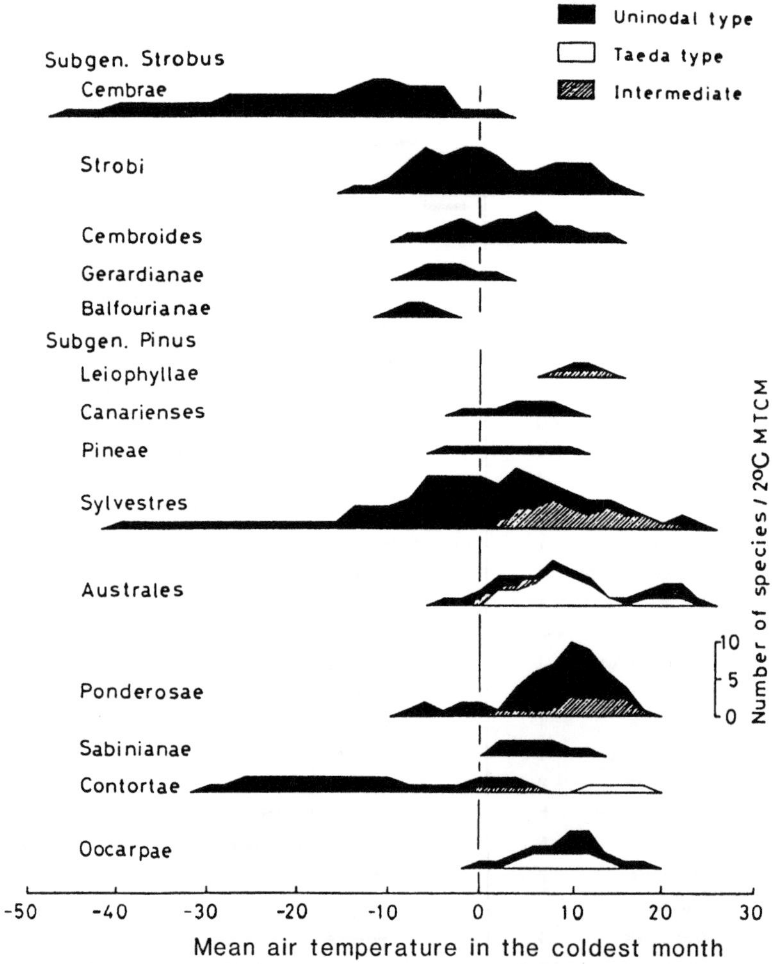

Fig. 7.23. Frequency distribution of each subsection of the two subgenera *Strobus* and *Pinus* in relation to the thermal gradient of the mean air temperature in the coldest month (MTCM). Distance on the y-axis: number of species included in each interval of 2 °C MTCM. For explanation of shoot growth types, see p. 104. (From Oohata and Sakai 1982)

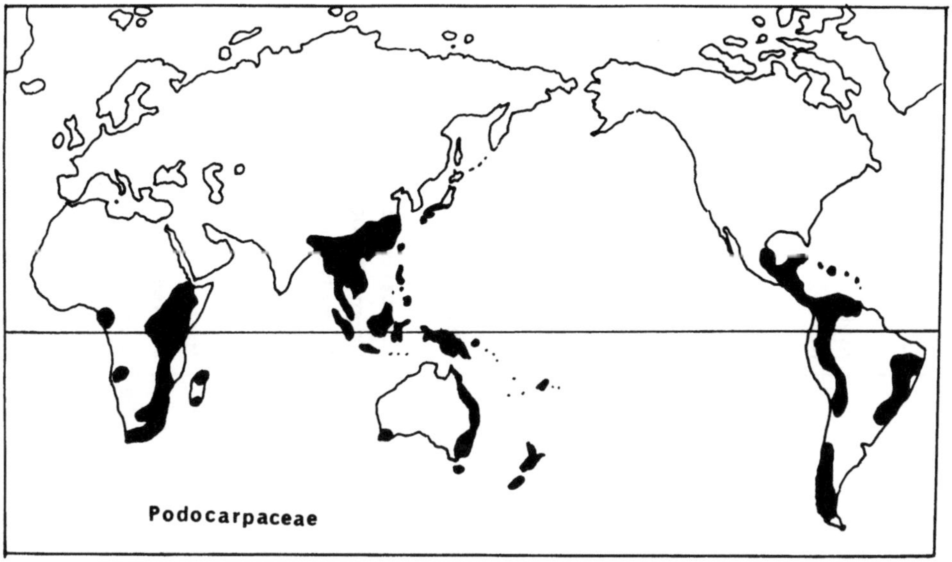

Fig. 7.24. Geographic distribution of Podocarpaceae. (From Florin 1963)

the subsections which are confined at present to a narrow discontinuous area, such as the subsections Strobi, Cembroides, Australes, Ponderosae and Oocarpae, have not evolved species with high frost resistance.

The family of Podocarpaceae ranges from low to high altitudes in the Southern Hemisphere and extends to the subtropics and tropics in the Northern Hemisphere (Fig. 7.24). Podocarpacea include three widespread genera, *Podocarpus, Dacrydium* and *Phyllocladus*. The subalpine scrub species, such as *Podocarpus nivalis, P. lawrencii, Dacrydium bidwillii* and *Phyllocladus asplenifolius* var. *alpinus* are the hardiest conifers in New Zealand and Australia, surviving only -20° to -23 °C (Table 7.7). Tree species that withstand freezing below -30 °C seem not to have evolved in the Southern Hemisphere (Sakai and Wardle 1978; Sakai et al. 1981); conversely, hardy northern tree genera, with minor exceptions, have been unable to cross the barrier formed by the tropics.

7.4.2 Characteristics of Freezing Resistance of Conifers

The potential frost resistance of leaves, buds and stem tissues of conifers originating from regions with different winter climates is presented in Fig. 7.25. In warm-temperate conifers (species 1-12) which are marginally hardy to -15° to -25 °C, only a slight difference exists between the freezing resistance of the primordia, leaves and twigs. In most of the conifers, except the genus *Pinus*, native to subboreal or boreal regions in the Northern Hemisphere (species 17-21) the freezing resistance is characterized by the relative susceptibility of the shoot and flower primordia. The leaves, cortex and xylem in twigs, and bud tissues other than the primordial shoots, tolerate freezing to -50 °C or below, while, except in conifers from Alaska *(Picea glauca, Picea mariana,*

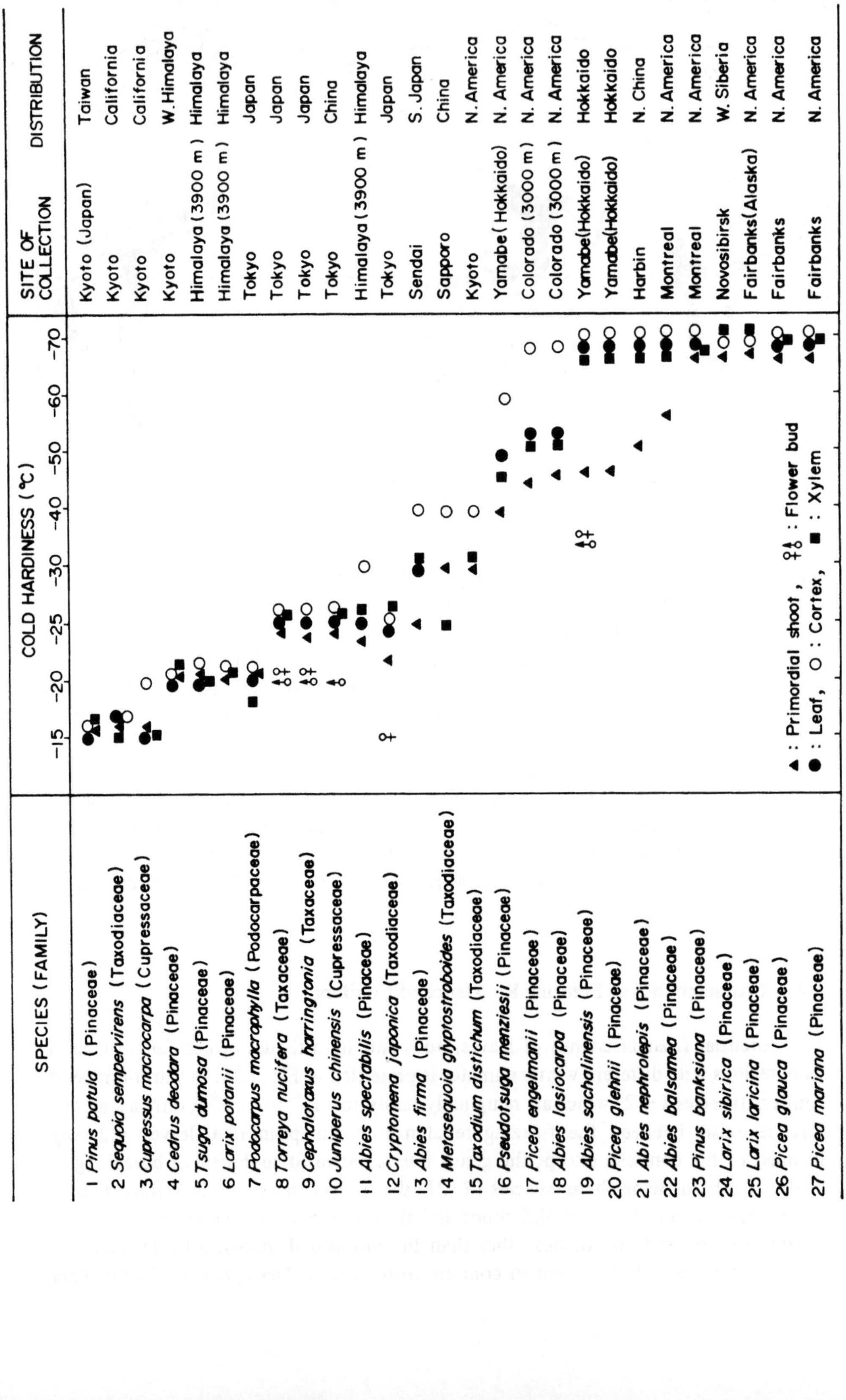
SPECIES (FAMILY)
COLD HARDINESS (℃)
-15 -20 -25 -30 -40 -50 -60 -70
SITE OF COLLECTION
DISTRIBUTION
1 Pinus patula (Pinaceae) Kyoto (Japan) Taiwan
2 Sequoia sempervirens (Taxodiaceae) Kyoto California
3 Cupressus macrocarpa (Cupressaceae) Kyoto California
4 Cedrus deodara (Pinaceae) Kyoto W. Himalaya
5 Tsuga dumosa (Pinaceae) Himalaya (3900 m) Himalaya
6 Larix potanii (Pinaceae) Himalaya (3900 m) Himalaya
7 Podocarpus macrophylla (Podocarpaceae) Tokyo Japan
8 Torreya nucifera (Taxaceae) Tokyo Japan
9 Cephalotaxus harringtonia (Taxaceae) Tokyo Japan
10 Juniperus chinensis (Cupressaceae) Tokyo China
11 Abies spectabilis (Pinaceae) Himalaya (3900 m) Himalaya
12 Cryptomeria japonica (Taxodiaceae) Tokyo Japan
13 Abies firma (Pinaceae) Sendai S. Japan
14 Metasequoia glyptostroboides (Taxodiaceae) Sapporo China
15 Taxodium distichum (Taxodiaceae) Kyoto N. America
16 Pseudotsuga menziesii (Pinaceae) Yamabe (Hokkaido) N. America
17 Picea engelmanii (Pinaceae) Colorado (3000 m) N. America
18 Abies lasiocarpa (Pinaceae) Colorado (3000 m) N. America
19 Abies sachalinensis (Pinaceae) Yamabe (Hokkaido) Hokkaido
20 Picea glehnii (Pinaceae) Yamabe (Hokkaido) Hokkaido
21 Abies nephrolepis (Pinaceae) Harbin N. China
22 Abies balsamea (Pinaceae) Montreal N. America
23 Pinus banksiana (Pinaceae) Montreal N. America
24 Larix sibirica (Pinaceae) Novosibirsk W. Siberia
25 Larix laricina (Pinaceae) Fairbanks (Alaska) N. America
26 Picea glauca (Pinaceae) Fairbanks N. America
27 Picea mariana (Pinaceae) Fairbanks N. America
▲ : Primordial shoot, ♀♂ : Flower bud
● : Leaf, ○ : Cortex, ■ : Xylem

Fig. 7.25. Potential frost resistance of leaves, shoots, flower primordia of winter buds and twigs (cortex and xylem) of conifers originating from different climatic zones. The twigs were cooled very slowly to −30 °C and in daily decrements of 5° daily to successively lower temperatures. (From Sakai 1983)

Larix laricina), N. Canada *(Pinus banksiana)* and Siberia *(Larix sibirica, Larix dahurica, Picea obovata, Pinus sylvestris)*, the frost resistance of the primordia usually lies between −40° and −60 °C. Thus, the freezing resistance of primordia and their survival mechanism are key problems in the understanding of cold adaptation in boreal conifers, especially in the subfamily Abietoideae.

In Fig. 7.26 the frost resistance of winter buds is related to thermal indices. Most of the samples, except those of Alaskan and Siberian conifers, were collected near the timberline. The gradation in winter hardiness of buds corresponds well with the average winter temperature and the continentality (in terms of the annual temperature fluctuation) of their natural distribution area. *Dacrycarpus compactus* (Podocarpaceae) grow-

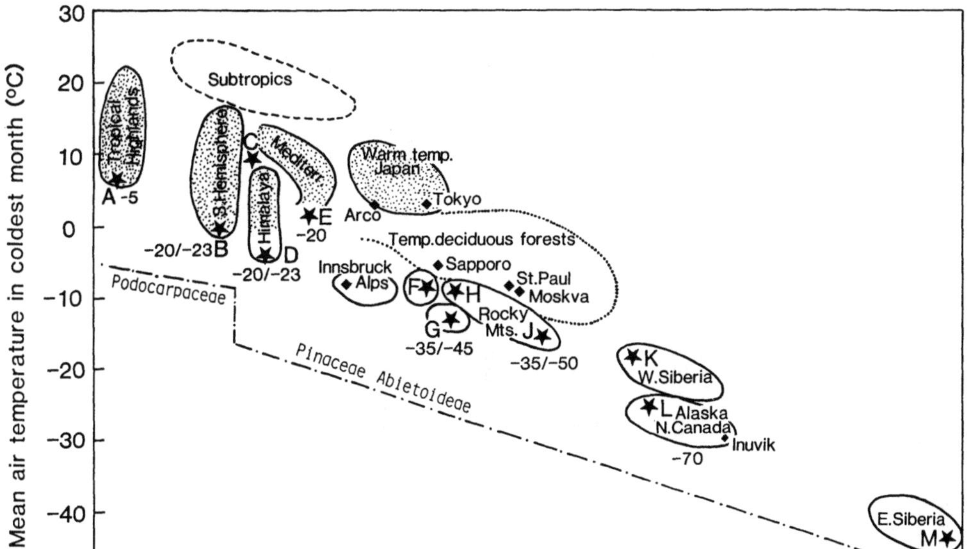

Fig. 7.26. Potential frost resistance after hardening (numbers as °C) of winter buds of conifers from different climatic zones. Collection sites (*star symbols*): *A* Mt. Wilhelm, Papua New Guinea (4°40′S, 3500 m a.s.l.); *B* Craigieburn, New Zealand (43°08′S, 1500 m); *C* Kunmin, Yunnan (25°02′N, 1893 m); *D* Shyangboche, Nepal (27°49′N, 3900 m); *E* Seattle, Washington (49°50′N, 38 m); *F* MT. Asama, Honshu (36°30′N, 2500 m); *G* Mt. Tokachi, Hokkaido (43°50′N, 1200 m); *H* Boulder, Colorado (39°46′N, 3300 m); *J* Summit Lake, Canada (54°17′N, 700 m); *K* Novosibirsk, W. Siberia (56°50′N, 107 m); *L* Fairbanks, Alaska (64°49′N, 124 m); *M* Yakutsk, E. Siberia (62°05′N, 102 m). *Dotted area:* occurrence of warm-temperate evergreen forests. The climates is represented by the mean air temperature of the coldest month and the difference between the mean temperatures of the coldest and the warmest months. (From Sakai 1983, revised)

Table 7.17. Freezing patterns of winter buds among conifer taxa. (From Sakai 1983)

Family, genus	Existence of a crow in the bud	Length of primordial shoot	Freezing pattern of the primordial shoot	Location of ice segregation	Development of bud scales
Pinaceae					
Pinoideae (*Pinus*)	No	1–5 cm	Extracellular freezing	Extracellular space	Very good. The bud surface is covered with resin.
Abietoideae (*Abies, Picea, Tsuga, Larix, Pseudotsuga, Cedrus*)	Yes	Less than about 3 mm	Extraorgan freezing	Beneath crown and inside scales surrounding basal primordial shoot	Very good. The bud surface is covered with resin (most species of boreal or subboreal conifers)
Taxodiaceae (*Metasequoia, Cryptomeria, Taxodium, Taiwania*)	No	Less than about 3 mm	Extraorgan freezing	Inside scales	Poor. Shoot and flower primordia are covered with a few scales
Taxaceae (*Taxus, Torreya*)	No	Less than about 3 mm	Extraorgan freezing	Inside scales	Poor. Shoot and flower primordia are covered with a few scales
Cephalotaxaceae (*Cephalotaxus*)	No	Less than about 3 mm	Extraorgan freezing	Inside scales	Poor. Shoot and flower primordia are covered with a few scales

ing near the forest limit at about 3500 m on Mount Wilhelm, Papua New Guinea, was damaged by frost already at -5 °C (see Table 7.2). The same order of potential frost resistance was assessed in the trees planted in Christchurch, New Zealand (Sakai and Wardle 1978). The subalpine species *Podocarpus nivalis* and *Dacrydium bidwillii* which occur in about 1550 m altitude in New Zealand, were resistant to only between -20° and -23 °C (Sakai et al. 1981). Most of the conifers growing in warm-temperate or temperate regions of both Northern and Southern Hemisphere were hardy to -25 °C. Among conifers from cold regions the hardiest were found in the regions with high continentality (difference between the mean air temperature of the warmest and the coldest month greater than 40 K).

To clarify factors governing the cold adaptation of conifers, it was necessary to elucidate the survival mechanism of shoot and flower primordia which are the least hardy, but the most important tissues for growth and propagation of conifers. The primordia of winter buds of extremely hardy taiga spruces survive freeze dehydration, even at -70 °C, by extraorgan freezing (see Table 4.14). The buds of *Abies sachalinensis* survive freezing to -45 °C, but not below -50 °C, even when cooled slowly at 5 °C decrements daily. Over 10 days from February 1st to 10th, 1982, in the coldest area of Hokkaido, temperatures around -40 °C were recorded three times. Yet, no freezing injury of shoot primordia was detected in *Abies sachalinensis* near the thermometer screen. In Abietoideae of more temperate regions, shoot and flower primordia become also more or less dehydrated during slow cooling by extraorgan freezing, and the survival temperature of primordia shifts to lower temperatures (Sakai 1979a, 1982b). However, freezable water still remains in these organs even if cooled slowly, and the primordia are killed by freezing when cooled below -25° or -30 °C. In milder temperate climates, temperatures below -25 °C seldom occur. Thus, winter buds of both boreal and temperate conifers tolerate freeze-induced dehydration by extraorgan freezing under natural conditions.

In the genus *Pinus* the shoot primordia are characteristically large (1 to 10 cm long), lack a crown and connect directly with the twig axis, unlike the conifers of the subfamily Abietoideae. The shoot primordia of pines freeze extracellularly (Table 7.17). In general, shoot primordia of hardier conifers, especially of taiga species, are much smaller than those of less hardy species, which permits faster segregation of ice from the shoot primordia. These facts suggest that the flower or shoot primordia of boreal conifers have enhanced their hardiness by reducing organ size and by increasing the rate of water flow and the ability to tolerate freeze dehydration. Many coniferous taxa other than Pinaceae lack crowns in their small shoot and flower primordia; ice segregation occurs inside the bud scales, which act as an ice sink.

Winter desiccation is also an important factor limiting survival of conifers in frozen-soil areas (Sakai 1970b; see Sect. 8.1). The number of bud scales in *Abies sachalinensis* increases with winter cold over the natural distribution range (Okada et al. 1973). In conifers belonging to the Taxodiaceae, Taxaceae, Podocarpaceae and Cupressaceae, the shoot and flower primordia are surrounded by only a few green or brown bud scales, whereas in boreal conifers of Pinaceae, shoot and flower primordia are enclosed by 40 or more scales and the bud surface is covered with resin, which prevents water loss. In the Pinaceae, these bud scales might have developed through modification of primordial leaves, making wintering in northern continental climates possible.

7.5 Mountain Plants

High mountain sites are stress-dominated habitats characterized by low air and soil temperatures, night frost during the growing season, strong winds, high evaporation and intensive daytime radiation (Larcher 1983b). In *humid* regions of high and intermediate latitudes, the decrease in temperature with increasing elevation means frequent night frost in summer down to $-5°$ to -10 °C at altitudes where vascular plants can thrive (cf. Fig. 1.4) and a longer winter (cf. Table 1.4). In *arid* regions the daily temperature fluctuations are greater, the nocturnal minima in the growing season fall several degrees below -10 °C (Pamir: Tyurina 1957), and in winter the absence of snow cover allows frost to penetrate deep into the ground. In *tropical* mountains of the equatorial zone, especially in the months with little rainfall, night frosts of $-5°$ to -10 °C are not infrequent at greater altitudes and at particularly unfavourable sites. High mountains, therefore, are unique environments favouring selection and adaptive radiation among regional floras in response to steep temperature gradients (Billings 1973, 1974; Larcher 1980a).

7.5.1 Dwarf Shrubs and Herbs of Temperate Mountains

Characteristic of mountainous plants is the very prompt adjustment of their resistance level to the changing habitat temperatures. During the autumn, dwarf shrubs and herbaceous plants above the alpine tree line very quickly develop their winter frost resistance (Fig. 7.27). Following several successive nights with temperatures below -10 °C their hardening is more or less complete. If the temperature rises, even for a brief period, there is a reduction in resistance after only a few days; the return of sharper frost temperatures, however, elicits an increase in frost resistance of 5–10 K within 1 day or even hours, according to species.

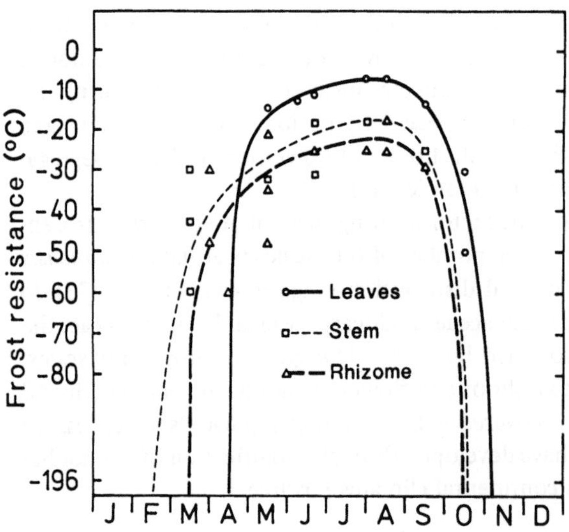

Fig. 7.27. Seasonal variation of the potential frost resistance (LT_{50}) of *Saxifraga oppositifolia*. Period with frost below -10 °C at the collection site (2200 m a.s.l.): December to March; flowering: April, May; sprouting: July. (From Kainmüller 1975)

Table 7.18. Potential frost resistance (LT_0 at °C) of ericaceous dwarf shrubs from temperate mountains. (From Ulmer 1937; Pisek and Schiessl 1947; Sakai and Otsuka 1970; Larcher 1977; and unpublished data; data for flower buds from Table 4.11)

Species	Collection site[a]	Leaves	Shoot bud	Flower bud	Stem	Below-ground stems
Arctostaphylos uva ursi	P	−30	−30		−30	
Calluna vulgaris	P	−35	−30		−30	−25
Cassiope lycopodioides	K	−40		−25	−30	−30
Diapensia lapponica	K	−70 (LN_2)			−70 (LN_2)	−70 (LN_2)
Empetrum nigrum	K	−70			−30	−30
Empetrum hermaphroditum	P	−30			−30	−20
Harrimanella stelleriana	K	−70			−30	−20
Loiseleuria procumbens	P	−50	−50		−60	−25
Loiseleuria procumbens	K	−70	−40	−20	−40	−30
Rhododendron aureum	K	−50	−40		−30	−30
Rhododendron ferrugineum	P	−25	−30	−25	−30	
Vaccinium myrtillus	P		−35		−35	−30
Vaccinium uliginosum	P, K		−40		−50	−35
Vaccinium vitis idaea	K	−70	−30	−30	−30	−20

[a] K = Mt. Kurodake, 1900 m a.s.l., Hokkaido; P = Mt. Patscherkofel, 2000 m a.s.l., Central Alps. (LN_2) = survives immersion in liquid nitrogen.

The potential frost resistance *in winter* is obviously connected with the geographic origin of the lineage and with the characteristic microsite preference of the species. Of the ericaceous dwarf shrubs of the alpine belt, species of predominantly oceanic distribution, like *Calluna vulgaris* and *Vaccinium myrtillus*, are only safe from serious frost damage down to about −30 °C (Larcher 1977; Table 7.18); *Rhododendron ferrugineum* and *R. hirsutum*, as Tertiary relicts, are equally sensitive (Ulmer 1937; Pisek and Schiessl 1947). Species originally of arctic distribution, such as *Loiseleuria procumbens* (Fig. 7.28) and *Vaccinium uliginosum* survive air temperatures to −60 °C and ground temperatures of −25° to −35 °C. Shoots of *Salix pauciflora* and *Diapensia lapponica* (circumpolar arctic plant), collected from a wind-swept mountain plateau in central Hokkaido where the lowest winter temperatures were −43 °C, were found by Sakai and Otsuka (1970) to be resistant to −70 °C. *Cassiope lycopodioides, Harrimanella stelleriana* and *Rhododendron aureum*, which are mainly distributed in coastal areas and on islands, attain much less winter hardiness (cf. Table 7.18). Of the herbs of high mountains, arctic-alpine species, such as *Silene acaulis* (found in the Alps at 1500–3600 m a.s.l.) and *Saxifraga oppositifolia* (1800–3500 m a.s.l.) become absolutely frost-tolerant (Kainmüller 1975); species of *Soldanella* and *Sempervivum*, descendants of S. European ancestors, survive only −20° to −25 °C (Fig. 7.29).

Genetically determined resistance properties reflect on microsite distribution patterns of alpine plants: *Loiseleuria, Silene* and *Saxifraga* can occupy wind-swept sites with little snow cover due to their extremely high levels of frost resistance, whereas

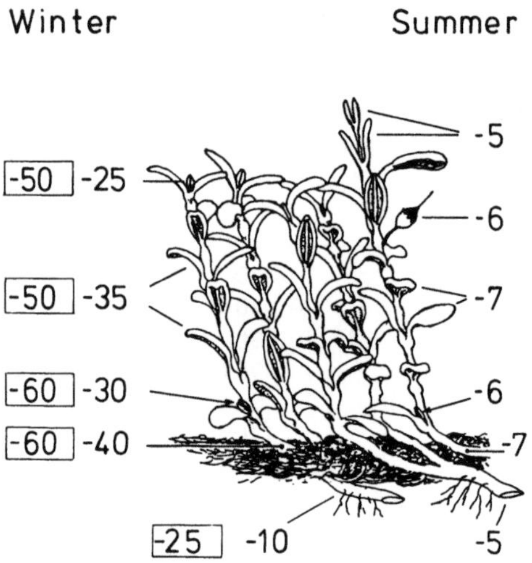

Fig. 7.28. Frost resistance (LT_{50}) of *Loiseleuria procumbens* in winter (*left*) and during the growing season (*right*). *Framed numbers:* maximal resistance; *open numbers:* minimum resistance after dehardening at 15 °C for 3 days. (From Larcher 1977)

Rhododendron, Calluna, Soldanella and *Sempervivum* are obligatory chionophytes, requiring the protection of a continuous covering of snow in winter. An intermediate position is occupied by *Carex firma*, an indigenous geoelement of the Alps found mainly on rocky sites with little snow (up to 3000 m a.s.l.). In the Rocky Mountains, *Cassiope mertensiana* suffers frost damage at -26 °C and remains restricted to snow beds, while *C. tetragona*, which survives -36 °C, grows on sites with shallow snow accumula-

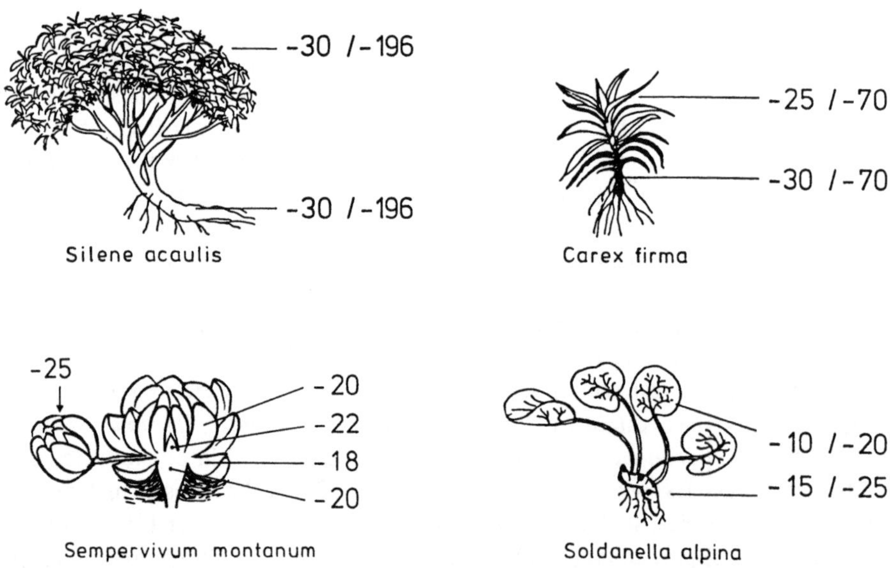

Fig. 7.29. Winter frost resistance (TL_{50}) of *Silene acaulis* and *Carex firma* on shallow snow sites, and of *Sempervivum montanum* and *Soldanella alpina* on snow-protected sites at 2000–2300 m altitude in the Alps. *Numbers* indicate minimal resistance after deacclimation at 20 °C for 3 days (*left*) and potential resistance after hardening by slow cooling at a rate of 1 K h^{-1} (*right*). *Sempervivum:* highest frost resistance observed on samples after collection from the habitat. (From Kainmüller 1974; Larcher and Wagner 1983)

Table 7.19. Summer frost resistance of leaves of mountainous plants from humid-temperate regions. (From Pisek et al. 1967; Sakai and Otsuka 1970; Kainmüller 1974; Larcher and Wagner 1976, 1983; Larcher 1977)

Plant species	LT_0 (°C)	LT_{100} (°C)
Dwarf shrubs (1900–2200 m a.s.l.)		
Vaccinium myrtillus	–4	
Vaccinium uliginosum	–4	
Vaccinium vitis idaea	–5	
Calluna vulgaris	–5	
Empetrum nigrum	–5	
Loiseleuria procumbens	–5	– 8
Arctostaphylos uva ursi	–7	– 9
Diapensia lapponica	–7	–12
Herbaceous plants (2200–3100 m a.s.l.)		
Primula minima	–3	– 4
Senecio incanus	–4	– 5
Geum reptans	–4	– 6
Soldanella pusilla	–4	– 6
Sempervivum montanum	–4	– 8
Saxifraga oppositifolia	–5	– 8
Silene acaulis	–6	
Carex firma	–6	– 8
Oxyria digyna	–6	– 8
Ranunculus glacialis	–7	– 8

Table 7.20. Frost resistance (surviving at °C) of leaves of subshrubs and perrenial herbs from an arid mountainous region (E. Pamir) in July and early August after previous night frosts of –1.5° to –3.5 °C; the samples were exposed to various test temperatures for ca. 12 h on the natural habitat. (From Tyurina 1957)

Upper alpine level (4300–4800 m a.s.l.)		**Lower alpine level** (3800–3900 m a.s.l.)	
Carex melanantha	–14.5	*Stipa glareosa*	–16.0
Androsace akbaitalensis	–14.0	*Carex duriusculiformis*	–15.0
Dracocephalum discolor	–13.0	*Artemisia skorniakowi*	–13.0
Potentilla pamiroalaica	–13.0	*Artemisia pamirica*	–12.5
Swertia marginata	–13.0	*Poa densissima*	–12.5
Artemisia pamirica	–12.5	*Ranunculus pseudohirculus*	–12.0
Neogaya simplex	–12.5	*Astragalus borodinii*	–12.0
Saussurea pamirica	–12.0	*Astragalus chadjanensis*	–12.0
Aster heterochaeta	–11.5	*Eurotia ceratoides*	–11.5
Tanacetum xylorrhizum	–11.5	*Oxytropis chiliophylla*	–11.5
Parrya excapa	–11.5	*Polygonum pamiricum*	–11.5
Waldheimia tridactylites	–11.5	*Zygophyllum rosovii*	–11.0
Sibbaldia tetrandra	–11.0	*Dracocephalum heterophyllum*	– 9.5
Ranunculus glacialis	–11.0	*Gypsophila capituliflora*	– 7.5
Saxifraga hirculus	–10.0		
Primula pamirica	–10.0		
Leontopodium ochroleucum	–10.0		
Rhodiola pamiroalaica	– 9.5		

tion (J.E. Harter, cited by Bliss 1985). Alpine dwarf shrubs and many herbaceous mountain plants can thus be regarded as indicators for the pattern of frost severity in mountainous areas (Larcher 1980a).

In *summer*, dwarf shrubs and herbaceous plants of high mountains of the humid regions resist episodic frosts of $-5°$ to -8 °C (Table 7.19; Figs. 7.27 and 7.28); mountain plants of arid regions survive $-10°$ to -15 °C (Table 7.20). At moderate frost temperatures leaf damage is prevented essentially by a low freezing threshold temperature. Where frost of below -10 °C may occur in the growing season, at least certain plants are capable of maintaining tolerance to equilibrium freezing even when metabolically active. This is the only possible explanation for the relatively high summer frost resistance of arid mountain plants (cf. Table 7.20) and of those which can be found, at any time in the growing season, to be frozen stiff in the morning, but to be entirely undamaged after thawing.

During the growing season, cold acclimation takes place at temperatures slightly above 0 °C within 1 day and results in depression of T_{sc} and T_f by 1–5 K (Tyurina 1957; Sakai and Otsuka 1970; Larcher 1980a). The degree of frost hardiness in summer, therefore, is readily adjusted to the temperature minima at the vegetation layer (Fig. 7.30). In addition, an inverse relationship to growth intensity can be recognized. Night frosts that arrest growth elicit a particularly large increase in resistance: in *Eurotia ceratoides*, for example, night frost in early June brought about a growth reduction of 50% and a rise in resistance from $-13°$ to -17 °C. After completion of shoot growth and until the beginning of autumn, the modulative adjustments of frost resistance were smaller.

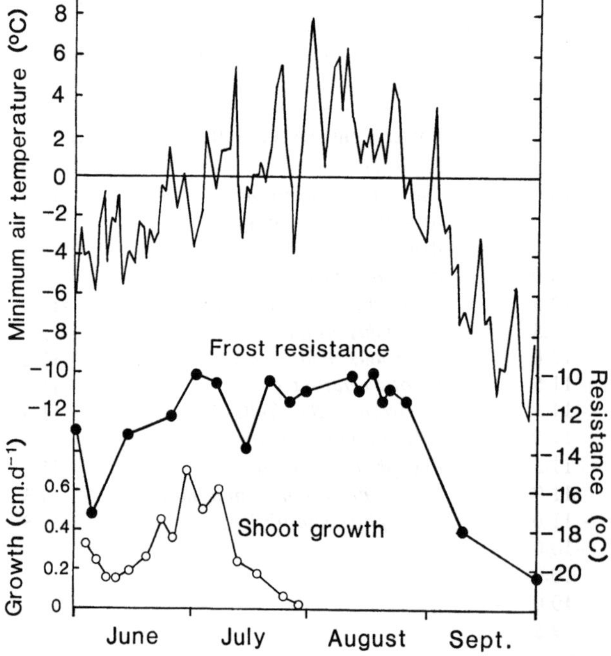

Fig. 7.30. Response of (•) leaf frost resistance and (○) shoot growth of *Eurotia ceratoides* at 4300 m a.s.l. on Pamir to temperature fluctuations (minimum air temperatures: *upper curve*) during the growing season of 1949. (From Tyurina 1957)

In rosette plants, which continuously produce new leaves, the ability to harden is connected with the climatic conditions prevailing at the time of leaf formation (Tyurina 1957). Leaves of *Potentilla multifida* and *Artemisia skorniakowii* which developed in spring at night temperatures of down to -20 °C and daytime temperatures of up to 30°-40 °C were, even after warm summer nights, distinctly more resistant to frost than leaves which had developed in July at a time when the night temperatures did not drop below -4 °C. Thus, it appears that rosette plants in high mountainous regions are capable of maintaining a persistent cold acclimation.

If foliage and shoot tips are damaged, the perennating parts of the shoot of high-alpine plants can still survive the lowest temperatures experienced in temperate mountains during the summer. The rhizomes of cushion plants of the European Alps, such as *Saxifraga oppositifolia* and *Silene acaulis*, tolerate temperatures to -20 °C (see Fig. 7.27). Cushion plants exhibit conspicuously large differences in resistance from one individual to another, a phenomenon indicative of great adaptive plasticity. An example is seen in *Saxifraga oppositifolia*: in some individuals the rhizomes retain nearly full freezing tolerance in summer, whereas the rhizomes of other individuals are totally destroyed at temperatures between -20° and -25 °C (Kainmüller 1975).

Little is known about the frost susceptibility of *seedlings* during germination and establishment. The germination of the seeds and growth of the young seedlings, which takes place during the short period of 2 to a maximum of 4 months favourable to growth, may be repeatedly interrupted by cold weather. When this happens the seedlings, even at this most tender life phase, achieve sufficient frost resistance to survive the frost temperatures that can occur at this time of year (Table 7.21). Seedlings of *Dryas octopetala* (6-days-old) and of *Loiseleuria procumbens* (18 days after germination) sustained temperatures to -4 °C without damage. At -6 °C the primary roots were killed in nearly all of the *Dryas* seedlings and in part of the *Loiseleuria* seedlings, while injury to the hypocotyls and cotyledons was slight. *Loiseleuria* seedlings were even observed to freeze solid without suffering lethal damage (M.Th. Eccher, pers. commun.). As long as the shoot axis remained intact, adventitious roots soon grew from the hypocotyl basis and most of the seedlings recovered (see Fig. 3.13).

Establishment and winter survival of seedlings depend on the completion of growth and the size acquired during the first summer after germination. The proportion of seedlings of *Polygonum cuspidatum* which survived the winter on Mt. Fuji was significantly higher at 1400 m than at 2500 m, approaching the upper distribution limit

Table 7.21. Frost resistance (°C) of imbibed seeds and young seedlings of alpine dwarf shrubs. (From M.Th. Eccher, unpubl.; cf. Fig. 3.13)

Plant species	LT_i	LT_{50}
Imbibed seeds ready for germination		
Dryas octopetala	-11	-20
Loiseleuria procumbens	-13	-21
Seedlings 1-3 weeks after germination		
Dryas octopetala	- 4	- 6
Loiseleuria procumbens	- 4.5	- 8

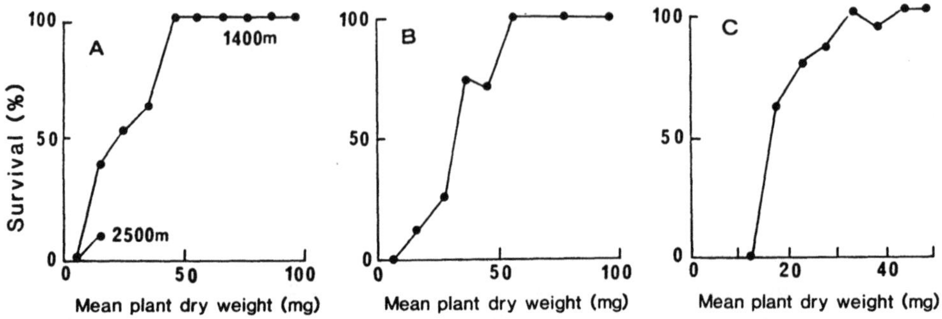

Fig. 7.31A–C. Survival of seedlings of *Polygonum cuspidatum* related to size. **A** Winter survival of seedlings in their natural habitats at 1400 m and 2500 m altitude on Mt. Fuji. **B** Winter survival of current-year seedlings grown at 1400 m and transferred to 2500 m a.s.l. where the lowest temperature at the soil surface was −19 °C. **C** Frost survival of pot-grown seedlings, brought to laboratory from field on November 25, after a freezing test involving exposure to −15 °C for 16 h. (From Maruta 1983)

(Maruta 1983). The climatic difference between the two altitudes did not affect the relative growth rate of the seedlings, but is reflected in dry matter accumulation which depends on the length of the growing season (130 days at 1400 m, and 70 days at 2500 m). The lowest winter temperature was −14 °C at 1400 m and −19 °C at 2500 m in the year of the investigation. Seedlings of less than 10 mg dry weight could not survive winter at any altitude. The survival rate increased with increasing dry weight up to 100% in seedlings of more than 40 mg dry weight (Fig. 7.31). At 2500 m, seedlings of less than 2 mg dry weight did not form perennation buds and died before winter; seedlings in the range of 2–20 mg dry weight, even if they had buds, did not survive −15 °C. Large seedlings (40–100 mg dry weight), grown in pots at 1400 m and transferred to 2500 m, survived the winter just as well as those at 1400 m. In *Polygonum cuspidatum* the size-dependent mortality involves two factors: (1) very small seedlings cannot produce perennating buds, since this requires the accumulation of a critical amount of photosynthates; (2) smaller seedlings show lower freezing resistance during winter both in laboratory tests and in field experiments (Fig. 7.31b,c).

A reduction in the growth period can also result in a lower frost resistance of mature plants in the following year. *Ribes nigrum* survives winter frost to −60 °C after a growing season of 5 months, which is the optimum duration, but only to −50 °C if the vegetative period is restricted to 4 months. If growth is arrested artificially after 3 months, severe injuries are suffered even at −20 °C, and if it is further limited to 2 months, at −10 °C (Tumanov and Krasavtsev 1975). This is an observation of considerable significance for mountain plants whose assimilatory period is progressively shortened by heat deficiency with increasing altitude, or by especially long snow cover.

7.5.2 Caulescent Giant Rosette Plants of Tropical High Mountains

Among the wide variety of growth forms seen in mountain plants a special place is occupied by the megaphytic rosette plants. They occur solely in the high mountains of

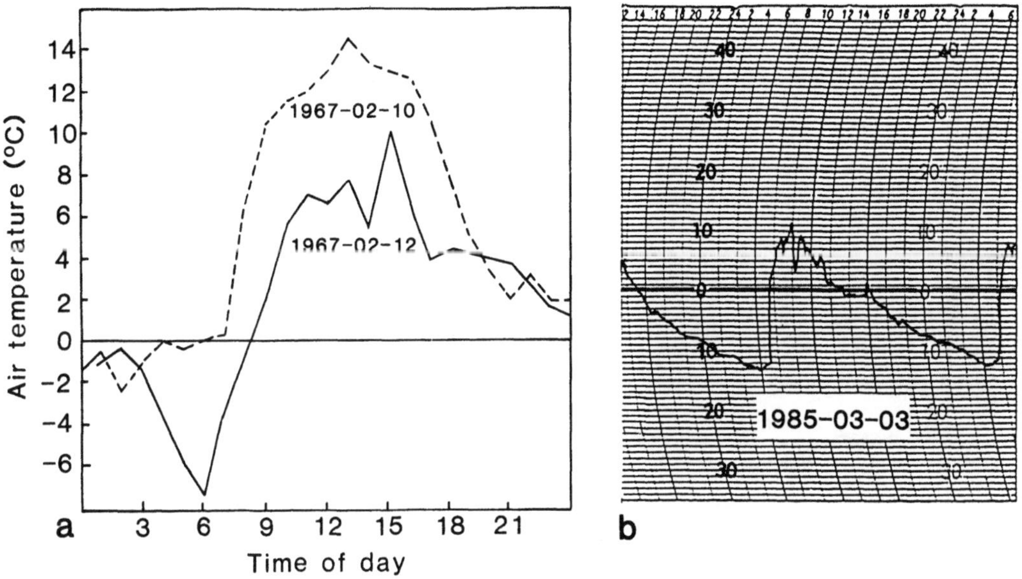

Fig. 7.32a,b. Night frost on tropical mountains. **a** Large diurnal temperature changes with radiation cooling during clear nights at Mucubaji, Venezuelan Andes, 3600 m a.s.l. (From Walter and Medina 1969). **b** Recording of extremely low night temperatures at 15 cm above soil surface in the Teleki Valley on Mt. Kenya, 4200 m a.s.l. (From Bodner 1985)

the equatorial zones of S. America and E. Africa, mainly at altitudes between 3500 m and 4500 m (Troll 1960; Hedberg 1964, 1973; Vareschi 1970; Cuatrecasas 1979). A comparison with the extratropical high mountain vegetation prompts the question as to how rosette plants, with stems several meters tall, can exist in a distinctly alpine milieu. At altitudes at which vascular plants thrive in tropical mountains there is of course not the persistent snow cover which in temperate mountains exerts a levelling effect on the plants during winter (Larcher 1975, 1980a). On the other hand, night frosts recur at any time of year in tropical high mountains. In the Paramos of the humid to semi-humid equatorial Andes, which is the chosen environment of various megaphytic species of *Espeletia* (Asteraceae), the temperature minima drop to −8° to −11 °C (Fig. 7.32a; Walter and Medina 1969; Goldstein et al. 1985a; Sturm and Rangel 1985). In the E. African high mountains, at altitudinal levels in which giant rosette plants of the genera *Dendrosenecio* and *Lobelia* occur, night frosts to −13 °C are experienced (Fig. 7.32b; Beck 1986; Bodner 1985). This means that the plants have to sustain frost whilst in a vegetatively active state, just like the plants of temperate and arid mountains during the short growing season.

The most suitable mechanism of frost mitigation and for avoidance of tissue freezing for protecting metabolically active and growing plants from injury in a climate where, after a few hours of night frost, the temperature suddenly rises to 5° to 10 °C at dawn, are these that can be activated at short notice. Freezing of the stem parenchyma of *Espeletia*, which would occur at −4° to −6 °C, is prevented by a thick insulating cover of attached dead leaves (Hedberg and Hedberg 1979; Goldstein and Meinzer 1983; Rada et al. 1985b; Figs. 7.33, 7.35). This parenchyma also acts as an internal water reservoir

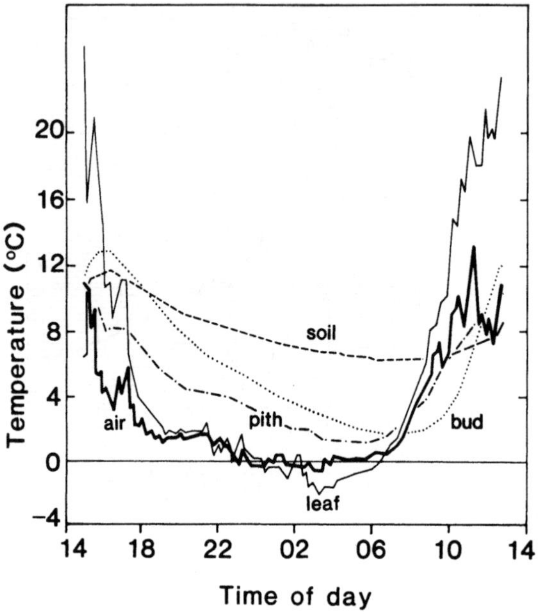

Fig. 7.33. Frost migration through insulation in *Espeletia timotensis*. Typical temperature decline in air (150 cm above ground), soil (15 cm below surface) and of various plant parts. (From Rada et al. 1985b)

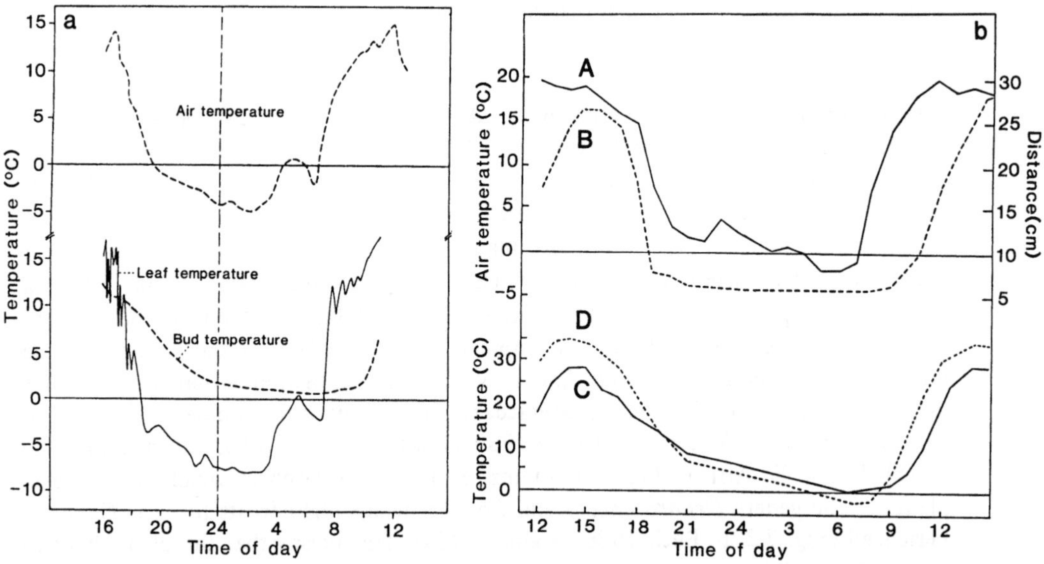

Fig. 7.34a,b. Frost mitigation through nyctinastic leaf movements. **a** *Dendrosenecio brassica*, Mt. Kenya. Temperatures of a rosette leaf and interior of the leaf bud, compared with air temperature. (From Beck et al. 1982). **b** *Espeletia schulzii*, Mucubaji. *A* Air temperature at ground level; *B* distance between tips of two opposing rosette leaves (average of seven leaf pairs); *C* bud-core temperature of plants with normal leaf folding; *D* bud-core temperature of plants when rosettes held open. (From A. Smith 1974)

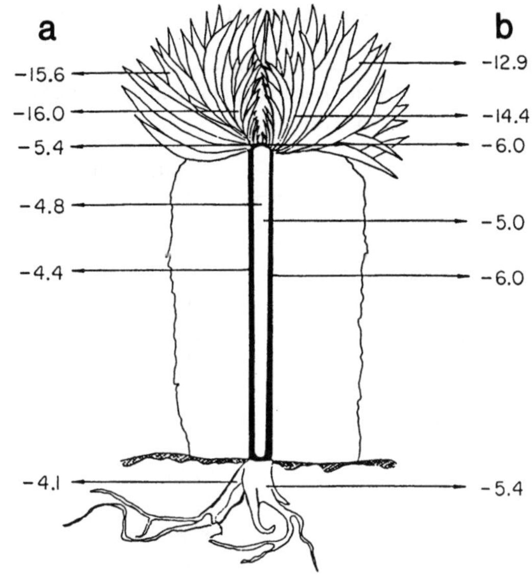

Fig. 7.35a,b. Supercooling thresholds for adult and young leaves, apical bud, stem pith, phloem and periderm tissues, and roots of (**a**) *Espeletia spicata* from depressions and (**b**) *Espeletia timotensis* growing on well-drained slopes at 3800–4100 m a.s.l. in the Venezuelan Andes. (From Rada et al. 1985b)

when cold root temperatures affect water uptake (Hedberg 1964; Goldstein et al. 1985b). Nocturnal heat loss from the apical bud, and in the case of the Asteraceae the flowers as well, is reduced by inward bending of densely packed leaf layers due to nyctinastic movement (Fig. 7.34). As a consequence, subzero temperatures were never observed in 'night buds' of *Espeletia* (A.P. Smith 1974; Hedberg and Hedberg 1979) and afroalpine megaphytes (Hedberg 1968; Beck et al. 1982). *Lobelia* species are able to exploit crystallization heat released by the freezing of water and so prevent too great a drop in tissue temperature: the hollow inflorescence axes of *L. telekii* are half-filled with water which begins to freeze soon after the temperature falls below 0 °C, thus keeping the organ temperature at zero point until the morning, even if the outer temperature drops to -5 °C (Hedberg 1964; Krog et al. 1979). The imbricate leaf bases of *L. keniensis* which are filled with an aqueous fluid act similarly as a thermal buffer for the apical bud of the rosette (Beck et al. 1982).

The temperature of the mature leaves drops well below 0 °C on clear nights (Fig. 7.35). Since subzero temperatures prevail only for several hours, a supercooling mechanism could provide sufficient protection from night frost in tropical mountain environments. Larcher and Wagner (1976), Goldstein et al. (1985a,b) and Rada et al. (1985b) found supercooling thresholds of *Espeletia* species in the range of -6° to -16 °C, i.e. adequate for avoidance of tissue freezing under the climatic conditions of the Paramos (Table 7.22). The leaves of *Lobelia* and *Dendrosenecio* species supercool to -4° to -7 °C (Table 7.22; Fig. 7.36); as they are not damaged until -10° to -20 °C, they can be considered as freezing-tolerant. Leaves which were frozen stiff by the morning regained photosynthetic activity immediately after thawing (Schulze et al. 1985; Bodner and Beck 1987). By microscopic observation of "frost plasmolysis" and by measuring the water potential of the frozen tissues, Beck et al. (1984) and Beck (1986) were able to provide evidence for equilibrium freezing in the leaves.

Table 7.22. Frost resistance (TL_{50} at °C) and supercooling ability (T_{SC}) of the leaves of megaphytes of tropical high mountains

(a) *Espeletia* species occurring along an altitudinal gradient on the Venezuelan Andes. (From Goldstein et al. 1985a,b)

Species	Elevation (m)	T_{SC}	TL_{50}
E. lindenii	2850	− 7.5	− 6.5
E. angustifolia	2850	− 6.6	− 6.1
E. atropurpurea	2850	− 6.4	− 5.9
E. marcana	3100	− 9.1	− 8.0
E. atropurpurea	3100	− 7.3	− 8.1
E. jahnii	3100	− 5.7	− 5.6
E. schultzii	3560	−10.8	−10.0
E. floccosa	3560	− 8.5	− 9.3
E. schultzii	4200	−10.0	−11.2
E. moritziana	4200	−10.6	−11.3
E. spicata	4200	−10.0	− 9.5
E. lutescens	4200	−10.5	−10.2
E. timotensis	4200		
Juvenile (15 cm tall)		−11.0	
Adult (90 cm tall)		− 8.1	
Adult (120 cm tall)		− 6.1	

(b) Giant rosette plants on Mt. Kenya. (From Beck et al. 1982, 1984; Beck 1986)

Species	Elevation (m)	T_{SC}	TL_{50}
Dendrosenecio keniodendron			
Seedling	4200		− 5
Adult	4200	−7	−14
Adult	4500		below −15
Dendrosenecio brassica			
Seedling	4100		− 8
Adult	4100	−5.5	− 5 to −10
Lobelia telekii			
Seedling	4200–4500		−10 to −14
Adult	4200–4500	−5	−14 to −20
Lobelia keniensis			
Seedling	4100		−10
Adult	4100	−4	−10 to −20

It appears that mountain plants during growth can employ two resistance mechanisms in adapting to the rising frequency and severity of frost with increasing altitude: (1) by improving their ability to supercool and (2) by retaining a permanent tolerance to equilibrium freezing. The first of these mechanisms suffices where temperatures do not drop much below −10 °C, which appears to be near the lowest limit for avoidance of freezing. Where temperatures below this critical threshold are experienced the mechanism of freezing tolerance has to be brought into play. This is more effective and can be extended to lower temperatures; it requires, however, specialized structures and biochemical measures for which a metabolic price has to be paid (Beck 1986).

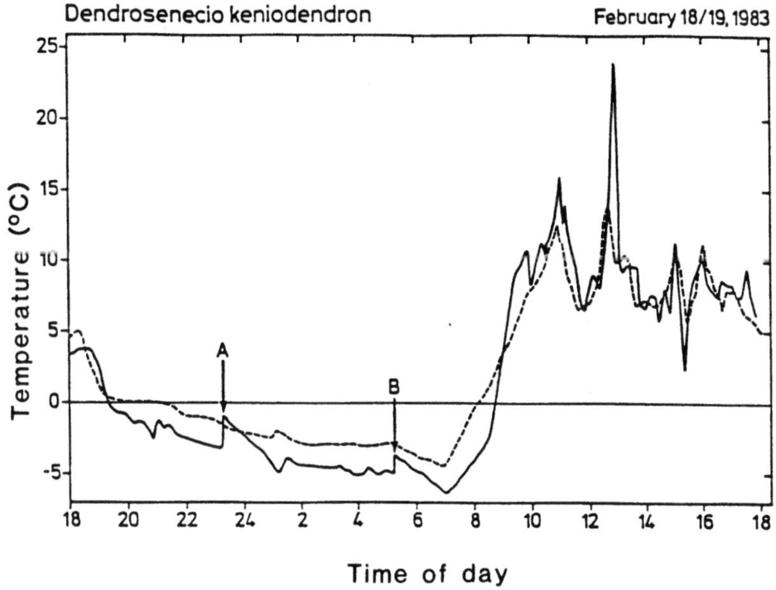

Fig. 7.36. In situ measurement of freezing exotherms on leaves of a tropical high mountain megaphyte (Mt. Kenya). *Solid line:* leaf temperature with two exotherms; *A* resulting from freezing of adhering water droplets; *B* from freezing of tissue water. *Broken line:* air temperature. (From Beck et al. 1984)

7.6 Trends of Adaptive Improvement of Low Temperature Resistance in Vascular Plants

Climatic stress promotes evolutionary adaptations and accelerates the differentiation of ecotypes and species (Stebbins 1952; Axelrod 1972). Progression in the cold resistance of plants along latitudinal and altitudinal gradients, and transitions between the various categories of cold resistance have been hypothesized by Larcher (1971, 1980a, 1981b) and Sakai (1978b–d) as being steps of evolutionary adaptation to low temperature constraints during the geological periods in which colder climates prevailed as well as in connection with the spread of plants to colder regions.

The *first step* in cold adaptation must have been that chilling-sensitive plants became resistant to temperatures above freezing by lowering of the critical phase transition temperature of their biomembranes. Evidence of a step of this nature can be seen in the modest, but nevertheless undeniable successful adaptation, following cold conditioning, of a variety of plants of tropical provenance, and in a seasonal displacement of the phase transition point to values below 0 °C (for references, see Lyons 1973; Graham and Patterson 1982; Larcher 1985a). Evolutionary adaptation to chilling resistance may have been brought up to the required levels in the border regions of the tropics (as in the case of mangroves: Sect. 7.1.1) and especially along altitudinal gradients. Among the chilling-sensitive tropical plants gradual progressions of cold adaptation have been found in species of *Passiflora* of differing altitudinal distribution (Patterson et al. 1976) and in altitudinal ecotypes and subspecies of *Lycopersicon hirsutum* (Patterson et al. 1978): in neither species is the transition to chilling tolerance complete. On Mt. Wilhelm, Papua New Guinea, Earnshaw et al. (1987) found a decreasing chilling sensitivity of tropical grasses from 1550 m to 2600 m a.s.l. (Fig. 7.37). All grass species from 3280 m to 4350 m a.s.l. were clearly chilling-resistant. Populations of *Miscanthus floridulus* were chilling-sensitive at 2600 m a.s.l. (mean minimum air temperature 9.4 °C), and resistant at 3280 m a.s.l. (mean minimum temperature 5.6 °C). It seems quite feasible that a transition might also be present in tropical plants with chilling-sensitive juvenile stages but resistant adult individuals, or with chilling-sensitive leaves and resistant buds or stems. That individual cells and tissues (e.g. pollen, epidermis) of chilling-susceptible plants can be resistant to chilling and are only killed by freezing has been repeatedly demonstrated (review: Larcher 1985a).

The *second step* in cold adaptation must have been the lowering of the threshold freezing temperature and improvement of the supercooling capacity, particularly the development of persistent supercooling in the freezing sensitive plants. This step, too, may have originated in modulative adaptations to short-term frost stress, comparable to the situation in the subtropics or in highlands. Tropical mountains have undoubted-

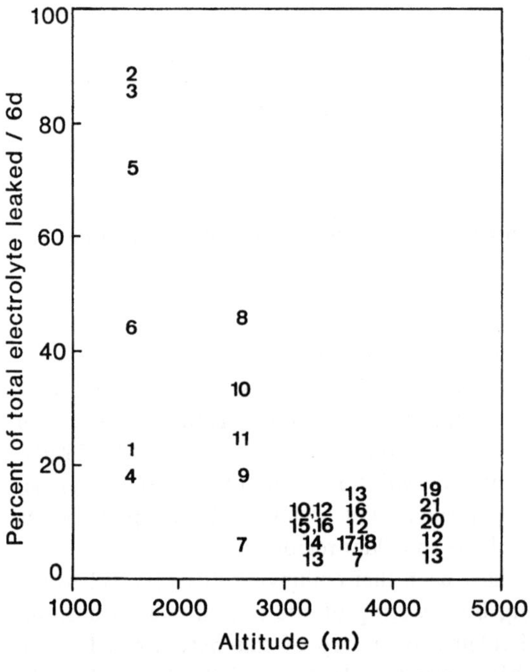

Fig. 7.37. Chilling susceptibility of leaves of tropical grasses along an altitudinal transect on Mt. Wilhelm, Papua New Guinea. The rate of electrolyte leakage during a 6-day period at 0 °C declines with decreasing susceptibility; specimens with values below 20% of total electrolytes are assumed to be chilling-tolerant. 1 *Coix lacryma-jobi*; 2 *Ischaemum polystachyum*; 3 *Paspalum conjugatum*; 4 *Saccharum officinarum*; 5 *Saccharum robustum*; 6 *Setaria palmifolia*; 7 *Agrostis avenacea* (different altitudinal populations); 8 *Athraxon ciliaris*; 9 *Imperata conferta*; 10 *Miscanthus floridulus* (different populations); 11 *Setaria montana*; 12 *Anthoxanthum redolens* (different populations); 13 *Deschampsia klossii* (different populations); 14 *Dichelachne rara*; 15 *Microlaena stipoides*; 16 *Poa keysseri*; 17 *Anthoxanthum angustum*; 18 *Deyeuxia brassii*; 19 *Danthonia vestita*; 20 *Festuca papuana*; 21 *Poa crassicaulis*. (From Earnshaw et al. 1987)

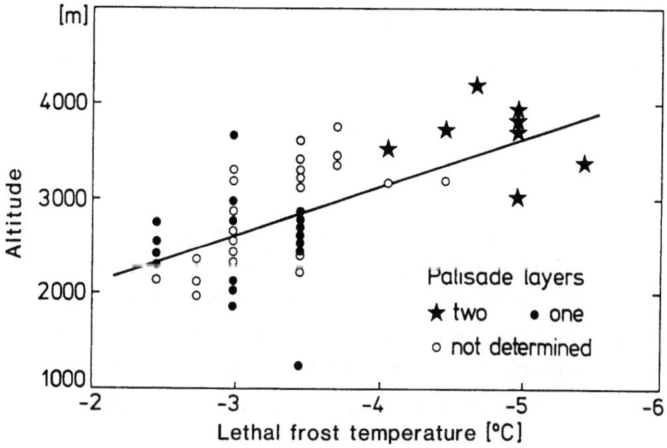

Fig. 7.38. Frost resistance of leaves of tuber-bearing *Solanum* species from Mexico, Guatemala and the Andes as far as 35° S, related to the altitude at which they were collected. (From Palta and Li 1979)

ly played an important role in eliciting adaptation to episodic frost. The biochemical basis of this adaptation lies in osmotic adjustments and in increased cryostability of the plasma membrane, which precludes nucleation of the supercooled intracellular solution. Mountain plants appear to be especially capable of adjusting their cell sap concentration and lowering their T_{sc} and T_f (see Fig. 4.7 and Sect. 7.5). Altitudinal resistance gradients of freezing-sensitive species (cf. Figs. 6.1 and 7.38; Table 7.22) and of woody species with deep supercooling xylem (Fig. 7.39) are reflected in a change in floristic composition and in ecotype differentiation with increasing altitude. An interesting case of complex adaptation to the climate at high altitudes has been reported by Palta and Li (1979): the leaves of *Solanum* species adapted to high altitudes froze at lower temperatures than those from lower sites and were characterized by a multilayered palisade parenchyma and a higher stomatal density, so that a close correlation exists

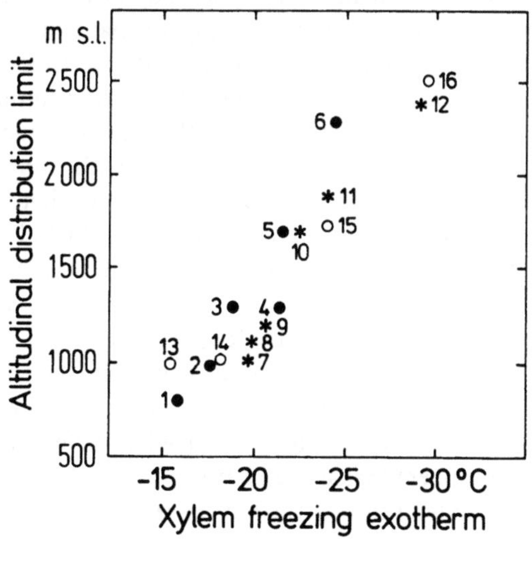

Fig. 7.39. Relation between the position of the low temperature exotherm of deep-supercooling xylem and the altitudinal distribution limit of various woody plant species. 1 *Quercus phillyreoides*; 2 *Qu. glauca*; 3 *Qu. acuta*; 4 *Qu. acutissima*; 5 *Qu. serrata*; 6 *Qu. mongolica* var. *grosseserrata*; 7 *Ilex rotunda*; 8 *I. integra*; 9 *I. chinensis*; 10 *I. pedunculosa*; 11 *I. pedunculosa* var. *senjoensis*; 12 *I. macropoda*; 13 *Viburum japonicum*; 14 *V. odoratissimum*; 15 *V. phlebotrichum*; 16 *V. furcatum*. (From Kaku and Iwaya 1979)

between elevated frost resistance and the occurrence of the above-mentioned histological characters (see Fig. 7.38). This correlation warrants the conclusion that adaptation to frost and to other climatic factors specific to these high altitudes went hand in hand. A most efficient factor is strong irradiation, the more so since is known to favour the development of this type of anatomical structure (Körner et al. 1983; Larcher 1983b).

The *third step* towards achieving freezing tolerance was the adaptive acquisition of increased dehydration tolerance necessary for resistance to equilibrium freezing. Such preadaptation (or co-adaptation) may have taken place in regions with a dry season. In cormophytes, water deficiency favours osmotic adjustments and elevated desiccation tolerance. It is a striking fact that plants of arid high mountains (Pamir), whose lowland ancestors were already selected for survival of low water potentials due to regular summer drought, develop a freezing tolerance in the vegetative state greater than those of humid mountains (European Alps). The wild, tuber-bearing species of *Solanum*, too, from both the colder and the drier subtropical regions of the Andes are more resistant to frost than those from warmer or wetter regions (Palta and Li 1979).

The adaptive improvement of frost hardiness comes to perfection in cormophytes in which the freezing tolerance is associated with an *activity rhythm*. This suggests that the cellular and ultrastructural basis for seasonal freezing tolerance and desiccation resistance evolved in step with the development of growth rhythmicity in arid subtropical regions and in intermediate latitudes with a seasonal climate. Where dry and rainy periods recur in definite seasons it would be possible for a synchronization between climatic and vegetational rhythms to have developed, which, in turn, became coupled with a rhythmicity in resistance. In response to different degrees of drought connected with the transition from an almost aseasonal to a strongly seasonal climate, certain tropical tree species changed from evergreen to deciduous (Borchert 1980). That the adaptive response of many tropical woody plants to drought has been met by the evolution of the deciduous habit is apparent from gradients across vegetation zones (Axelrod 1966; cf. Sect. 7.3). The fact that the rhythmicity of growth and resistance became genetically anchored, was decisive for penetration to higher latitudes: control of the growth and development by means of photoperiodic and thermoperiodic signals ensures that the plant is ready for hardening at the correct time, so that the complex and slow, stepwise acquisition of tolerance is activated before serious danger from winter cold can be expected. Adaptation to the conditions of existence in regions with cold winters thus involved both a precise timing of growth to coincide with a frost-free period that became steadily shorter with rising latitude, and the acquisition of a greater degree of freezing tolerance. In plants that spend the winter in a state of inherent, genetically determined dormancy, e.g. boreal conifers, freezing tolerance is linked with an increased desiccation tolerance (Pisek and Larcher 1954). Deep and protracted dormancy, however, means cessation of metabolic activity and a considerable reduction in the time available for productivity. Freezing tolerance coupled with rhythmicity is thus only an advantage for woody evergreen plants if it is exploited to a maximum.

In the Southern Hemisphere adaptive differentiation with respect to frost tolerance appears to have taken a somewhat different course from that seen in the Northern Hemisphere (Sakai et al. 1981; cf. Sect. 7.3 and 7.4). Many originally pantropic families that extend into extratropical regions contributed to both northern and southern temperate floras. However, they are usually represented in the north and south by different genera. As a counterpart to floral evolution in the north, there are a number of families that are principally tropical and austral (Proteaceae, Myrtaceae, Winteraceae, Pittosporaceae). At present, the temperate zone tree genera are basically either northern or southern, although several overlap at high altitudes on tropical mountains (e.g. *Quercus, Nothofagus, Podocarpus, Pinus* in southeast Asia; *Pinus* and *Podocarpus* in Mexico). *Podocarpus* is the only southern genus that reaches the northern temperate zone. Two northern deciduous woody genera with four species, all pioneering trees along streams and rivers, have crossed the tropics and reached the southern temperate zone: *Alnus jorullensis* and *Salix bonplandiana* in South America, and *Salix safsaf* and *Salix mucronata* in Africa. *Salix safsaf*, a primitive willow belonging to the subgenus *Protitea*, is widely distributed in Africa, from the Nile lands of Egypt and Sudan to South Africa through Kenya and Tanzania (Fig. 7.40). *Salix mucronata*, which is very closely related to *Salix safsaf*, ranges widely in South Africa, extending as far south as Cape Town. Winter hardiness may reflect a very long history of adaptation to climate with cold winters. Such hardiness, once acquired, can persist in mild climates where it is no longer appropriate. This was demonstrated for *Salix safsaf* and *S. tetrasperma* (Sakai 1978c; cf. p. 75).

The foregoing considerations are confined to the terrestrial cormophytes which, as revealed by research on phylogenetic evolution (Stebbins 1965; Cronquist 1968; Takhtajan 1970), may have developed under tropical-montane conditions and only later settled in periodically dry and cold zones. An interesting question is whether the phylogenetically older thallophytes, particularly the algae, achieved any form of resistance, and if so by what means. Transitions from chilling susceptibility to chilling tolerance have been shown in algae (Biebl 1962b). This seems to be a direct reaction to progressive cooling and can take place relatively easily. Among the algae there are also species with unlimited freezing tolerance, both in the anabiotic state and in hydrated, active

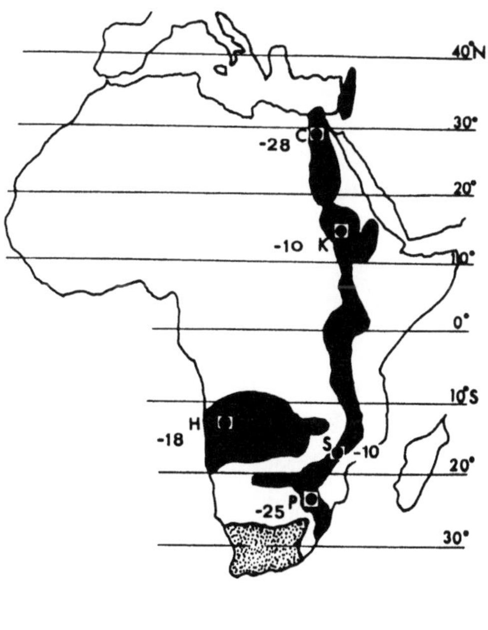

Fig. 7.40. Geographic distribution of *Salix safsaf* (*black area*) and *Salix mucronata* (*stippled area*). *Numerals* indicate frost resistance of collections from Cairo (*C*), Khartoum (*K*), Salisbury (*S*), Huambo (*H*) and Pretoria (*P*). (From Sakai 1978c)

cells (see Table 6.1). In all probability this is a consequence of the advantageous position of unicellular organisms and monolayer tissues with respect to extracellular freezing. Phasic or even rhythmicity-linked hardening processes, however, are an unique achievement of the woody cormophytes made necessary by their higher level of morphological organization and histological specialization.

8. Winter Damage as the Result of a Complexity of Constraints

Overwintering plants are subject to various constraints. The danger from winter cold is not only due to the primary effect of ice formation in the tissues. Additional threats are presented by the freezing of the water in and on the ground, and by snow accumulation. Especially for plants growing on mountains, the combinations of constraints to which they are exposed in winter vary over only very short distances due to steepness of slope and exposure to radiation and wind (Table 8.1; Figs. 8.1 and 8.2). Plants on sunny habitats are exposed to high radiation in the dormant state, to large, short-term fluctuation of shoot and soil surface temperatures, and to winter desiccation. Plants on

Table 8.1. Complex winter stress above timberline on mid-latitudinal mountains. (After Turner 1980, modified by Larcher 1985b)

Site exposure	Topoclimate[a]	Winter stress factors[b]
Sunny windward	R△W△ S▼ T_P▼ T_{CH}▼ T_R ▼ ~ TA△	LT, FT, FD W, R
Shaded windward	R ▼W△ S ▼ T_P ▼ T_{CH}▼ T_R▼ TA▼	LT, W (FD)
Sunny lee	R△W ▼ S△~ T_P~T_{CH}△ T_R△ TA△	(FT), (FD), (S)
Shaded lee	R ▼ W▼ S△ T_P~ T_{CH}△ T_R△ TA▼	S

[a] R = radiation (duration, potential intensity); W = wind (velocity, frequency); S = snow cover (depth, duration); T = temperature (T_P shoot temperature of phanerophytes; T_{CH} shoot temperature of chamaephytes and hemicryptophytes; T_R rhizosphere temperature); TA = daily temperature amplitude; △ high, ▼ low, ~ medium. S△ means > 180 days of continuous snow cover; S▼ means frequent snow-free periods during winter. Shoot temperatures are considered as low (▼) if night temperatures during mid-winter approach −20 °C or less; shoot temperatures below snow cover are usually in the range of 0° to −10 °C (△). Rhizosphere temperatures (soil temperatures at 10–15 cm depth) are indicated as ▼ at −5° to −10 °C, as △ at 0° to −2 °C. TA△ means frequent freeze-thaw cycles and maximum shoot temperatures > 10° to 15 °C.

[b] LT = low temperature stress; FT = freeze-thawing stress; FD = frost drought, winter desiccation; W = wind stress; S = stress induced by snow cover; R = irradiation stresses (high light, UV). In parenthesis: only temporarily and/or locally important.

Fig. 8.1. Typical distribution of snow on a wind-swept ridge above timerline on Mt. Patscherkofel (2200 m a.s.l.) near Innsbruck in late winter. Viewed towards the west. The sunlit slope is also the most wind-exposed site; the depression on the lee side is filled with several metres of snow. (From Larcher 1977)

windward habitats are likely to suffer from low-temperature stress, mechanical, cooling and desiccation effects of wind and frost drought due to frozen soil, whereas plants in wind-protected habitats where snow accumulates, especially on shaded sites, are safe from very low temperatures and desiccation; winter stress on such habitats depends on snow pressure and on the duration of the snow cover.

Winter survival is a highly complex problem, both with regard to stress factors as well as the plants' response. It was early recognized, particularly in the fields of agricultural and forestry science, that winter injuries are often indirect effects of frost rather than being due to inadequate freezing tolerance (in trees: Ebermayer 1873; in herbaceous crops: Vasilyev 1961). In specific cases it is frequently impossible to identify

Fig. 8.2a,b. Gradients of environmental factors, winter stress effects and plant cover along microtopographical transects above timberline. **a** Central Alps: A graminoids; B *Rhododendron ferrugineum*; C *Vaccinium uliginosum*; D *Loiseleuria procumbens*; E lichens; F *Juncus trifidus*; G *Arctostaphylos uva ursi, Empetrum hermaphroditum*; H *Calluna vulgaris*; I *Vaccinium vitis idaea; Juniperus nana*. Winter stress factors: *sp* snow pressure; *sm* infection by snow moulds; *wd* winter desiccation; *r* strong irradiation; *ft* frequent freeze-thaw cycles. (From Aulitzky 1963, modified). **b** Mt. Taisetsu, Hokkaido. Topography and plant cover: P *Pinus pumila* community; A alpine dwarf shrubs; L lichens; B bare ground; S snow bed; plant height refers to *Pinus pumila*. (From Okitsu and Ito 1983)

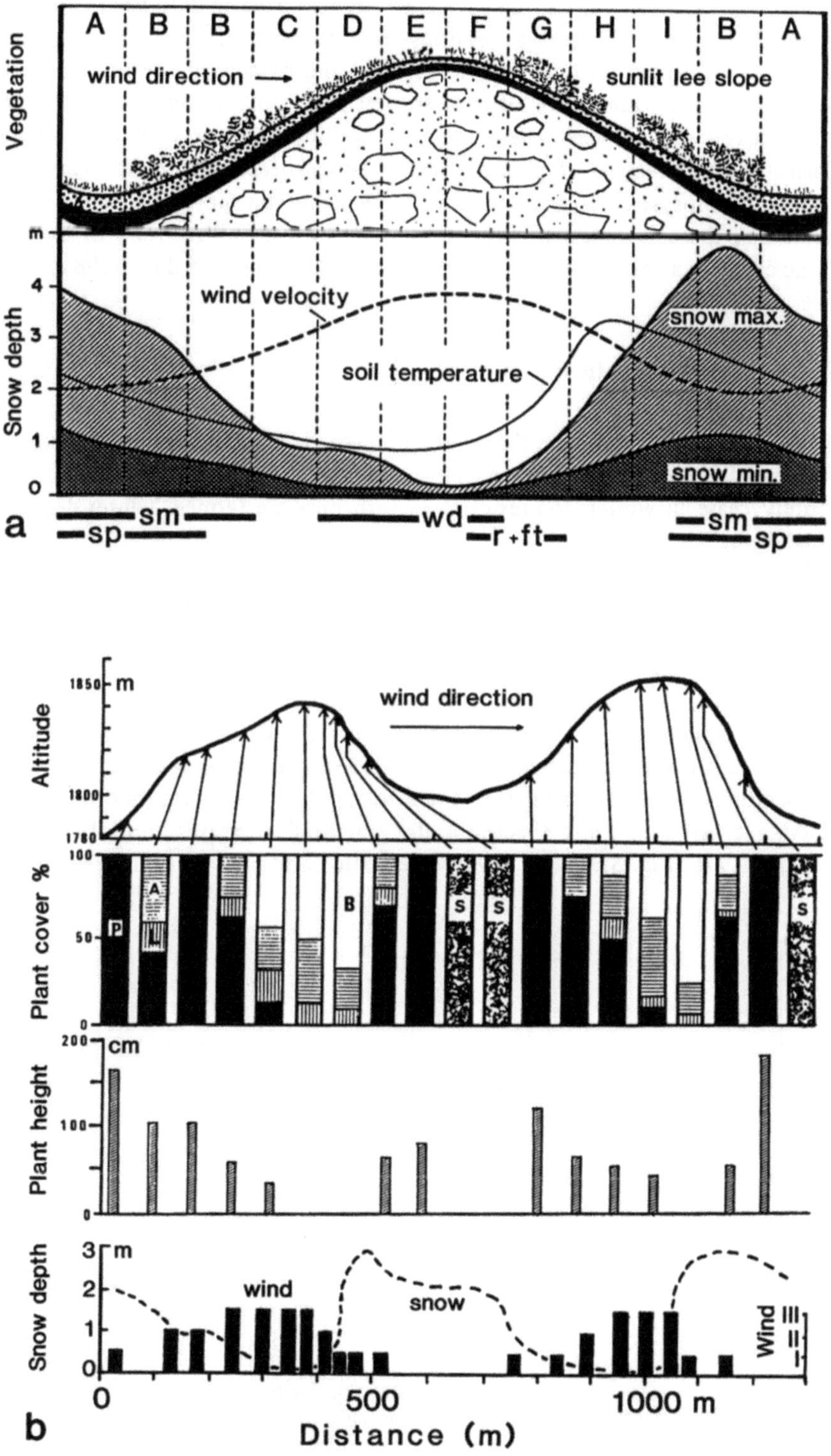
A B B C D E F G H I B A
Vegetation
wind direction
sunlit lee slope
m
Snow depth
wind velocity
soil temperature
snow max.
snow min.
a
sm
sp
wd
r+ft
sm
sp
b
Altitude
1850 m
1800
1780
wind direction
Plant cover %
100
50
0
A
P
L
B
S
S
S
Plant height
200 cm
100
0
Snow depth
3 m
2
1
0
wind
snow
Wind III II I
0
500
1000 m
Distance (m)

the proper cause of the damage. Detailed studies have mainly been concerned with winter desiccation and damage resulting from covering of snow and ice. Still little is known to date about interactions of primary and secondary stresses to which plants are subjected in winter. One of such stresses is photoinhibition of photosynthesis in frozen leaves of cold-acclimated evergreen conifers (Martin et al. 1978; Öquist and Martin 1980). Freezing of pine needles in light, in contrast to freezing in darkness, damages photosystem II and inhibits the electron transport linking the two photosystems of the chloroplast (Strand and Öquist 1985). Full recovery of the photosynthetic capacity after winter stress requires weeks (Larcher and Bauer 1981).

8.1 Winter Desiccation

If the water in the soil or in the conducting vessels of stems remains frozen, the water balance of the plant becomes negative due to the fact that transpiration does not entirely cease in winter; the resulting disturbances are termed 'winter desiccation' or 'frost drought'.

Winter desiccation is a widespread phenomenon, most frequently observed in woody plants, especially in young evergreen trees, but also in herbaceous plants (for references, see Vasilyev 1961; Larcher 1972, 1985a; Kozlowski 1976; Tranquillini 1976, 1982). Frost drought is a regular occurrence in mountains (Michaelis 1934a,b; Henson 1952; Larcher 1957, 1977; Wardle 1971, 1985; Tranquillini 1979b).

Winter desiccation has also been observed even when the ground was not frozen, e.g. in heath plants in Scotland (Bannister 1964). In *Citrus* species, below 10 °C soil temperature, the root resistance to water uptake becomes twice that of normal uptake (Elfving et al. 1972) and the leaf water potential decreases abruptly (Maotani and Machida 1976; Fig. 8.3). The lower water influx and transport rates reflect changes in

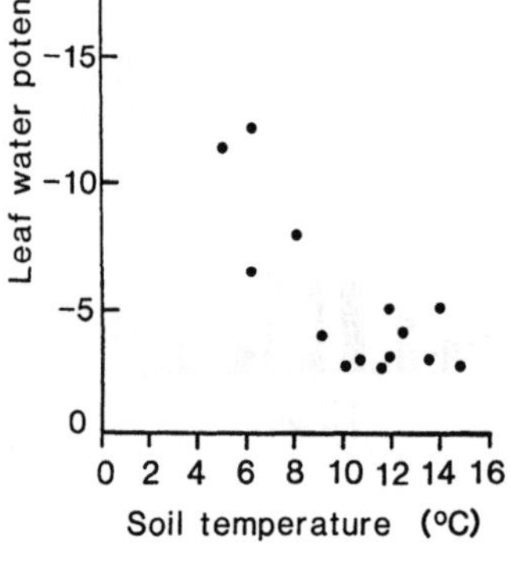

Fig. 8.3. Relationship between soil temperature and leaf water potential at dawn of Satsuma mandarin trees during the period November through April in Japan. (From Maotani and Machida 1976)

the biomembrane structure at low temperatures (Kramer 1942; Kaufmann 1975; Dalton and Gardner 1978; Running and Reid 1980). If citrus trees are subjected to strong cold wind at soil temperatures of about 4 °C, defoliation occurs within a few days after formation of fissures in the petioles (Konakahara 1975).

Attempts have been made to distinguish between damage by tissue freezing and winter desiccation on the basis of the appearance and localization of the injuries. The visual symptoms of winter desiccation (Fig. 8.4) are similar to freezing injuries. For an unequivocal diagnosis the water relations of the plants have to be monitored throughout the winter (Table 8.2a). In the case of frost drought a gradual decrease of the relative water content would *precede* the appearance of shrinking, discolouration and other signs of injury, which usually do not occur until spring. However, there are also cases as in *Picea* species where colour changes on needles and defoliation due to desiccation become visible already during winter. In contrast, after lethal freezing or mechanical damage a drastic water loss would *follow* the development of necrosis (Larcher 1985b). Without monitoring water content, a clear-cut distinction is in any case difficult since both events, freezing and desiccation, may have taken place (Wardle 1981). Visual signs of winter desiccation are: a greater degree of damage at the sunside of tree crowns and in sites more exposed to sunshine; the water loss in leaves, buds and shoot tips greater than in basal stems and roots (Table 8.2b and Fig. 8.5); damaged

Table 8.2. Phenomenology of winter desiccation

(a) Development of winter desiccation at and above timberline in the European Alps. The degree of desiccation is expressed as the average relative drought index (% RDI) of samples taken from numerous single plants on Mt. Patscherkofel in different years by various authors. (From Larcher 1985b)

Plant species	Nov.	Dec.	Jan.	Feb.	March	Apr.
Pinus cembra	12	15	40	45	45	25
Picea abies	12	15	20	50	55	62
Larix decidua	30	40	42	50	65	55
Rhododendron ferrugineum	5	30	45	45	50	10
Loiseleuria procumbens	10	20	70	60	65	21

$$RDI = \frac{WSD_{act}}{WSD_{crit}} \cdot 100$$

WSD_{act} = Actual water saturation deficit
WSD_{crit} = Critical WSD at first appearance of drought injury

(b) Water content (% fresh weight) of various plant organs of saplings of *Picea glehnii* in Hokkaido in relation to the degree of desiccation injury. (From Sakai 1968b)

Degree of damage	Water content at end of February				
	Leaves	Twigs	Stem cortex	Stem xylem	Main root
Serious to dead	23.3	17.5	30.7	24.4	48.2
Slight to medium	28.8	23.5	34.0	25.3	50.7
Uninjured	33.3	27.5	33.5	25.7	51.9

Damage was found in leaves, twigs and upper stems; no damage occurred in the middle and basal stems and roots. Data for stem tissues refer to the middle part of the stem.

Fig. 8.4a,b. Symptoms of winter desiccation above the alpine timberline (ca. 2100 m a.s.l.): a stunted *Picea abies*; b dieback and flagged growth form of young *Pinus cembra*. (Photos: W. Larcher)

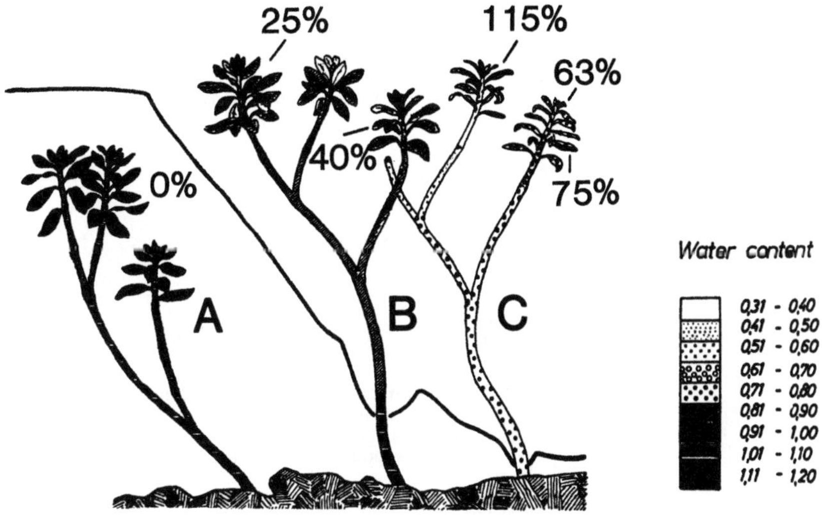

Fig. 8.5. Decrease in water content (g H_2O g^{-1} dry weight) of leaves and stems of *Rhododendron ferrugineum* at the alpine timberline during winter, depending on snow cover. *A* permanently below snow; *B* terminal twigs mostly without snow cover during winter, water status in February; *C* plants without snow cover on frozen soil in March. *Percentages* indicate relative drought indices (RDI; for explanation, see Table 8.2a) of leaves. (From Larcher 1963b)

plants found mainly at sites with shallow snow cover; the occurrence of damage even in winters in which prolonged frost but no extremely low temperatures were observed.

Among woody plants two types of winter desiccation can be distinguished (Larcher 1985a,b):

1. *Chronic* winter desiccation develops slowly due to gradual water loss by cuticular and peridermal transpiration after 2-3 months of soil frost. The chronic development of winter desiccation is common for conifers of boreal origin, which keep their stomata closed in the dormant state, and for deciduous woody plants which transpire peridermally. Resistance to this kind of frost drought depends essentially on the specific intensity of the cuticular and peridermal transpiration, on the water storage capacity of wood, bark and needles and on the protoplasmatic desiccation tolerance of the individual species (Pisek and Larcher 1954; Larcher 1957, 1963b, 1972; Baig and Tranquillini 1976; Tranquillini and Platter 1983; Hadley and Smith 1986; Richards and Bliss 1986). Since the performance of cuticular and peridermal layers depends on shoot and needle maturation, woody plants near their altitudinal distribution limit will be predisposed to greater injuries by frost drought after cool summers and in habitats with a short growing period (Michaelis 1934b; Tranquillini 1979a, 1982; Fig. 8.6).

2. *Acute* frost drought arises from a breakdown of water balance within a few days. Evergreen angiosperm plants, especially from warm-temperate regions, after being released from winter inactivation by sunny weather, open their stomata and therefore dry out strongly if the water supply is interrupted due to soil and stem frost. This type of winter desiccation, favoured by dry wind, is well known on tea plantations in Japan (Fuchinoue 1982, 1985; Fig. 8.7). Adult tea plants sustained serious damage within several days at subzero air temperatures if the soil remained frozen (Fig. 8.8). Thus,

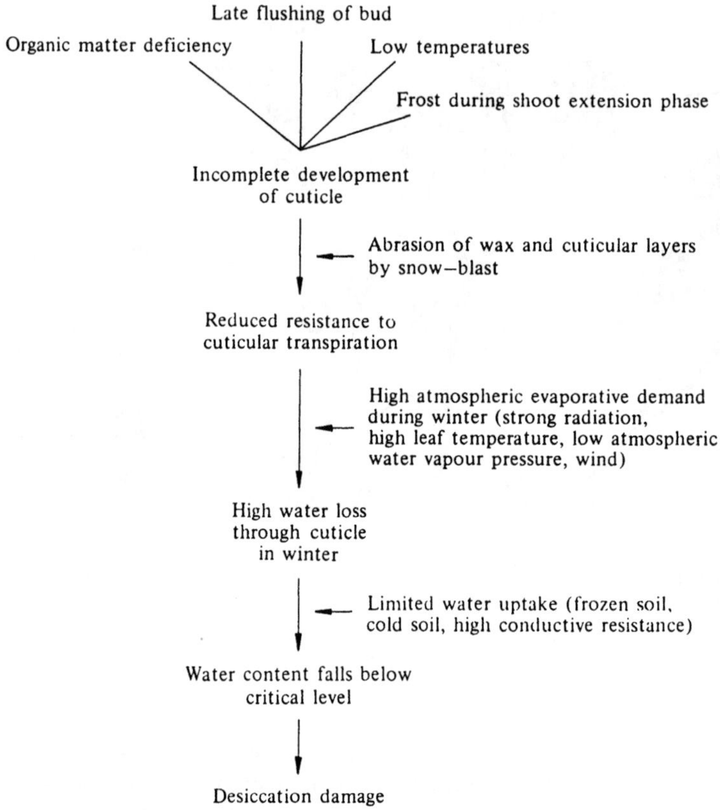

Fig. 8.6. Factors leading to chronic winter desiccation of conifers at the timberline in the Alps. (From Tranquillini 1979b)

Fig. 8.7. Winter desiccation in a tea plantation near Tokyo. (From Fuchinoue 1985)

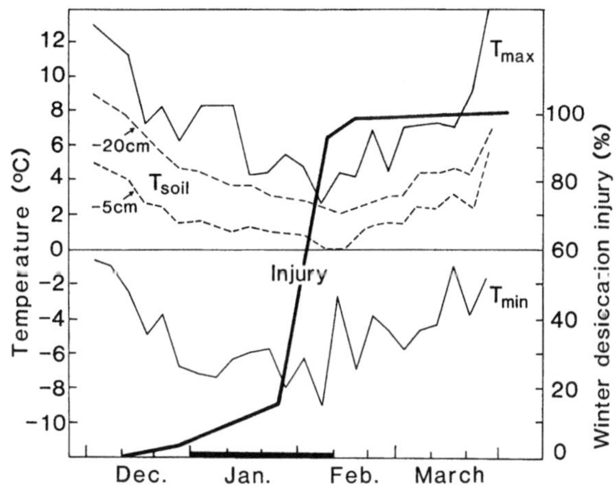

Fig. 8.8. Development of winter desiccation damages in leaves of adult tea plants in relation to air and soil temperature. The soil surface remained frozen from beginning of January until end of the first decade of February (*black bar*), the soil temperature below 5 cm depth remained above 0 °C throughout winter, the lowest air temperature was −9 °C. (From Fuchinoue 1985). Leaves of *Camellia sinensis* are frost resistant (LT_i) to −13 °C (Sakai and Hakoda 1979)

winter drought seems to be the most important factor setting the northern economical growth limit of tea plants in Japan. In mountain areas, acute frost drought has been found in *Rhododendron ferrugineum* (Larcher and Siegwolf 1985; Fig. 8.9). The acute type of winter desiccation depends primarily on the readiness of stomatal opening during winter and on a xylem structure which facilitates the formation of cavitation embolies (for conductive properties of wood and cavitation risk, see Zimmermann 1983; for critical sap tensions, see Crombie et al. 1985).

Meteorological factors which cause desiccation damage differ according to the particular habitat in which the plants are overwintering: Hokkaido is divided by central mountain ranges as high as 1500 to 2000 m altitude; the western and the eastern parts of the island contrast strikingly with regard to winter climate. W. Hokkaido, especially the Japan Sea coastal area, is characterized by deep snow accumulation. On the other hand, E. Hokkaido, especially the Pacific coast and its inlands, is very cold; a dry state

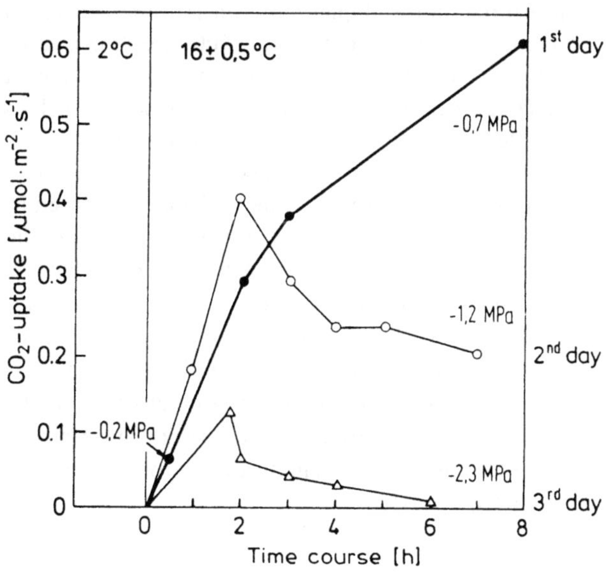

Fig. 8.9. Development of acute frost drought in *Rhododendron ferrugineum* in the course of 3 days after removal of the snow cover which had protected the twigs from strong light, elevated air temperatures and dry air. During the experiment the basal stems and roots remained frozen. CO_2 uptake reflects metabolic activity as well as stomatal conductivity for CO_2 and water vapour. Leaf water potential (MPa) was measured at the beginning of the experiment and at the end of each light period. (From Larcher and Siegwolf 1985)

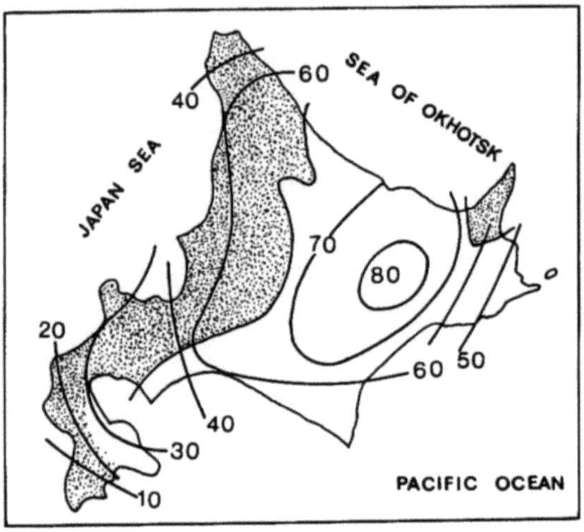

Fig. 8.10. Maximum depth (cm) of ground freezing during snow-free periods in 1974/75 in Japan. The *dotted area* indicates regions where snow accumulation is usually high (Nogami et al. 1980; Fig. 8.19)

prevails during winter and the soil usually remains frozen down to 60 to 80 cm (Sakai 1970b; Fig. 8.10).

In E. Hokkaido unusual frost drought prevailed especially throughout the winter of 1966 to 1967. The snow cover was less than 5 to 10 cm and the soil remained frozen even on the southern slopes for 4 months with minimum soil temperatures 10 cm below the ground surface of −5 °C (Sakai 1970b; cf. Fig. 1.24). The temperature of stems and leaves of young conifers in winter rose to about 17 °C and 9 °C at midday on the southern and northern slopes respectively. They remained unfrozen for 6 or 2 h, respectively, during daytime. Under these conditions, young conifers planted the previous spring, such as *Abies sachalinensis, Picea glehnii, Pinus strobus* and *Thuja occidentalis* suffered damage; those on sunny slopes and flatlands suffered greatest injury, whereas those on the northern slopes and sheltered by trees (even on the southern slope) suffered little or no damage (Fig. 8.11a).

In central and E. Japan between latitudes of 34°–36° N, on the northern slopes of mountain areas higher than about 400 to 600 m above sea level, the soil usually freezes for 1 to 2 months to the depth of 10 to 30 cm. Under such conditions saplings of *Cryptomeria japonica* often suffer desiccation damage in dry winters. In contrast, on southern and eastern slopes where the soil remains unfrozen, desiccation damage is hardly observed (Fig. 8.11b). On Mt. Tsukuba (876 m, 36°26′N), located near the boundary of the warm-temperate forest zone in Japan, evergreen hardwoods occur on the southern slopes where the soil remains unfrozen even near the summit, while only deciduous hardwoods are found on the northern slopes which are swept by northwesterly winds throughout the winter and where the soil usually remains frozen for 1 month or more. Similarly, in the Himalayan mountains, an uppermost evergreen oak forest of *Quercus semicarpifolia* occurs only on the southern slopes at around 3000 m altitude where the soil does not freeze; while at the same altitude on the northern slopes, where the soil freezes, deciduous hardwoods occur (Sakai and Malla 1981). These factors indicate that winter desiccation is an important factor limiting the natural range of many broad-leaved evergreen trees.

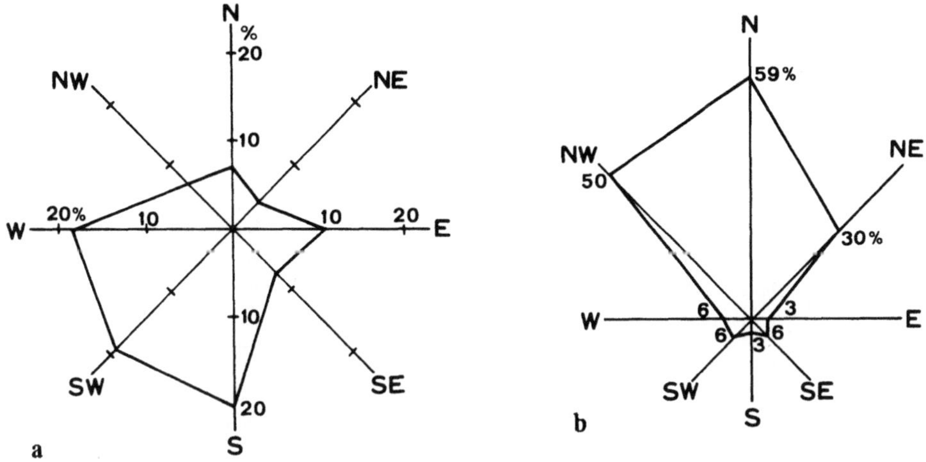

Fig. 8.11a,b. Relationships between winter desiccation of trees and slope direction in various regions of Japan. **a** *Abies sachalinensis* in E. Hokkaido. (From Rep. Forest Bureau Obihiro 1968). **b** *Cryptomeria japonica* in SW. Honshu. (From Ando and Saito 1969)

In the northern part of Honshu and in Hokkaido the soil freezes to a great depth even in the lowlands and on the southern slopes, if the snow cover does not exceed 30 to 60 cm. In these areas, therefore, winter desiccation damage to conifers can be observed in all wind-swept areas regardless of the direction of the slope (Sakai 1968b). At the Pacific Ocean side of Honshu, strong northwesterly winds sweep the country during winter. Here, dry wind is the decisive factor. Figure 8.12 shows that almost all of the young Ezo spruce *(Picea glehnii)*, planted in Hokkaido on the lee side of a wind break (60 m length, 30 m width and 4 to 10 m height) within a distance of about 60 m from the break, were uninjured, while those in the wind-swept area were seriously damaged during a test winter. In this area the time course of the desiccation damage differs considerably from year to year, depending on the environmental conditions, such as the time of soil freezing, depth of snow cover, humidity and strength of winds.

In the snowy wind-swept areas of Japan in which the mean air temperature in January is below about -3 °C, the soil generally remains unfrozen if the snow cover exceeds 40 to 60 cm. However, the stems 5 to 30 cm below the snow level usually remain in a frozen state during winter. Under the latter conditions, the leaves, small twigs and upper stems protruding above the snow level often sustain desiccation damage. The mechanical damage on the surface of stems and twigs caused by ice crystal abrasion in wind-swept areas may accelerate desiccation damage.

The situation at the timberline in the European Alps is dealt with in detail by Tranquillini (1979b), and that for the N. American mountains by Marchand and Chabot (1978), Kincaid and Lyons (1981), Hadley and Smith (1983, 1986), W.K. Smith (1985b), Richards and Bliss (1986); Wardle (1971, 1985) presents an overview which also includes the timberlines of the Southern Hemisphere.

In the Canadian Rocky Mountains a phenomenon commonly known as "Red Belt" appears as a horizontal stripe on south and west facing slopes of middle and upper foothills almost every spring as a result of winter injury to conifer foliage. All native

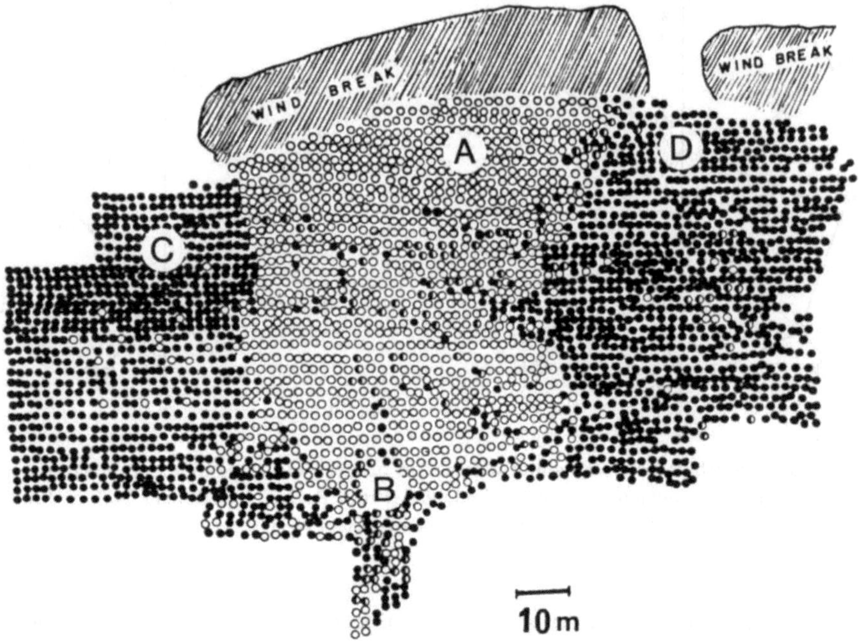

Fig. 8.12. Distribution of winter-desiccated trees in a plantation of young *Picea glehnii* on a windy plateau in Hokkaido (Lake Shikotsu, 250 m a.s.l.). The soil was continuously frozen from mid-November until late April down to a depth of 30 cm below a snow cover of 10–20 cm. Mean air temperature in January (1962) was −6 °C, the lowest air temperature was −20 °C. Wind velocity on March 5, 1963: *A* 0 m s^{-1}; *B* 2.5 m s^{-1}; *C* 4.5 m s^{-1}; *D* 5 m s^{-1}. Damages: ○ uninjured, ◐ partial injuries, ● dead. (From Sakai 1968b)

conifers may be affected, but lodgepole pine *(Pinus contorta)* is most susceptible. The most obvious damage is to the foliage which turns reddish-brown. The terminal buds and the bark remain uninjured and the affected trees usually recover, except in extreme cases. Typically, Red Belt occurs during late winter or in early spring; it has been associated with chinook winds which occur in the period between mid-December and February (Henson 1952; MacHattie 1963). Air of Pacific origin moves eastward and is dried and warmed after descent on the eastern slopes of the Rocky Mountains. Coming into the valleys which are filled with cold air, the warm winds do not mix with the air pool, but flow against the sunlit slopes of the valleys. Shaded slopes which are covered with cold air draining off the heights are not touched. In this way the chinook winds, which cause rapid warming and drying, move along a belt on the contour of the sunlight slopes. The altitude of the damaged zone is observed to be equivalent to the altitude of the top of the cold air pool in the valley which depends on the topography of each area. The mechanism of Red Belt damage is not definitely known. It is most likely climatological, winter drought (MacHattie 1963; Sakai 1970b), although little research has been undertaken to pinpoint the cause or to relate the visual damages to the impact (Robins and Sunset 1974). Since lodgepole pine foliage is extremely hardy in winter, freezing injury is unlikely to be the case. Whatever the cause, Red Belt is a natural phenomenon over which there is not likely to be any control.

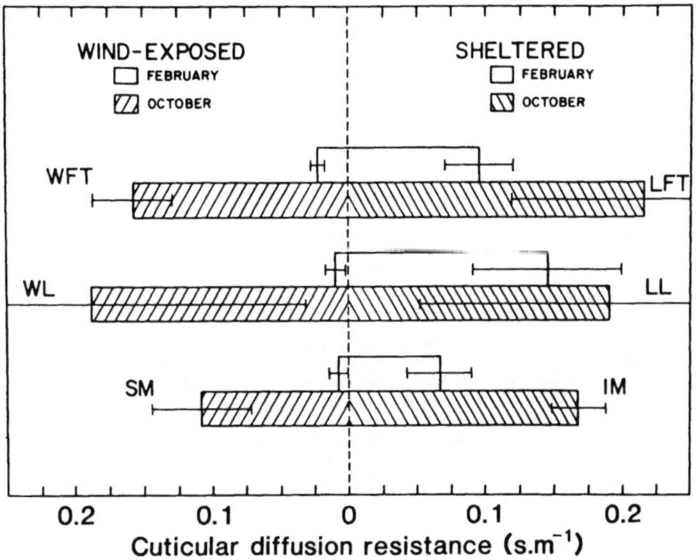

Fig. 8.13. Decline of the cuticular resistance to water loss of needles of *Picea engelmannii* during winter due to wind-borne abrasion. Wind-shielded shoots of krummholz spruces at 3200 m a.s.l. are compared with naturally wind-exposed shoots. *WFT* windward shoots on flagged trees; *LFT* leeward shoots of the same trees; *WL* windward side of leaders above mat; *LL* needles of the leeward side of leaders; *SM* shoot at the surface of the mat close to the surface of the snow cover; *IM* needles of shoots inside the mat. Bars indicate 95% confidence intervals of means for n = 4–6 shoots (October) and n = 8–12 shoots (February). (From Hadley and Smith 1986)

The mechanism of winter desiccation of wind-exposed needles of timberline conifers in the Rocky Mountains has recently been studied by Hadley and Smith (1986). Wind effects appeared to be more important than soil freezing for needle mortality of *Picea engelmannii*. Due to severe abrasion and removal of cuticular surface wax by wind, the resistance to cuticular transpiration decreased from 100–250 ks m^{-1} in autumn to 30 ks m^{-1} in mid-winter (Fig. 8.13). As a consequence, windward needles approached the 50% lethality level of 60% water content (on dry weight basis) already in early January, whereas the water content of leeward needles remained well above a critical level throughout winter.

Both *desiccation avoidance and tolerance* contribute to the resistance to frost drought. Dehydration of buds of *Larix lyallii* at the timberline in the Rocky Mountains of Canada is reduced by isolation from the apoplastic contact with the shoot xylem between October and February, thus preventing desiccation via transpirational water loss from the surface of the twig (Richards and Bliss 1986). At the same time the buds were tolerant of very low water potentials.

For an analysis and quantitative characterization of specific sensitivity to winter drought, data on the following features are required: (1) The efficiency of stomatal closure during dehydration stress in winter; (2) the specific intensity of cuticular transpiration under conditions of evaporation prevalent on the plant's habitat in winter, expressed as water lost per unit of time, or as the decrease in water content (or water potential) per unit of time; (3) the minimum water reserves of the various organs ne-

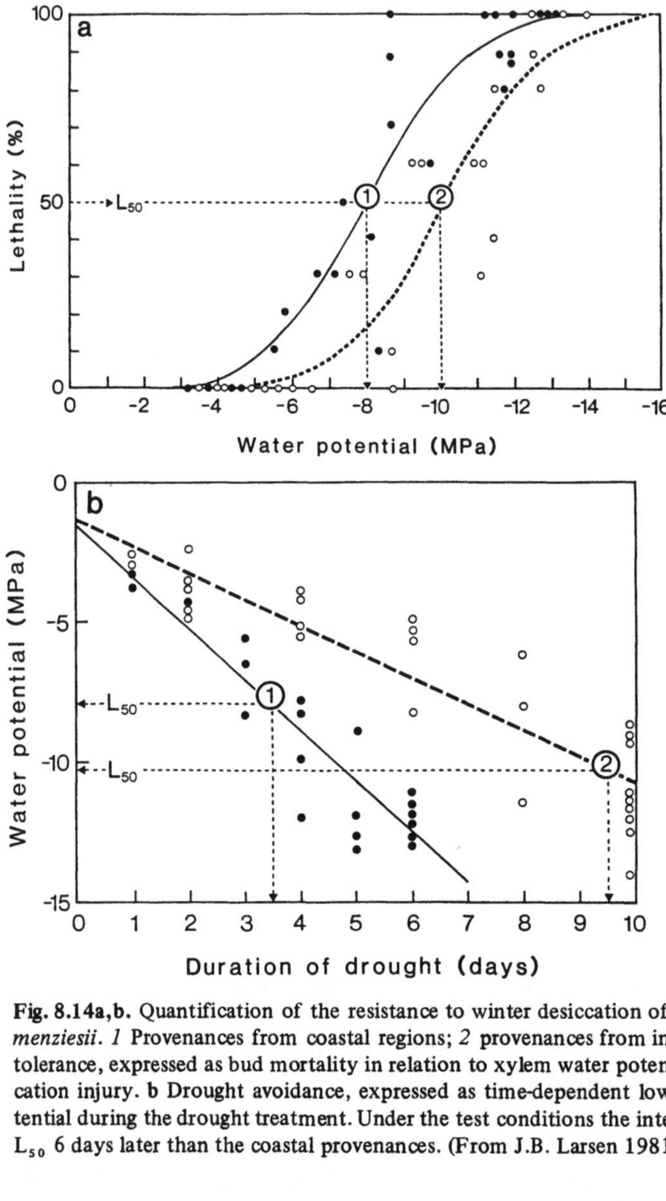

Fig. 8.14a,b. Quantification of the resistance to winter desiccation of young plants of *Pseudotsuga menziesii*. *1* Provenances from coastal regions; *2* provenances from interior N. America. **a** Drought tolerance, expressed as bud mortality in relation to xylem water potential. L_{50} indicates 50% desiccation injury. **b** Drought avoidance, expressed as time-dependent lowering of the xylem water potential during the drought treatment. Under the test conditions the interior provenances approached L_{50} 6 days later than the coastal provenances. (From J.B. Larsen 1981)

cessary for the maintenance of life, expressed as relative water content (or water potential) at 50% damage. With the help of graphs constructed from these data, comparisons can be made between species and varieties (Fig. 8.14).

In general, temperate conifers are more susceptible to winter drought than boreal conifers; however, *Pinus strobus* and *Thuja occidentalis*, which are frost-hardy to -70 °C and below, are sensitive to winter desiccation. Extremely frost-hardy conifers, including *Pinus sylvestris* and *Pinus banksiana*, suffered little or no damage even when subjected to unusual winter desiccation (Sakai 1970b). In the Rocky Mountains, *Larix lyallii* was less susceptible to winter desiccation as compared with sympatric evergreen

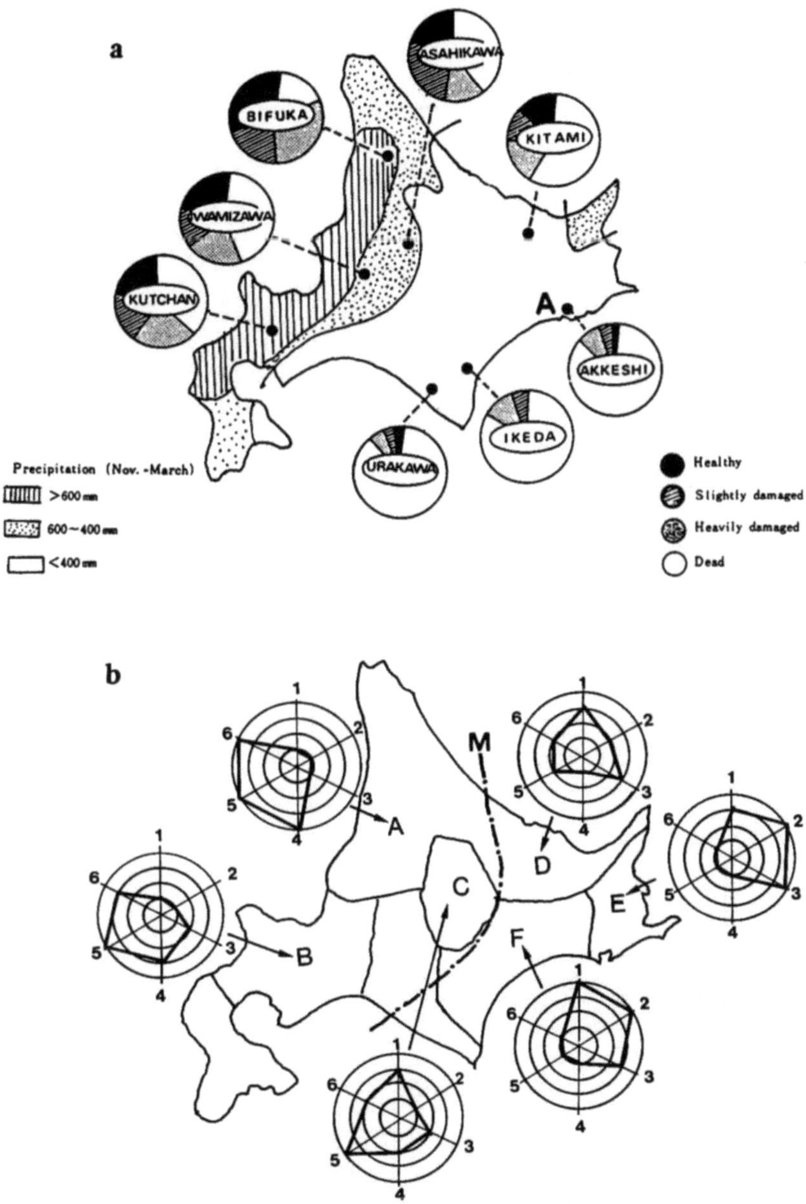

Fig. 8.15a,b. Provenance variation in response of *Abies sachalinensis* to winter stress factors. **a** Susceptibility to winter desiccation. *A* plantation site. (From Hatakeyama 1981). **b** Resistance to *1* freezing; *2* winter desiccation; *3* late frost (bud opening date); *4* snow pressure; *5* damages by field mice (related to the duration of snow cover); *6* tendency to early growth. Increasing resistance, late bud opening and fast growth are indicated by the greater radius of the circles. Climatic regions of Hokkaido: *A, B* heavy snow area; *C* central montainous area; *D* Okhotsk Sea coast; *E, F* Pacific coast; *M* climatic division between E. Hokkaido and central and W. Hokkaido. (From Eiga 1984; cf. Fig. 8.10)

Table 8.3. Multiple correlation, partial and standard partial regression coefficients of desiccation damage in winter of open-pollinated progenies of *Abies sachalinensis* on variables of climatic factors at the place of origin. (From Hatakeyama 1981)

Character (dependent variable)	Climatic factor (independent variable)	Partial regression coefficient[a]	SD	Standard partial regression coefficient	Multiple correlation coefficient[a]	Coefficient of determination
Degree of desiccation damage	Precipitation (Nov.–March)	0.0046	±0.0003	22.0138		
	Max. snow depth	0.276	±0.0013	2.8779	0.988	0.974
	Average temperature (Nov.–March)	–0.1379	±0.0144	–0.3541		
Mortality	Period of snow cover (250 cm)	0.1850	±0.0102	0.5540		
	Average temperature (Apr.–June)	2.1540	±0.3175	0.1577	0.999	0.999
	Sum of insolation time (Nov.–March)	–0.1710	±0.0112	–0.4499		

[a] Statistical significance at the 1% level.

conifers, among them *Abies lasiocarpa* appeared to be more resistant than *Picea engelmannii* and *Pinus contorta* (Richards and Bliss 1986). Among European trees and shrubs, *Pinus cembra, Larix decidua, Acer pseudoplatanus* and species of *Sorbus* are less sensitive to frost drought than *Picea abies, Pinus mugo* and *Fagus sylvatica* (Tranquillini 1982; Barclay and Crawford 1982; Larcher 1985a). The relationship of winter transpiration of different species of *Populus, Viburnum* and *Lonicera* to the northern border of their distribution area has been reported on by Bylinska (1975).

Provenance variation of winter desiccation damage in *Abies sachalinensis* is shown in Fig. 8.15. Seedlings of 29 families from 3 to 7 mother trees from 7 provenances were planted in eastern Hokkaido (Fig. 8.15a). A remarkable difference in the resistance to winter desiccation was observed (Hatakeyama 1981). The progenies from provenances along the Japan Sea coast showed considerably greater damage than those from the Pacific coast and the southern part of the Okhotsk Sea coast where severe cold and winter desiccation prevails in winter. The variation within families was much less than that between provenances, which amounts to 69.5% of the total variance. The genetic contribution of provenance variation to winter desiccation damage was as high as 92%. Thus, the significant difference in variance is mainly associated with provenance. Regression analysis evaluating the effect of the climatic factors on the provenance variation of desiccation damage revealed a multiple correlation coefficient of 0.988 and showed that most of the provenance variation for desiccation damage can be explained by three climatic factors at the place of seed origin: the period with snow cover exceeding 50 cm, the maximum snow depth and the total insolation time. The standard partial regression coefficient revealed that the period of snow cover over 50 cm is the most important climatic factor of the site of seed origin (Table 8.3a). On the other hand, progenies from the Japan Sea coast showed higher resistance to *Rhacodium* snow blight and to damage caused by snow pressure than those from along the Pacific coast (Hatakeyama 1981). A significant positive correlation between percentage of snow-damaged trees and number of branches per whorl for families within a provenance was observed, and also a significant negative correlation between percentage of snow-damaged trees and whorl stem length. These variation patterns probably resulted from regional gradients in the distribution of snow depth. Six characteristics of *Abies sachalinensis* related to winter survival are compiled in Fig. 8.15b: they demonstrate considerable differentiation corresponding to the climatic variation pattern of their natural ranges (Eiga 1984).

8.2 Damage Due to Ice Encasement and Compact Snow

A covering of ice or compact snow is especially dangerous for herbaceous plants and prostrate woody plants. Due to the low rates of gaseous diffusion through ice sheets and compressed snow, respiratory CO_2 attains concentrations as high as 44% (v/v), and O_2 drops to 3–5% (Freyman and Brink 1967; Rakitina 1970a; M.N. Smith and Olien 1981). Under such conditions, comparable to flooding-induced hypoxia, toxic substances accumulate via abnormal metabolic pathways. Especially injurious is the combination of high levels of ethanol and elevated CO_2 concentrations (Pomeroy and Andrews 1979). The plasma membrane is the primary site of injury, and the breakdown of the ion transport system is an early manifestation of damage (Pomeroy et al. 1983). The susceptibility to intoxication after ice encasement and snow compaction is dependent on species and variety (examples for crop plants: Smith and Olien 1981); it can be lessened by preconditioning (Andrews and Pomeroy 1983).

A further detrimental effect of hypoxic conditions is the inability of the weakened plants to achieve their full frost hardiness. This has been demonstrated in cereals by Rakitina (1965, 1967) and Andrews and Pomeroy (1981), in *Phleum pratense* by Andrews and Gudleifsson (1983) and in the roots and twigs of several conifer and angiosperm tree species by Rakitina (1968, 1970b).

8.3 Harmful Effects of Heavy and Long-Lasting Snow Cover

Snow is a viscoelastic material. Snow on the ground gradually subsides with time under the load of successively falling snow. The textures and properties of the layers in deposited snow change due to natural densification and metamorphism depending on thermal and mechanical conditions to which the layer is subjected. Deposited snow can be classified into three main types: (1) new snow: the crystals still have their original shape; the density ranges between 0.05 and 0.15 g cm^{-3}; (2) compact snow: ice-bound grains are connected to a fine network. The density of deposited snow increases gradually due to the weight of the overlying snow and thus the proportion of compact snow becomes larger with the lapse of time at subzero temperatures. As metamorphism and natural densification progress, the density is increased to 0.25–0.5 g cm^{-3}; (3) coarse-grained granular snow: by wet metamorphism, snow turns into coarse-grained granular snow which loses the network of ice-bound grains. Metamorphism into granular snow is a common tendency in the snow melt season.

The snow pressure is proportional to the mean density and depth of snow. A relation between maximum snow pressure (P_{max}) and maximum snow depth (H_{max}) can be formulated as shown in Fig. 8.16:

$$P_{max} = 162\, H_{max}^{1.89}\ [kg^{-1}\ m]\,.$$

The dimension of snow pressure also changes in proportion to the height, perimeter and form of the force-receiving surface of the underlaying object.

Snow damage to trees can be divided into four types (Shitei 1954): bending, breaking, splitting and falling or uprooting (Fig. 8.17). Damage can be done to tree tops, stems, branches and roots, of which the stem and root breaking or splitting are most serious. Bending moments and the stability of coniferous tree stems to breaking stress have been calculated by Petty and Worrell (1981). Buhl (1968) gave as load limits for conifers 50 kg m^{-2}, for broad-leaved trees 25 kg m^{-2}. Of the evergreen conifers, according to a survey by Frey (1977), pines, the types of spruce with widely spreading branches and firs are even more endangered from snow break, whereas *Pinus cembra*, slender-crowned spruces and the green varieties of Douglas fir are less vulnerable. In Japan, *Pinus densiflora, Pinus thunbergii, Chamaecyparis obtusa* and bamboos are most sensitive to snow pressure, whereas *Fagus crenata* and *Cryptomeria japonica* (from the Japan Sea side) are the most tolerant tree species. Ono and Inuma (1969) revealed

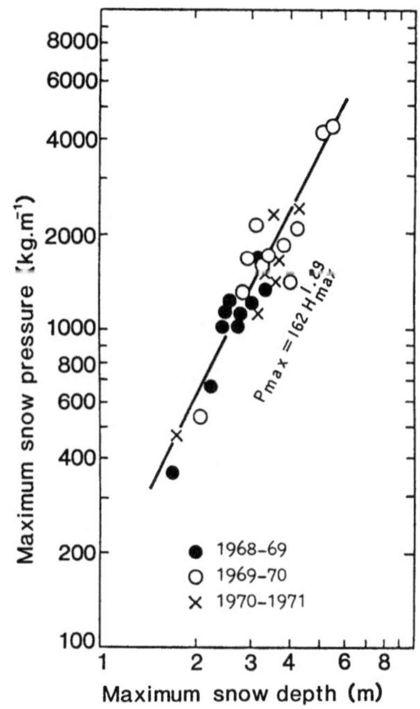

Fig. 8.16. Relationship between maximum snow depth and maximum snow pressure, determined by a spring gauge placed on a wooden bar of 1 m length and 10.5 cm width at 1 m above the ground. (From Ishikawa et al. 1974)

that damage done to *Cryptomeria japonica* increased with the depth of deposited snow, which parallels the increase in altitude from 250 to 750 m (Fig. 8.18). The flexible stems of young *Cryptomeria* trees less than 1 m in height are easily buried under snow every winter; they can quickly return to their upright position soon after the snow has melted. *Cryptomeria* trees grown to a height of 1.5 to 2 times of maximum snow depth usually are not buried, and thus this age class is very susceptible to damage by a heavy snow load on the branches.

Wet snowstorms often strike forests in early winter or spring. If snowfall continues under the conditions of less wind and slightly subzero temperatures (−0.5° to −1.0 °C), snow on the tree crowns grows remarkably large and heavy, resulting in severe damage, mostly in the form of stem breakage, often with stem split (W.H. Smith 1970; Nitta 1981).

The natural distribution of conifers in central Japan, where the precipitation in winter on the Sea of Japan side is very high (over 1500 mm), and is largely in the form of snow (Fig. 8.19), coincides approximately with the isogram for the mean yearly maximum snow depth of 50 to 100 cm (Takahashi 1960). *Cryptomeria japonica, Thuja standishi, Thujopsis dolobrata* and several dwarf coniferous species grow in areas with heavy snowfall (Table 8.4). On the other hand, *Larix leptolepis, Abies veichii, Picea jezoensis* var. *hondoensis, Pinus densiflora, Pinus koraiensis, Chamaecyparis obtusa* and *Tsuga diversifolia* are confined to less snowy areas. The greatest recorded snow depth (12 m) occurs in the mountainous region at the elevation of 2000 m altitude along the Sea of Japan. In such regions, a shrublike vegetation of *Betula ermanii, Fagus crenata, Quercus mongolica* and *Sasa kurilensis* prevails at the subalpine zone; the absence of tall conifers seems to be attributable to the deep snow cover (Shitei 1974). Attempts at afforestation of conifers in Japan have been unsuccessful in areas where the average yearly maximum snow depth exceeds 4 m (cf. Fig. 8.18).

Fig. 8.17a–c. Effects of snow pressure on woody plants. **a** Bending and splitting of bamboo stems. (Photo: S. Watanabe). **b** Stem deformation of *Betula platyphylla* planted in a heavy snow area. (Photo: A. Sakai). **c** Breaking and uprooting of *Larix leptolepis* planted in a heavy snow area. (Photo: S. Watanabe)

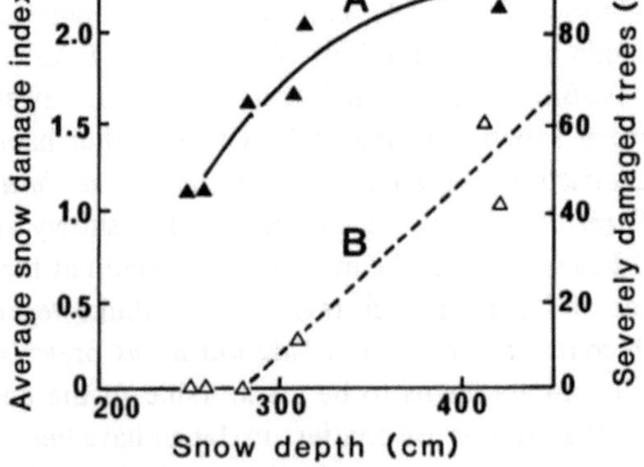

Fig. 8.18. Relationship between snow depth and snow damage in *Cryptomeria japonica*. *A* Mean snow damage index; the sum of damage ratings on individual trees is expressed as a mean value (0–5): *0* all trees undisturbed, *5* all trees with highest degree of snow damage. *B* Percent of severely damaged trees (i.e., with ratings 3 to 5) related to the total number of trees. (From Ono and Inuma 1969)

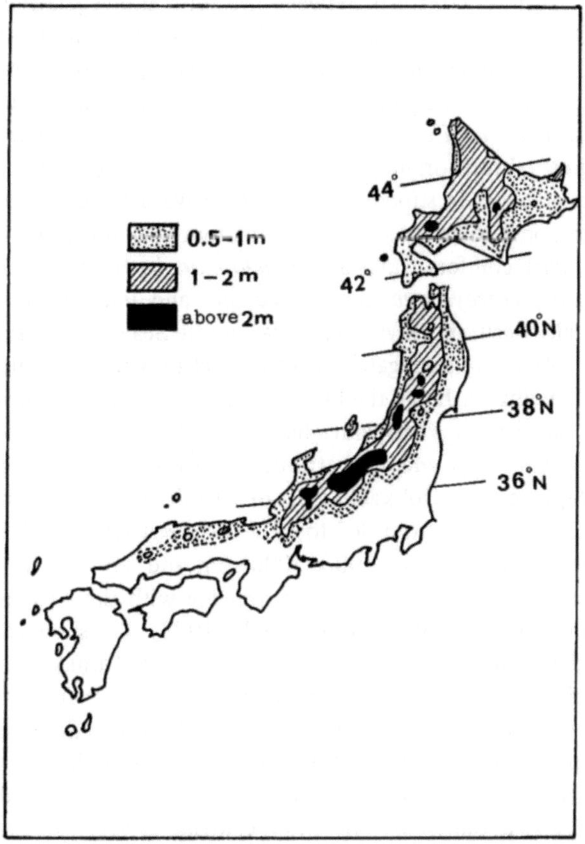

Fig. 8.19. Mean annual maximum snow depth in Japan. (From Department of Meteorology in Japan)

Table 8.4. Distribution of evergreen trees and shrubs depending on snow layer in Japan. (From Takahashi 1960; Usui 1961; Tuyama 1968; Shitei 1974)

	Genus	Districts with heavy snow	Districts without snow or shallow snow
Evergreen conifers	*Torreya*	*T. nucifera* var. *fruticosa*	*T. nucifera*
	Cephalotaxus	*C. harringtonia* var. *nana*	*C. harringtonia*
	Taxus	*T. cuspidata* var. *nana*	*T. cuspidata*
Broad-leaved evergreen trees and shrubs	*Aucuba*	*A. japonica* var. *borealis*	*A. japonica*
	Daphniphyllum	*D. macropodum* var. *humile*	*D. macropodum*
	Ilex	*I. leucoclada*	*I. integra*
	Ilex	*I. crenata* var. *paludosa*	*I. crenata*
	Camellia	*C. rusticana*	*C. japonica*
Dwarf bamboo	*Sasa*	*S. kurilensis*	*S. nipponica*
	Sasamorpha, Sasa	*Sasa palmata*	*Sasamorpha purpurascens*

Distinct contrasts between life form and vegetation on the Sea of Japan and Pacific Ocean sides can be attributed largely to the depth of snow cover. The beech forests on the Sea of Japan side are characterized by densely growing *Sasa kurilensis* in the shrub stratum and the presence of shrub species, such as *Aucuba japonica* var. *borealis, Camellia rusticana, Daphniphyllum macropodum* var. *humile* and *Ilex leucoclada*. These evergreen shrubs originate from regions with shallow snow cover where their growth form is erect. It is supposed that they have attained their prostrate form as an adaptive response to heavy snow accumulation. The same trend is observed in coniferous species. Dwarf conifers, such as *Cephalotaxus harringtonia* subsp. *nana, Taxus cuspidata* var. *nana, Torreya nucifera* var. *radicans* and *Juniperus communis* var. *nipponica* preferentially occur in heavy snow areas on the Sea of Japan side (see Table 8.4).

Adaptive segregation with respect to winter conditions is observed within the genus *Camellia*. In Japan the bush camellia *(Camellia japonica)* and the snow camellia *(C. rusticana)* occupy disparate habitats and are rarely found together (Fig. 8.20). *Camellia japonica* is a widely distributed species, mainly along the coastal regions throughout Japan except Hokkaido (Tuyama 1968). *C. rusticana* becomes predominant on the Sea of Japan side from 36° to 39° N between 100-1200 m a.s.l. There the average precipitation amounts to 2000-3000 mm and snow cover persists for 3 to 6 months. The eastern distributional ranges coincide with the 150 cm isogram for mean yearly maximum snow depth (Ishizawa 1973, 1978). The snow camellia can be distinguished from the bush camellia by its growth habit: the trunks and branches creep along the surface of the ground or are buried under humus; there is frequent budding from the base; even when the growth is erect, it is difficult to determine which is the main trunk. The branches are notably flexible to withstand the weight of deep snow accumulation (Hagiya and Ishizawa 1968). There is a conspicuous difference as to winter survival in the two camellias. The snow camellia is very sensitive to winter desiccation as compared with the bush camellia (Sakai and Hakoda 1979; Sakai, unpubl.), whereas the bush camellia cannot withstand the effects of heavy snow since its branches are very brittle. At low altitudes in the coastal region, the snow cover abruptly diminishes in depth from 3-4 m to 30-40 cm, even during cold winters, and the snow-free period becomes longer. At altitudes below about 100 m along the seashore, the bush camellia alternates with the snow camellia: the former does not grow below deep snow packs, and the latter does not flourish with an insufficient snow cover (Hagiya and Ishizawa 1961; Ishizawa 1978).

There are also intraspecific variations reflecting selection due to snow cover. *Cryptomeria japonica* segregates into two types differing in life form and mode of propagation. On the Sea of Japan side, the so-called Ura-sugi is dominant. It has dark green leaves, which project from the twig axis at a low angle and curl inwards. The form of the tree crown is narrow and sharp, the branches projecting from the stem at a large angle are flexible. Such cryptomerias readily propagate by vegetative means; they generate adventitious roots when their lower branches are pressed into the uppermost layers of the soil under heavy snow cover. Cryptomerias on the Sea of Japan side are

Fig. 8.20. Regional distribution of (●) Snow camellia *(C. rusticana)*; (○) Bush camellia *(C. japonica)* and the (△) intermediate form between the two species in Honshu. *Solid line:* isogram of 150 cm mean annual maximum snow depth. (From Ishizawa 1973, 1978)

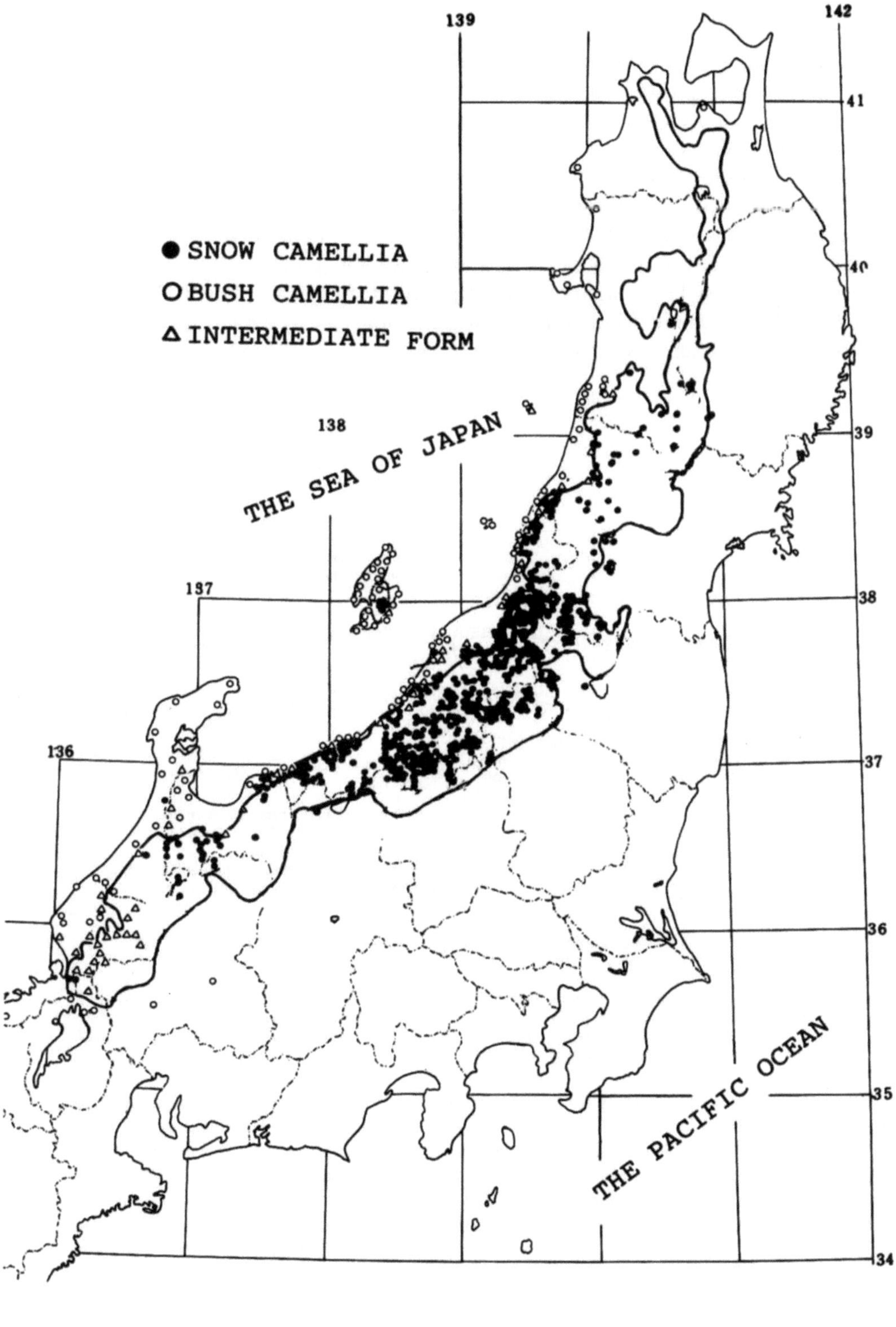
139
142
41
40
39
38
37
36
35
34
138
137
136
● SNOW CAMELLIA
○ BUSH CAMELLIA
△ INTERMEDIATE FORM
THE SEA OF JAPAN
THE PACIFIC OCEAN

more resistant to snow blight and to mechanical damage due to snow pressure than those on the Pacific side, but are sensitive to winter desiccation. The Pacific side is occupied primarily by a variety of cryptomeria with a wide and oblique tree crown. Their vegetative propagation is very difficult. Cryptomerias on the Pacific side, especially in northern districts, are much hardier than those on the Sea of Japan side (Muto and Horiuchi 1974).

Besides the sensational destruction of vegetation caused by the deformation of trees and shrubs, a permanent snow cover of 5–6 months' duration or more on lee sites and in snow beds weakens the plants to an extent which may result in decay (reviews: Frey 1977; Larcher 1985a). Plants buried below snow for a long time lose their frost resistance and can easily be damaged by low temperatures after being released. The needles of *Pinus cembra* shoots which were continuously protected by snow acquired 6–9 K less frost resistance than those of above-snow twigs (Tranquillini 1958).

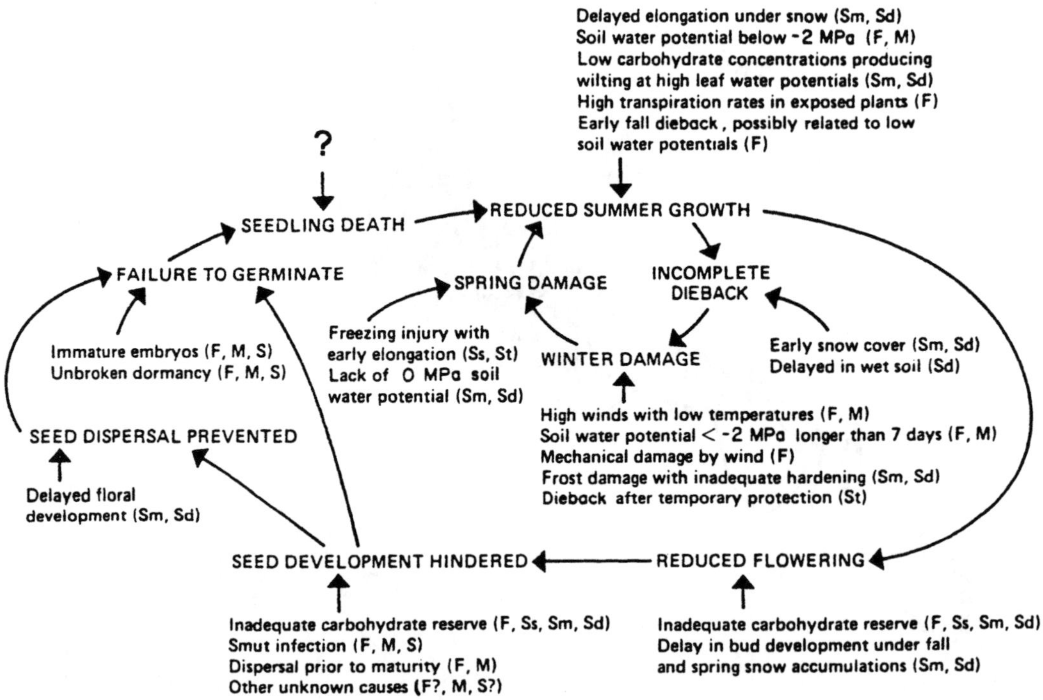

Fig. 8.21. Model of factors limiting small-scale distribution of *Kobresia bellardii*, a chionophobous alpine meadow sedge, as a result of snow accumulation in winter, based on transplant experiments. Microenvironments: *F* fellfield with little or no snow cover, abrasive wind effects, high summer evaporation rates and low summer soil moisture; *M Kobresia* meadow (control site); *Ss* shallow snow accumulation, early snow melt and premature spring growth accompanied by reduced frost resistance; *Sm* moderate snow accumulation; *Sd* deep snow accumulation associated with long duration of snow cover, restriction of the growing season and prevention of frost-hardening of the snow-protected plants; *St* temporary snow accumulation dispersing before spring, freezing risk for unhardened plants if snow cover is removed. *S* all conditions connected with snow accumulation. (From Bell and Bliss 1980)

The high air humidity below snow, together with the constant temperature of about 0 °C and darkness, facilitates the attack of the weakened plants by psychrophilous fungi, such as *Herpotrichia juniperi, Phacidium infestans* and *Lophodermium pinastri* on conifers and *Fusarium nivale, Sclerotinia borealis, Typhula* and *Pythium* species on grasses and cereals, which provoke widespread damage in winters with exceptional snow accumulation (Smith and Olien 1978; Aulitzky and Turner 1982; Tronsmo 1984).

The growth period of plants covered by snow is considerably shortened. Due to the attenuation of radiation in the snow and to the low temperatures little photosynthetic activity can be expected (Tranquillini 1957; Richardson and Salisbury 1977). Only especially well-adapted plants, e.g. snow-bed bryophytes and spring ephemerals, are able to make use of low light intensities at low temperatures in spite of small chlorophyll concentrations (Kimball and Salisbury 1974; Lösch et al. 1983).

8.4 Winter Survival: A Complex Response

Winter survival should not be regarded as a number of single incidents, but rather as a complex of interrelated events. Winter conditions additionally reflect on vegetative growth and reproduction during summer, and vice versa. An example on how the extent of low winter temperatures, soil frost and snow accumulation determine the microsite distribution of an alpine sedge is given in Fig. 8.21 (Bell and Bliss 1980); the transplant experiment shows that essentially all metabolic and developmental processes are implicated in the final response to the winter constraints and in the successful persistance of the plant. Such complexity makes it difficult to investigate winter survival and to evaluate and interprete observations. We are still far from a quantitative description and a causal analysis of the majority of known phenomena.

Appendix

Determination of Frost Resistance for Comparative Studies

Resistance research has two main goals: (1) to elucidate the causes underlying injury and survival and (2) to identify the properties of resistance peculiar to individual plant species or groups. This latter comparative approach provides us with the quantitative data needed for ecological purposes, especially for the analysis of the interaction of environmental hazards and the survival capacity of a plant in its habitat.

The basic mechanisms responsible for resistance can best be investigated in the laboratory, employing biophysical, biochemical and cell physiological procedures, as well as techniques for studying ultrastructure. The necessary information for comparative purposes is obtained by means of field observation following accidental severe frost, survival tests and artificial freezing tests.

Field Survival

The observations made following injurious frost events provide valuable information regarding the specific limits of resistance under naturally occurring stress conditions, as well as about later consequences and the chances of recovery. Unfortunately, only in rare cases have the temperature minima stated as being responsible for the injuries in question actually been measured in the immediate vicinity of the damage. Furthermore, the evaluated data sometimes lead to classifications that are neither easy to compare nor simple to interpret due to the fact that frost that is sufficiently severe to cause damage is mostly unexpected, and varies on each occasion in severity and duration. Depending on the time at which frost occurs, and on the preceding weather conditions, the plants may be at very different stages of maturity, activity and hardening, all factors which correspondingly influence their degree of sensitivity to frost. Nonetheless, every opportunity should be taken to record naturally occurring frost damage in order to verify experimental data.

The ability of plants to survive the severities of winter ("winter hardiness") can be measured by field survival tests. Agricultural and forestry research stations have extensive experience in this field and have established routine procedures. The general principles involved in field survival tests and a survey of the advantages and disadvantages of the various techniques are given, for example, in Fuchs and Rosenstiel

(1958), Fowler et al. (1981) and Marshall et al. (1981). In any case, a "test winter" is essential, i.e. a winter severe enough to kill the most tender plants and to cause varying degrees of damage to plants of intermediate hardiness (Horiuchi and Sakai 1978; Ikeda 1982). It is, therefore, important in planning field survival tests to choose planting sites in which test winters occur almost every year, or at least every few years. In evaluating the data obtained from such tests it has to be taken into account (especially in the case of woody plants) that the pattern of resistance may be highly atypical near and beyond the limits of the natural area of distribution. Although field survival tests may not give quick and sufficiently accurate results for physiological purposes, they are especially sutiable for studies on the genetic background of frost resistance and in programmes requiring a reliable estimation of recovery.

Artificial Freezing Tests

In order to provide a characterization of frost resistance, with quantitative data on resistance capacity and information on the components contributing to resistance, the use of experimental test procedures is indispensible. These must be such that the test material can be treated at any time with exact and reproducible degrees of frost, and that sufficiently large numbers of samples can be included to allow for individual scatter. For comparative and ecological purposes the determination of the resistance properties of a plant must include the following *parameters* (Larcher 1968, 1985a, 1987a):

1. Analysis of the freezing pattern: determination of T_f and T_{sc}, detection and possible quantification of extracellular and translocated ice formation.
2. Potential frost resistance at all stages of maturity and of every organ of the plant species. Complete hardening is necessary in order to reveal their inherent capacities.
3. The annual pattern of variability in resistance.
4. The acclimation amplitude of the various organs during different states of activity and at different phenological phases.

According to Scheumann (1965) and Scheumann and Schönbach (1968) the following aspects should be considered: the plant's readiness for hardening, which is important for its resistance to early frosts; the extent of its hardening potential, which is important for maximal resistance in winter; the stability of its hardiness, which is important during periods of widely fluctuating temperatures; the time of bud burst and flowering, which is important for tolerance of late frosts.

An exhaustive quantitative analysis of frost resistance involves considerable time and effort. Simple *screening tests* have therefore been developed for comparative studies, e.g. by Lapins (1962), Scheumann (1968), Mittelstädt and Murawski (1975), Warrington and Rook (1980), Holubowicz et al. (1982) for trees; and by A. Larsen (1978a,b), Prieur and Cousin (1978), Fowler et al. (1981), Marshall et al. (1981) and others, for herbaceous plants. For the purpose of screening of resistance levels, a graded series of injuries can be obtained by freezing the plants at one or more previously determined temperatures, depending on their hardiness. Numerical ratings were successfully used for the selection of hardy plants of many kinds (Manis and Knight 1967; Ikeda 1982; Toyao 1982; Eiga and Sakai 1984).

Artificial freezing tests have been found to give good agreement with winter survival in the field (Marshall 1965; Koch 1972; Pellett et al. 1981; Guinon et al. 1982). Nevertheless, differences between the two may be expected since field conditions are not completely simulated by freezing tests. In the latter, plants are exposed to a uniform temperature, whereas under natural conditions plants are exposed to a temperature gradient due to radiation cooling. A plant which is found to be less hardy in a laboratory test may nonetheless have a high degree of field hardiness since the freezing test measures only freezing tolerance, and neglects survival mechanisms which may mitigate the frost action. Another complication may arise where freezing injury in the field occurs mainly in early winter or in early spring, but not in mid-winter (Horiuchi and Sakai 1978).

Analysis of the Freezing Process

The onset and progress of freezing can be detected optically, by observation of the formation of ice in the tissues under the microscope, or thermometrically by measuring the temperature jump during crystallization of ice in the tissues.

Cryomicroscopic freezing analysis is employed in the case of single cells, tissue cultures, monolayer tissues and for sections. The construction of cooling microscopes has been described, e.g. by Biebl (1962a), Nei and Okada (1967), Diller (1982), Schwartz and Diller (1982). Ice formation in organs (buds, whole seeds) can be followed particularly well under a dissecting microscope equipped with a cooling device and using polarized light (Ishikawa and Sakai 1982).

Thermometric freezing analysis is used for determining T_f and T_{sc} of whole organs and large tissue portions. Measurement is based on recording of the temperature changes in the sample during cooling and freezing, and determination of the position of the exotherms on the directly recorded temperature curve (Müller-Thurgau 1886; Maximov 1914; Ullrich and Mäde 1940), or by means of differential thermal analysis (Tumanov and Krasavtsev 1959; Hudson and Idle 1962; Burke et al. 1976). For very small samples, or samples with low water content, or for material that undergoes deep supercooling, the DTA method is to be preferred, otherwise a special device has to be employed (e.g. Hayes et al. 1974; Cernusca and Vesco 1976). For the simultaneous registration of freezing events special procedures have been described (Quamme et al. 1975; Proebsting and Sakai 1979; Ashworth et al. 1981). The most accurate method for determining T_f is by means of an exotherm peak detection technique as developed by the Innsbruck group (Larcher 1963a; Pisek et al. 1967): in preliminary experiments the course of freezing is observed at a cooling rate of ca. 1 K min^{-1}. Freezing is provoked with as little supercooling as possible by means of ice inoculation. The verified T_f results from exposing samples for at least 6 h to temperatures upwards from the exotherm peak temperature of the first experiment until freezing no longer occurs.

Frost Treatment of the Samples

For the experimental determination of frost resistance the plants are exposed to a series of differing low temperatures. Numerous methods of freeze treatment are available for use in different situations (Larcher 1987a). An overview of the principles of construction and the requirements of suitable apparatus can be found, for example, in Kemmer and Schulz (1955), A. Larsen (1978b) and Warrington and Rook (1980). For field use mobile cooling cells are available (Loewel and Karnatz 1956; Murawski 1961; Scott and Spangelo 1964; Gill and Waister 1976; Reid et al. 1976; Sparks et al. (1976); a freezing device for field purposes has been described by Wiltbank and Rouse (1973). For laboratory measurements programmable refrigerators and climate chambers are commercially available. In principle, any cooling device that can be set and controlled with an accuracy of at least ± 1 K can be used. Small fluctuations in temperature can be eliminated by wrapping the plant in aluminium foil or by surrounding it with an insulating screen. This also prevents drying out of the samples. A one-directional heat loss which occurs in radiation frost can be obtained in cooling systems in which the ceiling lid of the freeze chamber functions as heat sink; constructional details can be found in Paton (1972), Warrington and Rook (1980) and Matsui et al. (1981).

For technical reasons it is impracticable to apply experimental tests of frost resistance of larger plants or of large numbers of individual plants in the field. Freezing experiments are therefore usually performed in the laboratory on potted plants or excised parts of plants. Herbaceous and smaller woody plants should always be frozen with their roots. In dealing with such plants it should be taken into consideration that the roots are less resistant than the shoot and, therefore, two test series should be used: in the one series the roots, in another the shoots, should be protected by heat-insulating material in order to avoid errors in the interpretation of injury. In the case of large plants, particularly trees, there is no alternative but to use detached parts of the plant. Experience has shown that experimental data concerning the resistance of excised parts of woody plants can be in good agreement with observations made in the field after severe frost (Cooper and Gorton 1954; Larcher 1954; Tyurina 1957; Lapins 1961; Pellett 1971).

For determining frost resistance, according to Levitt (1980), the following *principles* should be adhered to if comparable experimental results are to be obtained:

1. The plants must be inoculated to ensure freezing.
2. Cooling must be at a standard rate.
3. A single freeze must be used for a standard length of time.
4. Thawing must be made at a standard rate of rewarming.
5. Post-thawing conditions must be standardized.

The cooling method chosen depends on the nature of the desired data. A distinction is made between direct cooling, gradual cooling and simulation cooling. In *direct* cooling the object is brought to the test temperature quickly in order to avoid development of hardening during the process of cooling. A cooling speed of 5 to 10 K h^{-1} is usual in this case. The direct cooling method is used in determining R_{act} and R_{min}. In *gradual* cooling, the results of which indicate R_{pot}, the temperature is lowered stepwise and slowly enough to allow the plants to achieve full hardening. *Simulation* cooling is aimed at imitating naturally occurring frost, for which purpose the temperature is smoothly lowered at a speed of 1 to 2 K h^{-1}.

For ecological purposes (e.g. a comparison of frost resistance and frost stress at the plant's habitat) the most informative results are obtained by using a simulation method, whereas identification of the type of resistance is best achieved by direct cooling following controlled preconditioning. Samples on which the specific modulation amplitude of frost resistance is to be determined should, before the freezing test, be hardened or dehardened. The most effective temperature and duration for this conditioning vary according to species and season and should in every case be determined in preliminary experiments. Examples of conditioning programmes are given by Larcher (1985a, 1987a). Effective hardening in winter can be achieved by a two-step cold acclimation of 1-2 weeks, starting at 0° to -3 °C, then dropping to -5° to -10 °C (Sakai 1964, 1978; Scheumann and Hoffmann 1967; Tumanov 1967). Successful dehardening can be achieved by exposure for 1-3 days to temperatures above +15 °C in winter and above +20 °C in spring and autumn (Pisek and Schiessl 1947; Larcher 1987a).

The samples must be exposed to the test temperature long enough for thermodynamic equilibrium in the freezing process to become fully established. The required length of time has to be ascertained in experiments; i.e. the minimum length of frost treatment required for the production of a reproducible degree of damage. An exposure time of 4-6 h can be taken as a minimum (Larcher 1968). If the exposure is too brief, the samples may still be in a state of supercooling, and freezing may not have taken place at all, so that erroneous results are obtained. At temperatures below -30° to -40 °C ice formation proceeds very slowly and in some cases the full effect of frost only develops after 12-24 h (Christersson and Krasavtsev 1972).

At the Institute of Low Temperature Science in Sapporo the following procedure is employed for serial tests. Twig pieces (10 to 15 cm long) or leaves are taken from a number of plants which have been grown under identical conditions. After dipping the cut surfaces in water or wrapping them in wet cheesecloth to prevent supercooling, the twigs or leaves are put into small plastic bags. The samples are frozen at -5 °C and then cooled at the rate of 3 K h^{-1} or stepwise in 2 to 5 K decrements hourly or at 4 h intervals to successively colder test temperatures. Frozen twigs or leaves are rewarmed in air at 0 °C. After 16 to 24 h, they are transferred to room temperature.

Identification of Frost Injuries

Although a large number of tests are available for identifying damage, none of them is of general applicability. In every case the most suitable method has to be selected according to the object under investigation and the information being sought (for a comparison of available methods, see Palta et al. 1978; Duda and Kacperska 1983; Nelson et al. 1983; Samygin et al. 1985). In principle, a distinction has to be made between functional disturbances and necrotic phenomena (Larcher 1987b). In the former case it is important to know whether the disorders are reversible or will eventually lead to death.

Functional disturbances of diagnostic value include inhibition of photosynthesis (detected, for example, by chlorophyll-a-fluorescence techniques: Brown et al. 1977;

Klosson and Krause 1981; Bodner and Beck 1987), anomalies in respiratory processes (changes in respiratory intensity, metabolic pathways and adenylate energy charge; for references, see Larcher 1973; Bauer et al. 1975; Anderson 1978) and growth inhibition. A criterion that gives information about recovery from damage is the regrowth rate. In graminoids the measurements of leaf growth gives an indication of injury to the crown region and leaf bases (Kretschmer and Beger 1966). The regrowth rate of cultured cells can be estimated photometrically (Ericksson 1965; Sugawara and Sakai 1974).

Numerous methods are available for demonstrating events leading to *necrosis*, such as disruption of biomembranes and cessation of enzyme activity. Death of cells and tissues is usually recognizable by discolouration (e.g. from oxidation of polyphenols) and a characteristic odour (e.g. of coumarin or organic acids) due to decompartmentation and autolysis of the protoplasm. The loss of differential permeability of tonoplast and plasmalemma results in failure of plasmolysis, inability to accumulate vital stains and fluorochromes in the vacuole and leakage of cell solute constituents. Plasmolysis with balanced salt solutions may be combined advantageously with vital staining (Siminovitch and Briggs 1953; Sakai 1966a). Loss of ions can be detected by measuring the conductivity of the eluate (Stiles 1927; Dexter et al. 1932; Wilner 1960) or the electrical impedance of the tissue (Greenham 1966; Hayden et al. 1972; Wilner and Brach 1979), loss of soluble amino acids by a ninhydrin reaction (Siminovitch et al. 1962; Wiest et al. 1976), loss of secondary compounds such as anthocyanine and flavonoids by microscopic observation (autofluorescence: Larcher 1953; Oppenheimer and Jacoby 1961) and by colourimetry (Gäumann et al. 1952; Puth and Lüttge 1973) and loss of ability to maintain turgor by the resaturation test (Bornkamm 1958; Oppenheimer 1963; Rychnovská-Soudková 1963). Evidence for the activity of essential enzymes can be obtained by biochemical methods. A versatile method for demonstrating intactness of the respiratory electron transport system exploits the reducing capacity of dehydrogenases. A substance well-suited to this purpose is 2,3,5-triphenyltetrazolium chloride (TTC), whose formazan is red. The minute formazan crystals remain at their site of formation, so that the cells containing active dehydrogenases are coloured red (topographic TTC method: Lakon 1942; Parker 1953; Larcher 1969b). This method can be adapted to give quantitative results if the formazan is extracted from the samples by means of an appropriate solvent and the concentration measured (Steponkus and Lanphear 1967; Towill and Mazur 1976). The quantitative version is an integral procedure and only applicable to histologically uniform plant samples. It is especially useful for routine assessment of cell viability of parenchyma, callus, cell suspensions (Sugawara and Sakai 1974) and isolated protoplasts (Tao et al. 1983). As a vitality test for pollen grains, cultured cells and unilayer tissues the demonstration of esterase activity using fluorescein diacetate has proved useful (Heslop-Harrison and Heslop-Harrison 1970; Hölzl 1971; Widholm 1972; Nag and Street 1975; Withers 1978).

Whichever viability test is used, sufficient expression time must always be allowed for injuries to develop to the full before any procedure is begun. Following cell death the protoplasm may only become completely disorganized after some hours. Visible injuries usually attain their ultimate extent within 2–3 days, although this may take 1 week or even more in winter.

Quantification of Frost Resistance

In the case of tissues or organs that owe their resistance to supercooling or extraorgan freezing, a quantification can be derived from exotherm analysis (Hutcheson and Wiltbank 1972; Quamme et al. 1975; Eaton and Mahrt 1977). The beginning of injuries then corresponds to the average of the highest nucleation temperatures. Fifty percent damage to the material under investigation can be anticipated at the temperature at which in half of the samples the process of freezing is complete and in the other half has not yet begun (Proebsting and Sakai 1979).

Once damage has been recognized and localized, an attempt can be made to quantify its extent. This can best be defined as the injury rate (portion of sample killed as percent of total), or as its reciprocal value, the survival rate. Integral methods give direct, quantitative figures for the mass of tissue damaged in the total sample, whereas in the case of topographic methods an estimate has to be made. Given a sufficiently large number of replicates and sufficient practice it is possible to arrive at fairly accurate values. Valuable assistance in judging damage to leaves is provided by tables of estimates (Larcher 1987b); more objective results can be obtained by planimetric measurement of the damaged regions (Lange 1961) or by computer-assisted image analysis (Eguchi et al. 1982; Omasa et al. 1984).

The following injurious temperatures should be determined:

TL_0 = the lowest temperature sustained without necrotic damage;
TL_i = the incipient killing temperature at which initial injuries are revealed;
TL_{50} = the temperature at 50% lethality, as a standard measure for frost resistance;
TL_{100} = the temperature at which, at most, only a few cells survive.

The total sum of freezing injury to a quantity of plants can be expressed as the freezing injury index $\bar{d}$ according to Eiga and Sakai (1984) by $\Sigma^{n}_{d_i}$, where n is the number of plants, and d_i is the value of injury rating. For statistical analysis, $\bar{d}$ is transferred to $\sqrt{\bar{d}}$.

References

Alberdi M, Rios D (1983) Frost resistance of Embothrium coccineum Forst. and Gevuina avellana Mol. during development and aging. Acta Oecol (Oecol Plant) 4(18):3–9

Alberti M, Romero M, Rios D, Wenzel H (1985) Altitudinal gradients of seasonal frost resistance in Nothofagus communities of southern Chile. Acta Oecol (Oecol Plant) 6(20):21–30

Alden J, Hermann RK (1971) Aspects of the cold-hardiness mechanism in plants. Bot Rev 37: 37–142

Alexander NL, Flint HL, Hammer PA (1984) Variation in cold-hardiness of Fraxinus americana stem tissue according to geographic origin. Ecology 65:1087–1092

Alexandrov VYa, Lomagin AG, Feldman NL (1970) The responsive increase in thermostability of plant cells. Protoplasma 69:417–458

Anderson J (1978) Mitochondrial activity, adenosine triphosphate content and adenylate energy charge as a measure of plant cell and tissue viability. Cryobiology 15:700

Ando T, Saito A (1969) Winter desiccation of Cryptomeria in Shikoku in 1962. Kochi-Rinyu 507: 1–11

Andrew CJ, Gudleifsson BE (1983) A comparison of cold hardiness and ice encasement tolerance of timothy grass and winter wheat. Can J Plant Sci 63:429–435

Andrews CJ, Pomeroy MK (1978) The effect of anaerobic metabolites on survival and ultrastructure of winter wheat in relation to ice encasement. Plant Physiol 61 (Suppl 17)

Andrews CJ, Pomeroy MK (1981) The effect of flooding pretreatment on cold hardiness and survival of winter cereals in ice encasement. Can J Plant Sci 61:507–513

Andrews CJ, Pomeroy MK (1983) The influence of flooding pretreatment on metabolic changes in winter cereal seedlings during ice encasement. Can J Bot 61:142–147

Andrews CJ, Pomeroy MK, Roche de la IA (1974a) Changes in cold hardiness of overwintering winter wheat. Can J Plant Sci 54:9–15

Andrews CJ, Pomeroy MK, Roche de la IA (1974b) The influence of light and diurnal freezing temperature on cold hardiness of winter wheat seedlings. Can J Bot 52:2539–2546

Andrews JE, Horricks SJ, Roberts DWA (1960) Interrelationship between plant age, root-rot infection, and cold hardiness in winter wheat. Can J Bot 38:601–611

Andrews PK, Proebsting E (1985) The freezing properties and injury of sweet cherry (Prunus avium L.) and peach (Prunus persica (L.) Batsch) floral tissues. Ph D Thesis, Dep Hortic Landscape Architect, Washington State Univ, Pullman

Angelo E, Iversen VE, Brierley WG, Landon RH (1939) Studies on some factors relating to hardiness in the strawberry. Min Agric Exp Stn Tech Bull 135:1–36

Antropova TA (1974) Sezonnye izmeneniya kholodo- i teploistoichi-vosti kletok doukh vidov rihkov. (The seasonal changes of cold and heat resistance of cells in two moss species). Bot Zh 59: 117–122

Asahi T, Maeshima M, Matsuoka M, Uritani I (1982) Effect of temperature on the activity and stability of higher plant cytochrome c oxidase. In: Li PH, Sakai A (eds) Plant cold hardiness and freezing stress, vol II. Academic Press, London New York, pp 671–682

Asahina E (1956) The freezing process of plant cell. Contrib Inst Low Temp Sci Sapporo 10: 83–126

Asahina E (1978) Freezing processes and injury in plant cells. In: Li PH, Sakai A (eds) Plant cold hardiness and freezing stress. Academic Press, London New York, pp 17–38

Ashworth EN, Abeles FB (1984) Freezing behavior of water in small pores and the possible role in the freezing of plant tissues. Plant Physiol 76:201–204

Ashworth EN, Lightner GW, Rowse DJ (1981) Evaluation of apricot flower bud hardiness using a computer-assisted method of thermal analysis. Hort Sci 16:754–756

Aulitzky H (1961) Die Bodentemperaturverhältnisse an einer zentralalpinen Hanglage beiderseits der Waldgrenze. I. Die Bodentemperatur oberhalb der zentralalpinen Waldgrenze. Arch Meteor Geophys Bodenklimatol B10:445–532

Aulitzky H (1963) Grundlagen und Anwendung des vorläufigen Wind-Schnee-Ökogrammes. Mitt Forstl Bundes-Versuchsanst Mariabrunn 60:763–834

Aulitzky H, Turner H (1982) Bioklimatische Grundlagen einer standortgemäßen Bewirtschaftung des subalpinen Lärchen-Arvenwaldes. Mitt Eidgen Anst Forstl Versuchsw 58:327–580

Aulitzky H, Turner H, Mayer H (1982) Bioklimatische Grundlagen einer standortsgemäßen Bewirtschaftung des subalpinen Lärchen-Arvenwaldes. Mitt Eidgen Forstl Versuchsw 58:327–580

Axelrod DI (1966) Origin of deciduous and evergreen habits in temperate forest. Evolution 20:1–15

Axelrod DI (1972) Edaphic aridity as a factor in angiosperm evolution. Am Nat 106:311–320

Baig MN, Tranquillini W (1976) Studies on upper timberline: morphology and anatomy of Norway spruce (Piciea abies) and stone pine (Pinus cembra) needles from various habitat conditions. Can J Bot 54:1622–1632

Bannister P (1964) The water relations of certain heath plants with reference to their ecological amplitude. II. Field studies. J Ecol 52:481–497

Bannister P (1984a) Winter frost resistance of leaves of some plants from the Three Kings Islands, grown outdoors in Dunedin, New Zealand. N Z J Bot 22:303–306

Bannister P (1984b) The seasonal course of frost resistance in some New Zealand pteridophytes. N Z J Bot 22:557–563

Barclay AM, Crawford RMM (1982) Winter desiccation stress and resting bud viability in relation to high altitude survival in Sorbus aucuparia L. Flora 172:21–34

Baross JA, Morita RY (1978) Microbial life at low temperatures: ecological aspects. In: Kushner DJ (ed) Microbial life in extreme environments. Academic Press, London New York, pp 9–71

Barry RG, Hare FK (1974) Arctic climate. In: Ives JD, Barry RG (eds) Arctic and alpine environments. Methuen, London, pp 19–54

Bauch J, Klein P, Frühwald A, Brill H (1979) Alterations of wood characteristics in Abies alba Mill. due to 'fir-dying' and considerations concerning its origin. Eur J For Pathol 9:321–331

Bauer H, Harrasser J, Bendetta G, Larcher W (1971) Jahresgang der Temperaturresistenz junger Holzpflanzen im Zusammenhang mit ihrer jahreszeitlichen Entwicklung. Ber Dtsch Bot Ges 84:561–570

Bauer H, Larcher W, Walker RB (1975) Influence of temperature stress on CO_2-gas exchange. In: Cooper JP (ed) Photosynthesis and productivity in different environments, vol III. Cambridge Univ Press, pp 557–586

Beccari O (1933) Asiatic palms – Corypheae. Ann R Bot Garden Calcutta 13:1–356

Beck E (1986) Cold tolerance. In: Rundel PW (ed) Tropical alpine environments: plant form and function (in press)

Beck E, Senser M, Scheibe R, Steiger H-M, Pongratz P (1982) Frost avoidance and freezing tolerance in Afroalpine 'giant rosette' plants. Plant Cell Environ 5:215–222

Beck E, Schulze E-D, Senser M, Scheibe R (1984) Equilibrium freezing of leaf water and extracellular ice formation in Afroalpine 'giant rosette' plants. Planta 162:276–282

Becquerel P (1954) La cryosynérèse cytonucléoplasmique jusqu'aux confins du zéro absolu, son rôle pour la végétation polaire et la conservation de la vie. Rapp Commun 8. Congr Int Bot, Paris, Sect 11:269–270

Becwar MR, Rajashekar C, Briston KFH, Burke MJ (1981) Deep undercooling of tissue water and winter hardiness limitation. Plant Physiol 68:111–114

Bell KL, Bliss LC (1980) Autecology of Kobresia bellardii: Why winter snow accumulation limits local distribution. Ecol Monogr 49:377–402

Bervaes JCAM, Kuiper PJC, Kylin A (1972) Conversion of digalactosyl diglyceride (extra long chain conjugates) into monogalactosyl diglyceride of pine needle chloroplasts upon dehardening. Physiol Plant 27:231–235

Bervaes JCAM, Ketchie DO, Kuiper PJC (1978) Cold hardiness of pine needles and apple bark as affected by alteration of day length and temperature. Physiol Plant 44:365–368

Bialobok ST (1974) Variation of cold hardiness of woody plants. Pol Acad Sci Inst Dendr, Kórnik Arboretum, FG-Po-238/E-CR-68, Final Rep

Biebl R (1939) Über die Temperaturresistenz von Meeresalgen verschiedener Klimazonen und verschieden tiefer Standorte. Jahrb Wiss Bot 88:389–420

Biebl R (1958) Temperatur- und osmotische Resistenz von Meeresalgen der bretonischen Küste. Protoplasma 50:218–242

Biebl R (1962a) Protoplasmatische Ökologie der Pflanzen. Wasser und Temperatur. Protoplasmatologia, vol XII/1. Springer, Berlin Wien

Biebl R (1962b) Temperaturresistenz tropischer Meeresalgen. (Verglichen mit jener der Algen in temperierten Meeresgebieten). Bot Mar 4:241–254

Biebl R (1964) Temperaturresistenz tropischer Pflanzen auf Puerto Rico. Protoplasma 59:133–156

Biebl R (1967a) Temperaturresistenz einiger Grünalgen warmer Bäche auf Island. Botaniste 50: 33–42

Biebl R (1967b) Temperaturresistenz tropischer Urwaldmoose. Flora (Jena) 157:25–30

Biebl R (1968) Über Wärmehaushalt und Temperaturresistenz arktischer Pflanzen in Westgrönland. Flora (Jena) Abt B 157:327–354

Biebl R (1969) Untersuchungen zur Temperaturresistenz arktischer Süßwasseralgen im Raum von Barrow, Alaska. Mikroskopie 25:3–6

Biebl R (1970) Vergleichende Untersuchungen zur Temperaturresistenz von Meeresalgen entlang der pazifischen Küste Nordamerikas. Protoplasma 69:61–83

Biebl R (1972) Studien zur Temperaturresistenz der Gezeitenalge Ulva pertusa Kjellmann. Bot Mar 15:139–143

Billings WD (1973) Arctic and alpine vegetations: similarities, differences, and susceptibility to disturbance. Bio Sci 23:697–704

Billings WD (1974) Adaptations and origins of alpine plants. Arctic Alpine Res 6:129–142

Bittenbender BC, Howell GS (1976) Cold hardiness of flower buds from selected highbush blueberry cultivars (Vaccinium australe Small). J Am Soc Hortic Sci 101:135–139

Black RA, Bliss LC (1980) Reproductive ecology of Picea mariana (Mill.) at tree line near Inuvik, Northwest Territories, Canada. Ecol Monogr 50:331–354

Bliss LC (1971) Arctic and alpine plant life cycles. Annu Rev Ecol Syst 2:405–438

Bliss LC (1985) Alpine. In: Chabot BF, Mooney HA (eds) Physiological ecology of North American plant communities. Chapman & Hall, New York, pp 41–65

Bodner M (1985) Frostwirkungen auf Riesenrosettenpflanzen am Mt. Kenia. Sitzungsber Österr Akad Wiss Math-Naturwiss Kl Abt I 194:301–309

Bodner M, Beck E (1987) Effect of supercooling and freezing on chlorophyll fluorescence induction in freezing tolerant leaves of afroalpine 'giant rosette' plants. (in press)

Bodner M, Larcher W (1987) Chilling susceptibility of different organs and tissues in Saitpaulia ionantha and Coffea arabica. Angew Bot (in press)

Bootsma A (1976) Estimating minimum temperature and climatological freeze risk in hilly terrain. Agric Meteorol 16:425–443

Borchert R (1980) Phenology and ecophysiology of tropical trees: Erythrina poeppigiana O.F.COOK. Ecology 61:1065–1074

Borisov AA (1965) Climates of the USSR (English translation). Oliver & Boyd, Edinburgh

Bornkamm R (1958) Standortbedingungen und Wasserhaushalt von Trespen-Halbtrockenrasen (Mesobromion) im oberen Leinegebiet. Flora (Jena) 146:23–67

Boysen-Jensen P (1949) Causal plant-geography. K Dan Vidensk Selsk 21:1–19

Brewer M (1958) The thermal regime of an arctic lake. Trans Am Goephys Union 39:278–284

Briggs SP, Hang AR, Scheffer RP (1982) Localization of spin labels in oat leaf protoplasts. Plant Physiol 70:662–667

Brisse H, Grandjouan G (1974) Classification climatique des plantes. Oecol Plant 9:51–80

Brock TD (1978) Thermophilic microorganisms and life at high temperatures. Springer, Berlin Heidelberg New York

Brown GN, Bixby JA, Melcarek PK, Hinckley TM, Rogers R (1977) Xylem pressure potential and chlorophyll fluorescence as indicators of freezing survival in black locust and western hemlock seedlings. Cryobiology 14:94–99

Brown MS, Reuter FW (1974) Freezing of nonwoody plant tissues. III. Videotape micrography and the correlation between individual cellular freezing events and temperature changes in the surrounding tissue. Cryobiology 11:185–191

Brown MS, Pereira ESB, Finkle BJ (1974) Freezing of nonwoody plant tissues. II. Cell damage and the fine structure of freezing curves. Plant Physiol 53:709–711

Buhl C (1968) Wunden. In: Sorauer P (found) Handbuch der Pflanzenkrankheiten, vol I, pt 3. Parey, Berlin Hamburg, pp 1–96

Bünning E (1953) Entwicklungs- und Bewegungsphysiologie der Pflanze, 3rd edn. Springer, Berlin Wien

Burckhardt H (1963) Meteorologische Voraussetzungen der Nachtfröste. In: Schnelle F (ed) Frostschutz im Pflanzenbau, vol I. Bayr Landw Verlag, München, pp 13–81

Burke MJ, Stushnoff C (1979) Frost hardiness: A discussion of possible molecular causes of injury with particular reference to deep supercooling of water. In: Mussell H, Staples R (eds) Stress physiology in crop plants. Wiley & Son, New York, pp 197–225

Burke MJ, Gusta LV, Quamme HA, Weiser CJ, Li PH (1976) Freezing and injury in plants. Annu Rev Plant Physiol 27:507–528

Butin H, Volger Ch (1982) Untersuchungen über die Entstehung von Stammrissen („Frostrissen") an Eiche. Forstwiss Centralbl 101:295–303

Bylinska E (1975) The relationship of winter transpiration of selected species from the genera Viburnum, Populus, and Lonicera to the northern boundary of their geographical distribution. Bot Monogr 50:5–59

Candolle de A (1855) Géographie botanique. Masson, Paris

Carter JV, Brenner ML (1985) Plant growth regulators and low temperature stress. In: Pharis RP, Reid DM (eds) Hormonal regulation of development. III. Role of environmental factors. Encyclopedia of plant physiology, vol XI. Springer, Berlin Heidelberg New York, pp 418–443

Cary JW (1975) Factors affecting cold injury of sugarbeet seedlings. Agron J 67:258–262

Castri di F, Mooney HA (eds) (1973) Mediterranean type ecosystems. Ecol Stud 7. Springer, Berlin Heidelberg New York

Cernusca A, Vesco A (1976) A thermoelectric, heat-flux-controlled cooling stage for measuring parameters of small plants. Cryobiology 13:638–644

Chamber R, Hale HP (1932) The formation of ice in protoplasm. Proc R Soc London Ser B 110: 337–352

Chandler WH (1954) Cold resistance in horticultural plants: a review. Proc Am Soc Hortic Sci 64: 552–572

Chaney RW (1940) Tertiary forests and continental history. Bull Geol Soc Am 51:469–488

Chaney RW (1947) Tertiary centers and migration routes. Ecol Monogr 17:140–148

Chapin III FS, Shaver GR (1985) Arctic. In: Chabot BF, Mooney HA (eds) Physiological ecology of North American plant communities. Chapman & Hall, New York, pp 16–40

Chapman D (1975) Phase transition and fluidity characteristics of lipids and cell membrane. Rev Biophys 8:185–235

Chapman VP (1970) Mangrove phytosociology. Trop Ecol 11:1–19

Chauvin C (1984) Resistance au froid des Jacaranda. Rapp Dév Amél Vég Lab Biol Evol Univ Provence, Marseille

Chaw MW, Rubinsky B (1985) Cryomicroscopic observations on directional solidification in onion cells. Cryobiology 22:392–399

Chen HH, Li PH (1980a) Characteristics of cold acclimation and deacclimation in tuber-bearing Solanum species. Plant Physiol 65:1146–1148

Chen HH, Li PH (1980b) Biochemical changes in tuber-bearing Solanum species in relation to frost hardiness during cold acclimation. Plant Physiol 66:414–421

Chen HH, Li PH (1982) Potato cold acclimation. In: Li PH, Sakai A (eds) Plant cold hardiness and freezing stress, vol II. Academic Press, London New York, pp 5–22

Chen HH, Li PH, Bremner ML (1983) Involvement of abscisic acid in potato cold acclimation. Plant Physiol 71:362–365

Chen THH, Gusta LV (1983) Abscisic acid-induced freezing resistance in cultured plant cells. Plant Physiol 73:71–75

Chen THH, Kartha KK, Constabel F, Gusta LV (1984) Freezing characteristics of cultured Catharanthus roseus (L.). G. Don cells treated with dimethylsulfoxide and sorbitol in relation to cryopreservation. Plant Physiol 75:720–725

Chiba S (1965) Frost cracks of poplar cultivars and Alnus hirsuta, var. sibirica. Hokkaido For Breed 8:9–11

Chollet R, Anderson LL (1977) Conformational changes associated with the reversible cold inactivation of ribulose-1-5-biphosphate carboxylase-oxygenase. Biochem Biophys Acta 482:228–240

Christersson L, Krasavtsev OA (1972) Vlijanie skorosti i prodolžitelnosti ochlaždenija na morozostoikost chvoinych derevev. (Effect of the rate and the duration of cooling on frost resistance of coniferous trees). Fiziol Rast 19:638–642

Christophersen J (1973) Basic aspects of temperature action on microorganisms. In: Precht H, Christophersen J, Hensel H, Larcher W (eds) Temperature and life. Springer, Berlin Heidelberg New York, pp 3–59

Clausen E (1964) The tolerance of hepatics to desiccation and temperature. Bryologist 67:411–417

Clausen KE (1982) Variation in frost injury to white ash families in an Ontario plantation. Can J For Res 12:440–443

Clements RJ, Ludlow MM (1977) Frost avoidance and frost resistance in Centrosema virginianum. J Appl Ecol 14:551–556

Cloutier Y, Andrews CJ (1984) Efficiency of cold hardiness induction by desiccation stress in four winter cereals. Plant Physiol 76:595–598

Cloutier Y, Siminovitch D (1982) Correlation between cold- and drought-induced frost hardiness in winter wheat and rye varieties. Plant Physiol 69:256–258

Cooper WC, Gorton BS (1954) Freezing tests with small trees and detached leaves of grapefruit. Am Soc Hortic Sci 63:167–172

Cooper WC, Peynado A (1959) Winter temperatures of 3 citrus areas as related to dormancy and freeze injury of Citrus trees. Proc Am Soc Hortic Sci 74:333–347

Cooper WC, Gorton BS, Tayloe SD (1954) Freezing tests with small trees and detached leaves. Proc Am Soc Hortic Soc 63:167–172

Corbet PS (1972) The microclimate of arctic plants and animals, on land and in fresh water. Acta Arctica, Fasc 18, 43 pp. Kϕbenhavn: Munksgaard, Kϕbenhavn

Cottrell StF (1981) Yeast freeze-thawe survival rates as a function of different stages in the cell cycle. Cryobiology 18:506–510

Cox W, Levitt J (1972) An inproved leaf-disk method for determining the freeze-killing temperature of leaves. Cryobiology 9:251–256

Crombie DS, Milburn JA, Hipkins MF (1985) Maximum sustainable xylem sap tensions in Rhododendron and other species. Planta 163:27–33

Cronquist A (1968) The evolution and classification of flowering plants. Mifflin, Boston

Crowe JH, Crowe LM, Mouradian R (1983) Stabilization of biological membranes at low water activities. Cryobiology 20:346–356

Crowe LM, Crowe JH (1982) Hydration dependent hexagonal phase in a biological membrane. Arch Biochem Biophys 217:582–587

Cuatrecasas J (1979) Growth forms of the Espeletiinae and their correlation to vegetation types of high tropical Andes. In: Larsen K, Holm-Nielsen LB (eds) Tropical botany. Academic Press, London New York, pp 397–410

Daie J, Campbell WF (1981) Response to tomato plants to stressful temperatures. Plant Physiol 67:26–29

Dalton FN, Gardner WR (1978) Temperature dependence of water uptake by plant roots. Agron J 70:404–406

Darrall NM, Jäger HJ (1984) Biochemical diagnostic tests for the effect of air pollution on plants. In: Koziol MJ, Whatley FR (eds) Gaseous air pollutants and plant metabolism. Butterworths, London, pp 333–349

Daubenmire RF (1974) Plants and environment, 3rd edn. Wiley & Son, New York

Davies MD, Dickinson DB (1971) Effects of freeze-drying on permeability and respiration of germinating lily pollen. Plant Physiol 24:5–9

Day WR, Peace TR (1937) The influence of certain accessory factors on frost injury to forest trees. II: temperature conditions before freezing. III: time factor. Forestry 11:13–29

Deamer DW, Leonard R, Tardieu A, Branton D (1970) Lamellar and hexagonal lipid phases visualized by freezing-eching. Biochim Biophys Acta 219:47–60

Dereuddre J (1978) Effets de divers types de refroidissements sur la teneur en eau et sur la résistance au gel des bourgeons de rameaux d'Épicea en vie ralentie. Physiol Veg 16:469–489

Dereuddre J (1979) Étude comparative du comportement des bourgeons d'arbres en vie ralentie, pendant un refroidissement graduel des rameaux. Bull Soc Bot Fr 126:399–412

Devay M, Paldi E (1977) Cold induced tRNA synthesis in wheat cultivars during the hardening period. Plant Sci Lett 8:191–195

Dexter ST, Tottingham WE, Graber LF (1932) Investigations of the hardiness of plants by measurement of electrical conductivity. Plant Physiol 7:63–78

Diller KR (1982) Quantitative low temperature optical microscopy of biological systems. J Microsc 126:9–28

Dilley DR, Heggestad HE, Powers WL, Weiser CJ (1975) Environmental stress. In: Brown AWA, Byerly TC, Gibbs M, San Pietro A (eds) Crop productivity – research imperatives. Kettering Foundation, East Lansing, pp 309–355

Dircksen A (1964) Vergleichende Untersuchungen zur Frost-, Hitze- und Austrocknungsresistenz einheimischer Laub- und Lebermoose unter besonderer Berücksichtigung jahreszeitlicher Veränderungen. Diss, Univ Göttingen

Dowgert MF, Steponkus PL (1983) Effect of cold acclimation on intracellular ice formation in isolated protoplasts. Plant Physiol 72:978–988

Duda U, Kacperska A (1983) Frost tolerance estimation in callus derived from poplar and winter rape plants using three different methods. Z Pflanzenphysiol 111:69–73

Duthie HC (1964) The survival of desmids in ice. Br Phycol Bull 2:376–377

Dvorak J, Fowler DB (1978) Cold hardiness potential of triticale and tetraploid rye. Crop Sci 17: 477–478

Eamus D, Wilson JM (1983) ABA levels and effects in chilled and hardened Phaseolus vulgaris. J Exp Bot 34:1000–1006

Earnshaw MJ, Carver KA, Kerenga K, Harvey V, Harvey J, Vogt P, Gunn TC, Griffiths H, Lewis DH (1987) Photosynthetic mechanism, chilling susceptibility and cell sap osmotic potential of grasses along an altitudinal gradient in Papua New Guinea. Oecologia (Berlin) (in press)

Eaton GW, Mahrt BJ (1977) Cold hardiness testing of cranberry flower buds. Can J Plant Sci 57: 461–465

Ebermayer E (1873) Die physikalischen Einwirkungen des Waldes auf Luft und Boden und seine klimatologische und hygienische Bedeutung, begründet durch die Beobachtungen der forstlich-meteorologischen Stationen im Kgr. Bayern. Result Forstl Versuchsst Kgr Bayern, Aschaffenburg, Bd I: Krebs

Eckel O (1960) Bodentemperatur. In: Steinhauser F (ed) Klimatographie von Österreich. Denkschr Österr Akad Wiss 3(2):207–292

Edlich F (1936) Einwirkung von Temperatur und Wasser auf aerophile Algen. Arch Mikrobiol 7: 62–109

Eguchi H, Hamakoga M, Matsui T (1982) Digital image processing in polarized light for evaluation of foliar injury. Environ Exp Bot 22:277–283

Eiche V (1966) Cold damage and plant mortality in experimental provenance plantations with Scots pine in Northern Sweden. Stud Forest Suec 36:1–219

Eiche V, Andersson E (1974) Survival and growth in Scots pine (Pinus silvestris L.). Provenance experiments in Northern Sweden. Theor Appl Genet 44:49–57

Eiga S (1984) Ecogenetical study on the freezing resistance of Sakhalin fir (Abies sachalinensis Mast.) in Hokkaido. Bull Tree Breed Inst 2:61–107

Eiga S, Sakai A (1984) Altitudinal variation in freezing resistance of Sakhalin fir (Abies sachalinensis). Can J Bot 62:156–160

Elfving DC, Kaufmann MR, Hall AE (1972) Interpreting leaf water potential measurements with a model of a soil-plant-atmosphere continuum. Physiol Plant 27:161–168

Ellenberg H (1966) Leben und Kampf an den Baumgrenzen der Erde. Naturwiss Rundsch 19:133–139
Emberger L (1932) Sur une formule climatique et ses application en botanique. Météorologie 92: 1–10
Encyclopaedia Britannica (1963) vol XX. Encycl Br, London, p 959
Enright JT (1982) Sleep movements of leaves: In defense of Darwin's interpretation. Oecologia (Berlin) 54:253–259
Ericksson T (1965) Studies on the growth requirement and growth measurements of cell cultures of Haplopappus gracilis. Physiol Plant 18:976–993
Ernst WHO (1971) Zur Ökologie der Miombo-Wälder. Flora (Jena) 106:317–331
Esterbauer H, Grill D (1978) Seasonal variation of glutathione and glutathione reductase in needles of Picea abies. Plant Physiol (Lancaster) 61:119–121
Eunus AM, Johnson LPV, Aksel R (1962) Inheritance of winter hardiness in an eighteen-parent diallel cross of barley. Can J Genet Cytol 4:356–376
Eurola S (1975) Snow and ground frost conditions of some Finnish mire types. Ann Bot Fenn 12: 1–16
Farkas T, Deri-Hadlaczky E, Belea A (1975) Effects of temperature upon linolenic acid level in wheat and rye seedlings. Lipids 10:331–334
Farkas T, Singh B, Nemecz G (1985) Abscisic acid-related changes in composition and physical state of membranes in bean leaves. J Plant Physiol 118:373–379
Farrar DR (1967) Gametophytes of four tropical fern genera reproducing independently of their sporophytes in the Southern Appalachians. Science 155:1266–1267
Farrar DR (1978) Problem in the identity and origin of the Appalachian Vittaria gametophytes, a sporophyteless fern of the eastern United States. Am J Bot 65:1–12
Fennema OR (1973) Freezing injury and cryoprotectants. In: Fennema OR, Powrie WD, Marth EH (eds) Low temperature preservation of foods and living matter. Dekker, New York, pp 476–503
Fey RL, Workman M, Marcellos H, Burke MJ (1978) Electron spin resonance of 2.2.6.6-tetra-methyl-piperidine-1-oxyl (TEMPO)-labeled plant leaves. Plant Physiol 63:1220–1222
Fischer HW (1911) Gefrieren und Erfrieren, eine physikochemische Studie. Beitr Biol Pflanzen 10: 133–234
Fletcher NH (1970) The chemical physics of ice. Cambridge Univ Press
Flint HL (1972) Cold hardiness of twigs of Quercus rubra L. as a function of geographic origin. Ecology 53:1163–1170
Fliri F (1975) Das Klima der Alpen im Raume von Tirol. Wagner, Innsbruck
Florin R (1963) The distribution of conifer and taxod genera in time and space. Acta Hortic Bergiani 20:121–312
Fowler DB, Dvorak J, Gusta LV (1977) Comparative cold hardiness of several Triticum species and Secale cereale L. Crop Sci 17:941–943
Fowler DB, Gusta LV, Tyler NJ (1981) Selection for winterhardiness in wheat. III. Screening methods. Crop Sci 21:896–901
France RC, Cline ML, Reid CPP (1979) Recovery of ectomycorrhizal fungi after exposure to sub-freezing temperatures. Can J Bot 57:1845–1848
Franz H (1979) Ökologie der Hochgebirge. Ulmer, Stuttgart
Fraser JW, Farrar JL (1957) Frost hardiness of white spruce and red pine seedlings in relation to soil moisture. For Res Div Techn Note 59:1–5
Frey W (1977) Wechselseitige Beziehungen zwischen Schnee und Pflanze. Eine Zusammenstellung anhand von Literatur. Mitt Eidgen Schnee- Lawinenforsch 34, Birmensdorf
Freyman S, Brink VC (1967) Nature of ice-sheet injury to alfalfa. Agron J 59:557–560
Fuchigami LH, Evert DR, Weiser CJ (1970) A translocatable cold hardiness promotor. Plant Physiol 47:164–167
Fuchigami LH, Weiser CJ, Kobayashi K, Timis R, Gusta LV (1982) A degree growth stage (°GS) model and cold acclimation in temperate woody plants. In: Li PH, Sakai A (eds) Plant cold hardiness and freezing stress. Academic Press, London New York, pp 93–116
Fuchinoue H (1982) The winter desiccation damage of the tea plant in Japan. In: Li PH, Sakai A (eds) Plant cold hardiness and freezing stress, vol II. Academic Press, London New York, pp 499–510

Fuchinoue H (1985) Time course of cold injury of tea plants related to the climate. In: Fuchinoue H (ed) Special report on cold injury in tea plant in 1984. Saitama Prefectural Tea Exp Stn, Jpn, pp 24–28

Fuchs WH (1933) Zur Analyse des physiologischen Zustandes der Pflanzen in Zusammenhang mit Temperatureinflüssen. Kühn-Arch 38:232–286

Fuchs WH, Rosenstiel K (1958) Ertragssicherheit. In: Kappert H, Rudorf W (eds) Handbuch der Pflanzenzüchtung, vol I, 2nd edn. Parey, Berlin, pp 365–442

Fujikawa S, Miura K (1986) Plasma membrane ultrastructural changes caused by mechanical stress in the formation of extracellular ice as a primary cause of slow freezing in fruit-bodies of basidiomycetes [*Lyophyllum ulmarium* (Fr.) Kühner]. Cryobiology 23:371–382

Fuller MP, Eagles CF (1978) A seedling test for cold hardiness in Lolium perenne L. J Agric Sci 91: 217–222

Furst G (1983) Anatomičeskoe stroenie steblija različnych po zimostoikosti vidov roda Betula L (Anatomische Struktur von den in der Winterfestigkeit unterschiedlichen Baumstämmen der Gattung Betula L.). Folia Dendrol Arb (Mlyňany) 10:107–129

Futrell MC, Lyles WE, Pilgrim AJ (1962) Ascorbic acid and cold hardiness in oats. Plant Physiol Suppl 37:70

Garber MP, Steponkus PL (1976) Alterations in chloroplast thylakoids during an in vitro freeze-thaw cycle. Plant Physiol 57:673–680

Gäumann E, Naeff-Roth St, Reusser P, Amann A (1952) Über den Einfluß einiger Welketoxine und Antibiotica auf die osmotischen Eigenschaften pflanzlicher Zellen. Phytopathol Z 19:160–220

Gazeau C, Dereuddre J (1980) Effets de refroidissement prolongés sur la teneur en eau et la résistance au froid des tissus de la Troène (Ligustrum vulgare L. Oléacées). CR Acad Sci Paris Ser D 290:1443–1446

Geiger R (1957) The climate near the ground. Harvard Univ Press, Cambridge Ma

George MF, Burke MJ (1977a) Supercooling in overwintering Azalea flower buds. Plant Physiol 59:326–328

George MF, Burke MJ (1977b) Cold hardiness and deep supercooling in xylem of Shagbark Hickory. Plant Physiol 59:319–325

George MF, Burke MJ, Weiser CJ (1974a) Supercooling in overwintering Azalea flower buds. Plant Physiol 54:29–35

George MF, Burke MJ, Pellett HM, Johnson AG (1974b) Low temperature exotherm and woody plant distribution. HortSci 9:519–522

George MF, Becwar MR, Burke MJ (1982) Freezing avoidance by deep undercooling of tissue water in winter-hardy plants. Cryobiology 19:628–639

Gerloff ED, Richardson T, Stahmann MA (1966) Changes in fatty acid of alfalfa roots during cold hardening. Plant Physiol 41:1280–1284

Giertych M (ed) Definition of terms on dormancy. IUFRO Div 2 WP 2.01.4, Circular Lett 5, July 1974

Gill PA, Waister PD (1976) Design and performance of a portable frost chamber for the investigation of hardiness of strawberries. J Hortic Sci 51:509–513

Gola G (1929) Osservazioni sui danneggamenti alle piante legnose della regione veneta in seguito ai freddi del gennaio febbraio 1929. Atti Mem R Accad Sci Lett Arti, Padova 45:5–15

Goldstein G, Meinzer F (1983) Influence of insulating dead leaves and low temperatures on water balance in an Andean giant rosette plant. Plant Cell Environ 6:649–656

Goldstein G, Rada F, Azocar A (1985a) Cold hardiness and supercooling along an altitudinal gradient in Andean giant rosette species. Oecologia (Berlin) 68:147–152

Goldstein G, Meinzer F, Monasterio M (1985b) Physiological and mechanical factors in relation to size-dependent mortality in an Andean giant rosette species. Acta Oecol (Oecol Plant) 6(29): 263–275

Good R (1974) The geography of the flowering plants, 4th edn. Longman, London

Gordon-Kamm WJ, Steponkus PL (1984) Lamellar-to-hexagonal II phase transitions in the plasma membrane of isolated protoplasts after freeze-induced dehydration. Proc Natl Acad Sci USA 81:6373–6377

Goryshina TK (1969) Rannevesennie efemeroidy lesostepnykh dubrav. Izdat Leningrad Univ

Goryshina TK (1972) Recherches écophysiologiques sur les plantes éphéméroides printanières dans les chênaies de la zone forêt-steppe de la Russe centrale. Oecol Plant 7:241–258
Goujon C, Maia N, Doussinault G (1968) Resistance au froid chez le blé. 2. Réactions au stade coléoptile étudiées en conditions artificielles. Ann Amelior Plantes 18:49–57
Graham D, Patterson BD (1982) Responses of plants to low, nonfreezing temperatures: Proteins metabolism, and acclimation. Annu Rev Plant Physiol 33:347–372
Graham RP (1971) Cold injury and its determination in selected Rhododendron species. Master thesis, Univ Minn, St. Paul
Graham PR, Mullin R (1975) The determination of lethal freezing temperatures in buds and stems of deciduous azalea by a freezing curve method. J Am Soc Hortic Sci 100:3–17
Graham PR, Mullin R (1976) The determination of lethal freezing temperatures in buds and stems of deciduous azalea by a freezing curve method. J Am Soc Hortic Sci 101:3–7
Green DC, Ratzlaff CD (1975) An apparent relationship of soluble sugars with hardiness in winter wheat varieties. Can J Bot 53:2198–2201
Green JW (1969) Temperature response in altitudinal populations of Eucalyptus pauciflora, Sieb. Ex. Spreng. New Phytol 68:339–410
Greenham CG (1966) The stages at which frost injury occurs in alfalfa. Can J Bot 44:1471–1483
Grenier B, Willemot C (1975) Lipid phosphorus content and ^{33}Pi incorporation in roots of alfalfa varieties during frost hardening. Can J Bot 53:1473–1477
Grime JP (1979) Plant strategies and vegetation processes. Wiley and Son, New York Chichester
Grisebach A (1838) Über den Einfluß des Climas auf die Begrenzung der natürlichen Floren. Linnaea 12:159–200
Guinon M, Larsen JB, Spethmann W (1982) Frost resistance and early growth of Sequoidendron giganteum seedlings of different origins. Silvae Genet 31:173–178
Gusta LV, Fowler DB (1975) Effects of temperature on dehardening and rehardening of winter cereals. Can J Plant Sci 56:673–678
Gusta LV, Fowler DB (1976) Dehardening and rehardening of spring-collected winter wheats and a winter rye. Can J Plant Sci 56:775–779
Gusta LV, Fowler DB (1979) Cold resistance and injury in winter cereals. In: Mussell H, Staples RC (eds) Stress physiology in crop plants. Wiley and Son, New York, pp 159–178
Gusta LV, Weiser CJ (1972) Nucleic acid and protein changes in relation to cold acclimation and freezing injury of Korean boxwood leaves. Plant Physiol 49:91–96
Gusta LV, Burke MJ, Kapoor AC (1975) Determination of unfrozen water in winter cereals at subfreezing temperatures. Plant Physiol 56:707–709
Gusta LV, Butler JD, Rajashekar C, Burke MJ (1980) Freezing resistance of perennial turfgrasses. HortSci 15:494–496
Gusta LV, Tyler NJ, Chen HH (1983) Deep undercooling in woody taxa growing north of the -40 °C isotherm. Plant Physiol 72:122–128
Guy CL (1983) Changes in glutathione levels during cold acclimation and its role in freezing tolerance. PhD Thesis, Univ, St. Paul
Guy CL, Carter JV (1982) Effect of low temperature on the glutathione status of plant cells. In: Li PH, Sakai A (eds) Plant cold hardiness and freezing stress, vol II. Academic Press, London New York, pp 169–179
Häckel H (1985) Kälte und Frost als Belastungsfaktor. In: Sorauer P (found) Handbuch der Pflanzenkrankheiten, vol I, p 5, 7th edn. Parey, Berlin, pp 70–106
Hadley JL, Smith WK (1983) Influence of wind exposure on needle desiccation and mortality for timberline conifers in Wyoming, USA. Arctic Alpine Res 15:127–135
Hadley JL, Smith WK (1986) Wind effects on needles of timberline conifers: seasonal influence on mortality. Ecology 67:12–19
Hagiya K, Ishizawa S (1961) Studies on snow camellia (Camellia rusticana). I. On variation and distribution of native and domesticated camellia in Niigata prefecture. J Jpn Soc Hortic Sci 30: 270–290
Hagiya K, Ishizawa S (1968) On some physiological aspects of the bush and snow camellia. In: Tuyama T (ed) Camellias of Japan. Takeda Sci Found, Osaka, pp 27–32
Hagner M (1969) Thermal mapping of conifer seedlings – a useful method in pathological and physiological research. Z Pflanzenphysiol 61:322–331

Hamaya T, Kurahashi A, Takahashi N, Sakai A (1968) Studies in frost-hardiness of the Japanese and the Dahurian larch and their hybrids. Bull Tokyo Univ For 64:197–240

Harper JL (1982) After description. In: Newman EI (ed) The plant community as a working mechanism. Blackwell, Oxford, pp 11–25

Harvey RB (1923) Cambial temperatures of trees in winter and their relation to sunscald. Ecology 4:261–265

Harwood CE (1980) Frost resistance of subalpine Eucalyptus species. I. Experiments using a radiation frost room. Aust J Bot 28:587–599

Hasegawa K, Tsuboi A (1960) The effect of low temperature on the breaking of rest for winter bud in mulberry tree. Jpn Assoc Sericult 25:320–326

Hatakeyama I (1960) The relation between growth and cold hardiness of leaves of Camellia sinensis. Biol J Nara Women's Univ 10:65–69

Hatakeyama I (1961) Studies on the freezing of living and dead tissues of plants with special reference to the colloidally bound water in living state. Mem Coll Sci Univ Kyoto Ser B 28:401–429

Hatakeayama I, Kato J (1965) Studies on the water relation of Buxus leaves. Planta 65:259–268

Hatakeyama S (1981) Genetical and breeding studies on the regional differences of intraprovenance variation in Abies sachalinensis. Bull Hokkaido For Exp Stn 19:1–87

Hatano S (1978) Studies on frost hardiness in Chlorella ellipsoidea: Effects of antimetabolites, surfactants, hormones, and sugars on the hardening process in the light and dark. In: Li PH, Sakai A (eds) Plant cold hardiness and freezing stress. Academic Press, London New York, pp 175–196

Hatano S, Kabata K (1982) Transition of lipid metabolism in relation to frost hardiness in Chlorella ellipsoidea. In: Li PH, Sakai A (eds) Plant cold hardiness and freezing stress, vol II. Academic Press, London New York, pp 145–156

Hatano S, Sadakane H, Tutumi M, Watanabe T (1976) Studies on frost hardiness in Chlorella ellipsoidea II. Effects of inhibition of RNA and protein synthesis and surfactants on the process of hardening. Plant Cell Physiol 17:643–651

Hatano S, Sadakane H, Nagayama J, Watanabe T (1978) Studies on frost hardiness in Chlorella ellipsoidea III. Changes in O_2 uptake and evolving during hardening and after freeze-thawing. Plant Cell Physiol 19:917–926

Havis JR (1976) Root hardiness of woody ornamentals. HortSci 11:385–386

Hayden RE, Dionne L, Fenson DS (1972) Electrical impedance studies of stem tissue of Solanum clones during cooling. Can J Bot 50:1547–1554

Hayes AR, Pegg DE, Kingston RE (1974) A multirate small-volume cooling machine. Cryobiology 11:371–377

Heber U (1959) Beziehungen zwischen der Größe der Chloroplasten und ihrem Gehalt an löslichen Eiweißen und Zuckern im Zusammenhang mit dem Frostproblem. Protoplasma 51:284–298

Heber U (1968) Freezing injury in relation to loss of enzyme activities and protection against freezing. Cryobiology 5:188–201

Heber U, Santarius KA (1973) Cell death by cold and heat, and resistance to extreme temperatures. Mechanisms of hardening and dehardening. In: Precht H, Christophersen J, Hensel H, Larcher W (eds) Temperature and life. Springer, Berlin Heidelberg New York, pp 232–263

Heber U, Tyankova L, Santarius KA (1971) Stabilization and inactivation of biological membranes during freezing in the presence of amino acids. Biochim Biophys Acta 241:578–592

Heber U, Volger H, Overbeck V, Santarius KA (1979) Membrane damage and protection during freezing. In: Fennema O (ed) Proteins at low temperatures. Adv Chem 180:159–189

Hedberg I, Hedberg O (1979) Tropical-alpine life forms of vascular plants. Oikos 33:297–307

Hedberg O (1964) Features of Afro-alpine plant ecology. Acta Phytogeogr Suecica 49:1–144

Hedberg O (1968) Taxonomic and ecological studies on the afroalpine flora of Mt. Kenya. Hochgebirgsforschung 1:171–194

Hedberg O (1973) Adaptive evolution in a tropical-alpine environment. In: Heywood VN (ed) Taxonomy and ecology, vol V. Academic Press, London, pp 71–92

Heinze W, Schreiber D (1984) Eine neue Kartierung der Winterhärtezonen für Gehölze in Europa. Mitt Dtsch Dendrol Ges 75:11–56

Hellergren J, Lundborg T, Widell S (1984) Cold acclimation in Pinus sylvestris: Phospholipids in purified plasma membranes from needles of pine. Physiol Plant 62:162–166
Hendershott GH (1961) The effect of water content and surface moisture on the freezing of orange fruits. Proc Am Soc Hortic Sci 78:186–189
Hendershott GH (1962) The response of orange trees and fruits to freezing temperatures. Proc Am Soc Hortic Sci 80:247–254
Henson WR (1952) Chinook winds and red belt injury to lodgepole pine in the Rocky Mountain parks area of Canada. For Chron 28:62–64
Heslop-Harrison J, Heslop-Harrison Y (1970) Evaluation of pollen viability by enzymatically induced fluorescence; intracellular hydrolysis of fluorescein diacetate. Stain Technol 45:115–120
Hincha DK, Schmidt JE, Heber U, Schmitt JM (1984) Colligative and non-colligative freezing damage to thylakiod membranes. Biochim Biophys Acta 769:8–14
Hintikka V (1963) Über das Großklima einiger Pflanzenareale in zwei Klimakoordinatensystemen dargestellt. Ann Bot Soc Zool Bot Fenn Vanamo 35(5):1–64
Hirsh AG, Williams RJ, Meryman H (1985) A novel method of natural cryoprotection. Plant Physiol 79:41–56
Hoffmann G (1960) Die mittleren jährlichen und absoluten Extremtemperaturen der Erde. II. Met Abh 8/3. Reimer, Berlin
Hoffmann G (1963) Die höchsten und tiefsten Temperaturen auf der Erde. Umschau 1963:16–18
Holm-Hansen O (1963) Viability of blue-green and green algae after freezing. Phsiol Plant 16:530–540
Holt MA, Pellett NE (1981) Cold hardiness of leaf and stem organs of Rhododendron cultivars. J Am Soc Hortic Sci 106:608–612
Holtmeier FK (1973) Geoecological aspects of timberlines in northern and central Europe. Arctic Alpine Res 5:A45–A54
Holubowicz T, Cummins JN, Forsline PL (1982) Responses of Malus clones to programmed low-temperature stresses in late winter. J Am Soc Hortic Sci 107:492–496
Holzer K (1970) A late frost injury in an alpine Norway spruce (Picea abies L. Karst.) provenance test. 2nd FAO-IUFRO World Cons For Tree Breeding, Doc. vol I, 6/10
Hölzl J (1971) Fluorochromierungsversuche mit Fluoreszeinacetylester an Allium cepa-Epidermen. Biochem Physiol Pflanzen 162:357–362
Hong SG, Sucoff E (1980) Units of freezing of deep supercooled water in woody xylem. Plant Physiol 66:40–45
Horiguchi K (1979) Effect of the rate of heat removal on the rate of frost heaving. Eng Geol 13: 63–71
Horiuchi T, Sakai A (1978) Effect of solar radiation on frost damage to young cryptomerias. In: Li PH, Sakai A (eds) Plant cold hardiness and freezing stress, vol I. Academic Press, London New York, pp 281–295
Horváth I, Vigh L, Belea A, Farkas T (1979) Conversion of phosphatidyl cholin to phosphatidic acid in freeze injured rye and wheat cultivars. Physiol Plant 45:57–62
Horváth I, Vigh L, Farkas T (1981) The manipulation of polar head group composition of phospholipids in the wheat Miranovskaja 808 affects frost tolerance. Planta 151:103–108
Hosokawa T, Tagawa H, Chapman VJ (1977) Mangals of Micronesia, Taiwan, Japan, the Philippines and Oceania. In: Chapman VJ (ed) Wet coastal ecosystems. Elsevier, Amsterdam, pp 271–291
Howell GS, Weiser CJ (1970a) The environmental control of cold acclimation in apple. Plant Physiol 45:390–394
Howell GS, Weiser CJ (1970b) Fluctuations in the cold resistance of apple twigs during spring dehardening. J Am Soc Hortic Sci 95:190–192
Hudson MA, Brustkern P (1965) Resistance of young and mature leaves of Mnium undulatum (L.) to frost. Planta 66:135–155
Hudson MA, Idle DB (1962) The formation of ice in plant tissues. Planta 57:718–730
Hummel RL, Johnson CR (1985) Freezing tolerance in the genus Ficus. HortSci 20:287–289
Huner NPA, MacDowall FDH (1976a) Chloroplastic proteins of wheat and rye grown at warm and cold-hardening temperatures. Can J Biochem 54:848–853
Huner NPA, MacDowall FDH (1976b) Effect of cold adaptation of puma rye on properties of RuBPCase carboxylase. Biochem Biophys Res Commun 73:411–429

Huner NPA, MacDowall FDH (1979a) Changes in net charge and subunit properties of RubPCase during cold hardening of Puma rye. Can J Biochem 57:155–164

Huner NPA, MacDowall FDH (1979b) The effect of low temperature acclimation of winter rye on catalytic properties of its ribulose bisphosphate carboxylase oxygenase. Can J Biochem 57: 1036–1041

Huner NPA, Hopkins WG, Elfman B, Hayden DB, Griffith M (1982) Influence of growth at cold-hardening temperature on protein structure and function. In: Li PH, Sakai A (eds) Plant cold hardiness and freezing stress, vol II. Academic Press, London New York, pp 129–144

Hutcheson CE, Wiltbank WJ (1972) The freezing point of detached leaves as a measure of cold hardiness of young budded citrus plants. HortSci 7:27–28

Hwang S (1968) Investigation of ultra-low temperature for fungal cultures. I: An evaluation of liquid-nitrogen storage for preservation of selected fungal cultures. Mycologia 60:613–621

Ichikawa S, Shidei T (1971) Fundamental studies on deep freezing storage of tree pollen. Bull Kyoto Univ For 42:51–82

Ichiki S, Yamaya H (1982) Sorbitol in tracheal sap of dormant apple (Malus domestica Borkh) shoots as related to cold hardiness. In: Li PH, Sakai A (eds) Plant cold hardiness and freezing stress, vol II. Academic Press, London New York, pp 181–187

Idle DB (1966) The photography of ice formation in plant tissue. Ann Bot New Ser 30:199–205

Ikeda I (1982) Freezing injury and protection of citrus in Japan. In: Li PH, Sakai A (eds) Plant cold hardiness and freezing stress, vol II. Academic Press, London New York, pp 575–589

Ikeda I, Kobayashi S, Nakatani M (1980) Differences in cold resistance of various citrus varieties and cross-seedlings based on the data obtained from the freezes in 1977. Bull Fruit Tree Res Stn Akitsu Ser E 3:49–65

Ingold CT (1981) Flammula velutipes in relation to drying and freezing. Trans Br Mycol Soc 76: 150–152

Ingold CT (1982) Resistance of certain basidiomycetes to freezing. Trans Br Mycol Soc 79:554–556

Irmscher E (1912) Über die Resistenz der Laubmoose gegen Austrocknung und Kälte. Jahrb Wiss Bot 50:387–449

Irmscher E (1922) Pflanzenverbreitung und Entwicklung der Kontinente. Mitt Inst Allg Bot (Hamburg) 5:19–240

Irving RM, Lanphear FO (1967) The long-day leaf as a source of cold hardiness inhibitors. Plant Physiol 42:1384–1388

Ishida S (1963) On the development of frost cracks on "Todomatsu" trunks, Abies sachalinensis, especially in relation to their wetwood. Res Bull College Exp For, Hokkaido Univ 22:273–373

Ishikawa M (1984) Deep supercooling in most tissues of wintering Sasa senanensis and its mechanism in leaf blade tissues. Plant Physiol 75:196–202

Ishikawa M, Sakai A (1978) Freezing avoidance in rice and wheat seeds in relation to water content. Low Temp Sci Ser B 36:39–49

Ishikawa M, Sakai A (1981) Freezing avoidance mechanisms by supercooling in some Rhododendron flower buds with reference to water relations. Plant Cell Physiol 22:953–967

Ishikawa M, Sakai A (1982) Characteristics of freezing avoidance in comparison with freezing tolerance: A demonstration of extraorgan freezing. In: Li PH, Sakai A (eds) Plant cold hardiness and freezing stress, vol II. Academic Press, London New York, pp 325–340

Ishikawa M, Sakai A (1985) Extraorgan freezing in wintering flower buds of Cornus officinalis Sieb. et Zucc. Plant Cell Environ 8:333–338

Ishikawa M, Yoshida S (1985) Seasonal changes in plasma membranes and mitochondrial isolated from Jerusalem artichoke tubers. Possible relationship to cold hardiness. Plant Cell Physiol 26: 1331–1334

Ishikawa MY, Ono S, Kawaguchi M (1974) Determination of settling force of deposited snow in a heavy snow area in Yamagata prefecture. J Jpn For Soc 85:289–290

Ishikawa N (1977) Studies of radiative cooling at land basins in the snowy season. Contrib Inst Low Temp Sci Ser A 27:1–46

Ishizawa S (1973) The ecology of Camellia rusticana Honda. 1. On the habitat segregation between Camellia rusticana Honda (snow camellia) and Viburnum urceolatum Sieb. et Zucc. var. procumbens Nakai. Sci Rep, Niigata Univ Ser D 10:53–70

Ishizawa S (1978) Climatic factors affecting the distribution of snow camellia (C. rusticana Honda). In: Plant ecology to the memory of Dr. Kuniji Yoshioka, Soc Tohoku Plant Ecol, Sendai, pp 298–308

Iversen J (1944) Viscum, Hedera and Ilex as climate indicators. Geol Fören Förhandl 66:463–483

Ives JD (1974) Permafrost. In: Ives JD, Barry RG (eds) Arctic and alpine environments. Methuen, London, pp 159–194

Ivory DA, Whiteman PC (1978) Effects of environmental and plant factors on foliar freezing resistance in tropical grasses. II. Comparison of frost resistance between cultivars of Cenchrus ciliaris, Chloris gayana and Setaria anceps. Aust J Agric Res 29:261–266

Iwaya-Inoue M, Kaku S (1983) Cold hardiness in various organs and tissues of Rhododendron species and the supercooling ability of flower buds as the most susceptible organ. Cryobiology 20: 310–317

Jenkins G (1969) Transgressive segregation for frost resistance in hexaploid oats (Avena ssp.). J Agric Sci (Cambridge) 73:477–482

Jeremias K (1964) Über die jahresperiodisch bedingten Veränderungen der Ablagerungsform der Kohlenhydrate in vegetativen Pflanzenteilen. Fischer, Jena. Bot Stud 15:1–96

Jian LC, Sun LH, Dong HZ, Sun DL (1982) Changes in adenosine triphosphate activity associated with membranes and organelles during freezing stress and cold hardening. In: Li PH, Sakai A (eds) Plant cold hardiness and freezing stress, vol II. Academic Press, London New York, pp 243–259

Johnson JR, Havis JR (1977) Photoperiod and temperature effects on root cold acclimation. J Am Soc Hortic Sci 102:306–308

Johnson-Flanagan AM, Singh J (1986) Induction of freezing tolerance in suspension cultures of *Brassica napus* by ABA at 25 °C: Studies on protein and RNA changes. 3rd Plant Cold Hardiness Seminar, Abstr 38, Shanghai

Johnston WJ, Dickens R (1976) Centipedegrass cold tolerance as affected by environmental factors. Agron J 68:83–85

Jones HG (1983) Plant and microclimate. Cambridge Univ Press

Kacperska A (1985) Biochemical and physiological aspects of frost hardening in herbaceous plants. In: Kaurin Å, Juntila O, Nilsen J (eds) Plant production in the North. Norwegian Univ Press, Tromsø, pp 99–115

Kacperska-Palacz A (1978) Mechanism of cold acclimation in herbaceous plants. In: Li PH, Sakai A (eds) Plant cold hardiness and freezing stress, vol I. Academic Press, London New York, pp 139–152

Kacperska-Palacz A, Wcislinska B (1972) Electrophoretic pattern of soluble proteins in the rape leaves in relation to frost hardiness. Physiol Vég 10:19–25

Kacperska-Palacz A, Dlugokecka E, Breitenwald J, Wcislinska B (1977a) Physiological mechanisms of frost tolerance: possible role of protein in plant adaptation to cold. Biol Plant 10:10–17

Kacperska-Palacz A, Jasinska M, Sobczyk EA, Wcislinska B (1977b) Physiological mechanisms of frost tolerance: subcellular localization and some physical-chemical properties of protein fractions accumulated under cold treatment. Biol Plant 19:18–26

Kacperska-Palacz A, Zielinski K, Lewak St (1980) Frost tolerance of apple embryos as related to their dormancy. Physiol Veg 18:325–329

Kainmüller Ch (1974) Die Temperaturresistenz von Hochgebirgspflanzen. Diss Univ Innsbruck

Kainmüller Ch (1975) Temperaturresistenz von Hochgebirgspflanzen. Anz Math.-Naturwiss Kl Österr Akad Wiss 1975:67–75

Kaku S (1964) Undercooling points and frost resistance in mature and immature leaf tissue of some evergreen plants. Bot Mag (Tokyo) 77:283–289

Kaku S (1973) High ice nucleating ability in plant leaves. Plant Cell Physiol 14:1035–1038

Kaku S (1975) Analysis of freezing temperature distribution in plants. Cryobiology 12:154–159

Kaku S, Iwaya M (1978) Low temperature exotherms in xylems of evergreen and deciduous broad-leaved trees in Japan with references to freezing resistance and distribution range. In: Li PH, Sakai A (eds) Plant cold hardiness and freezing stress, vol I. Academic Press, London New York, pp 227–239

Kaku S, Iwaya M (1979) Deep supercooling in xylems and ecological distribution in the genera Ilex, Viburnum and Quercus in Japan. Oikos 33:402–411
Kaku S, Iwaya M, Kunishige M (1980) Supercooling ability of Rhododendron flower buds in relation to cooling rate and cold hardiness. Plant Cell Physiol 21:1205–1216
Kaku S, Iwaya-Inoue M, Gusta LV (1985) Estimation of the freezing injury in flower buds of evergreen azaleas by water proton nuclear magnetic resonance relaxation times. Plant Cell Physiol 26:1019–1025
Kappen L (1964) Untersuchungen über den Jahresverlauf der Frost-, Hitze- und Austrocknungsresistenz von Sporophyten einheimischer Polypodiaceen. Flora (Jena) 155:124–166
Kappen L (1965) Untersuchungen über die Widerstandsfähigkeit der Gametophyten einheimischer Polypodiaceen gegenüber Frost, Hitze und Trockenheit. Flora (Jena) 156:101–116
Kappen L (1966) Der Einfluß des Wassergehaltes auf die Widerstandsfähigkeit von Pflanzen gegenüber hohen und tiefen Temperaturen, untersucht an Blättern einiger Farne und Ramonda myconi. Flora (Jena) 156:427–445
Kappen L (1985) Vegetation and ecology of ice-free areas of northern Victoria Land, Antarctica. 2. Ecological conditions in typical microhabitats of lichens at Birthday Ridge. Polar Biol 4: 227–236
Kappen L, Lange OL (1970) The cold resistance of phycobionts from macrolichens of various habitats. Lichenologist 4:289–293
Kappen L, Lange OL (1972) Die Kälteresistenz einiger Makrolichenen. Flora (Jena) 161:1–29
Kappen L, Friedmann EI, Garty J (1981) Ecophysiology of lichens in the dry valleys of southern Victoria on the cryptoendolithic lichen habitat. Flora (Jena) 171:216–235
Kärcher H (1931) Über Kälteresistenz einiger Pilze und Algen. Planta 14:515–516
Karnatz H (1956) Untersuchungen über die Frostresistenz der Obstgehölze im Baumschulstadium. III. Über die relative Frosthärte unveredelter Kernobstunterlagen. Züchter 26:307–315
Karow AM, Webb WR (1965) A theory for injury and survival. Cryobiology 2:99–108
Kaufmann MR (1975) Leaf water stress in Engelmann spruce. Influence of the root and shoot environments. Plant Physiol 56:841–844
Kaurin Å, Stushnoff C, Junttila O (1982) Vegetative growth and frost hardiness of cloudberry (Rubus chamaemorus) as affected by temperature and photoperiod. Physiol Plant 55:76–81
Keefe PD, Moore KG (1981) Freeze desiccation: a second mechanism for the survival of hydrated lettuce (Lactuca sativa L.) seed at sub-zero temperatures. Ann Bot (London) 47:635–645
Kemmer E, Schulz F (1955) Das Frostproblem im Obstbau. Bayerischer Landwirtschaftsverlag, München
Kemmer E, Thiele I (1955) Frostresistenzprüfungen an keimenden Kernobstsamen. Züchter 25: 57–60
Ketchie DO, Beeman CH (1973) Cold acclimation in 'Red Delicious' apple trees under natural conditions during four winters. J Am Soc Hortic Sci 98:257–261
Kimball SL, Salisbury FB (1974) Plant development under snow. Bot Gaz 135:147–149
Kincaid DT, Lyons EE (1981) Winter water relations of red spruce on Mt. Monadnock, New Hampshire. Ecology 62:1155–1161
Kitaura K (1967a) Supercooling and ice formation in mulberry trees. In: Asahina E (ed) Cellular injury and resistance in freezing organisms. Inst Low Temp Sci, Hokkaido Univ, Sapporo, pp 143–156
Kitaura K (1967b) Freezing and injury of mulberry trees by late spring frost. Bull Sericult Exp Stn (Tokyo) 22:202–323
Kleemann W, McConnel HH (1976) Interactions of proteins and cholesterol with lipids in bilayer membranes. Biochim Biophys Acta 419:206–222
Klimov SV, Džanumov DA, Bočarov EA (1981) Mechanism povyšenija morozoustoičivosti i zimostoikosti pri cholodovom zakalivanii rastenii (Mechanism of the increase in frost and winter resistance in cold hardening of plants). Fiziol Rast 28:1230–1238
Klosson RJ, Krause GH (1981) Freezing injury in cold-acclimated and unhardened spinach leaves. II. Effects of freezing in chlorophyll fluorescence and light scattering reactions. Planta 151: 347–352

Knutson RM (1974) Heat production and temperature regulation in Eastern scunk cabbage. Science 186:746–747

Koch HD (1972) Genetic variability of frost hardiness in winter barley and some remarks on ecological aspects. In: Rajki S (ed) Proceedings of a colloquium on the winter hardiness of cereals. Agric Res Inst Hung Acad Sci, Martonvásár, pp 125–142

Koch KE, Kennedy RA (1980) Effects of seasonal changes in the Midwest on Crassulacean Acid Metabolism (CAM) in Opuntia humifusa Raf. Oecologia (Berlin) 45:390–395

Koga S, Echigo A, Nunomura K (1966) Physical properties of cell water in partially dried Saccharomyces cerevisiae. Biophys J 6:665–674

Kohn H (1959) Experimenteller Beitrag zur Kenntnis der Frostresistenz von Rinde, Winterknospen und Blüten verschiedener Apfelsorten. Gartenbauwissenschaft 24:315–329

Kojima K, Kobayashi D, Kobayashi S, Akitaya E, Narita H (1970) Report of pit-wall observations of snow cover in Sapporo, 1968–1969. Low Temp Sci Ser A 38, Data Rep:25–36

Konakahara M (1975) Experimental studies on the mechanisms of cold damage and its protection methods in Citrus trees. Shizuoka Pref Citrus Exp Stn Spec Bull 3:1–164

Konda K, Sasaki J (1959) Togai and Sougai. Hoppo Ringyo, Sapporo

Körner C, Allison A, Hilscher H (1983) Altitudinal variation of leaf diffusive conductance and leaf anatomy in heliophytes of montane New Guinea and their interrelation with microclimate. Flora (Jena) 174:91–135

Kozlowski TT (1971) Growth and development of trees, vol II: Cambial growth, root growth, and reproductive growth. Academic Press, London New York

Kozlowski TT (1976) Water supply and leaf shedding. In: Kozlowski TT (ed) Water deficits and plant growth, vol IV. Academic Press, London New York, pp 191–231

Kramer PJ (1937) Photoperiodic stimulation of growth by artificial light as a cause of winter killing. Plant Physiol 12:881–883

Kramer PJ (1942) Species differences with respect to water absorption at low soil temperatures. Am J Bot 29:828–832

Kramer PJ (1980) Drought, stress, and the origin of adaptations. In: Turner NC, Kramer PJ (eds) Adaptation of plants to water and high temperature stress. Wiley & Son, New York, pp 7–20

Kramer PJ, Kozlowski TT (1979) Physiology of woody plants. Academic Press, London New York

Kramer JP, Wetmore TH (1943) Effects of defoliation on cold resistance and diameter growth of broad leaved evergreens. Am J Bot 30:428–431

Krasavtsev OA (1960) Zakalivanie drevesnykh rastenii k morozu (Frost hardiness of woody plants). Tr Konf Fiziol Ustoichivost Rast Nauka, Moscow, pp 229–234

Krasavtsev OA (1967) The autofluorescence of cell of some northern forest plants with regard to their frost resistance. In: Troshim AS (ed) The cell and environmental temperature. Pergamon, Oxford New York, pp 35–43

Krasavtsev OA (1972) Kalorimetriya rastenii pri temperaturakh nizne nulya. Nauka, Moscow

Krasavtsev OA, Khvalin NM (1978) Ob osobennostiakh morozostoikosti i vymerzaniya parenkhymikh kletok drevesny yabloni. (Frost resistance and frost killing of apple wood parenchyma cells). Fiziol Rast 25:5–11

Krasavtsev OA, Khvalin NM (1982) Nezamerzayushchaya voda v zakalennykh zachatochnykh pobegakh ozimykh zlakov (Nonfreezing water in hardened primordial shoots of winter cereals). Fiziol Rast 29:437–446

Krasavtsev OA, Tutkevich GI (1970) Elektronnomikroskopicheskoe issledovanie zamerzaniya i vymerzaniya drevesnykh rastenii. (Microscopic and electron-microscopic investigations of freezing and frost death of woody plants). Fiziol Rast 17:385–393

Krasavtsev OA, Tutkevich GI (1971) Izmeneniya submikroskopicheskoi struktury kletok morozostoikykh rastenii vo vremya ottaivaniya. (Changes in the submicroscopical cell structure of frosthardy woody plants during thawing). Tsitologiya 13:1443–1447

Krause GH, Klosson RJ, Tröster V (1982) On the mechanism of freezing injury and cold acclimattion of spinach leaves. In: Li PH, Sakai A (eds) Plant cold hardiness and freezing stress, vol II. Academic Press, London New York, pp 55–76

Kretschmer G, Berger B (1966) Zur Torsomethode: Die Nachwuchslängen als Indikator für Frostschäden und Frostresistenz. Züchter 36:328–340

Kreutz W (1942) Das Eindringen des Frostes in den Boden unter gleichen und verschiedenen Witterungsbedingungen während des sehr kalten Winters 1939/40. Reichsamt Wetterdienst, Wiss. Abh (2) Berlin (cited in Franz 1979)

Krog JO, Zachariassen KE, Larsen B, Smidsrφd O (1979) Thermal buffering in afro-alpine plants due to nucleating agent-induced water freezing. Nature (London) 282:300

Kuhn MH, Riordan AJ, Wagner IA (1973) The climate of Plateau Station. In: Weller G, Bowling SM (eds) Climate of the arctis. Geophys Inst Univ Alaska, Fairbanks, pp 255–267

Kuntz ID, Brassfield TS, Law GD, Purcell GV (1969) Hydration of macromolecules. Science 163: 1329–1331

Kurahashi A, Hamaya T (1981) Variation of morphological characters and growth response of Saghalien fir (Abies sachalinensis) in different altitudes. Bull Tokyo Univ For 71:101–151

Kuwashima N, Singh S, Wildman SG (1971) Reversible cold inactivation and heat reactivation of RuDP carboxylase activity of crystallized tobacco fraction I proteins. Biochem Biophys Res Commun 42:664–668

Lakon G (1942) Topographischer Nachweis der Keimfähigkeit der Getreidefrüchte durch Tetrazoliumsalze. Ber Dtsch Bot Ges 60:299–305

Lalk I, Dörffling K (1985) Hardening, abscisic acid, proline and freezing resistance in two winter wheat varieties. Physiol Plant 63:287–292

Lambers H (1982) Cyanide-resistant respiration: A non-phosphorylatic electron transport pathway acting as an energy overflow. Physiol Plant 55:478–485

Lang GA, Early JD, Arroyave NJ, Darnell RL, Martin GC, Stutte GW (1985) Dormancy: Toward a reduced, universal terminology. Hort Science 20:809–812

Lange OL (1961) Die Hitzeresistenz einheimischer immer- und wintergrüner Pflanzen im Jahreslauf. Planta 56:666–683

Langlet O (1937) Studier över tallens fysiologiska variabilitet och dess samband med klimatet. Medd Stat Skogsförs Anst 29:421–470

Lapedes DN (ed) (1978) McGraw-Hill dictionary of scientific and technical terms, 2nd edn. McGraw-Hill, New York

Lapins K (1961) Cold hardiness of young apple trees originating from the juvenile and adult zones of seedlings. Can J Plant Sci 42:521–526

Lapins K (1962) Artificial freezing as a routine test of cold hardiness of young apple seedlings. Proc Am Soc Hortic Sci 81:26–34

Lapins K (1965) Cold hardiness of sweet cherries as determined by artifical freezing tests. Can J Plant Sci 45:429–435

Larcher W (1953) Schnellmethode zur Unterscheidung lebender von toten Zellen mit Hilfe der Eigenfluoreszenz pflanzlicher Zellsäfte. Mikroskopie 8:299–302

Larcher W (1954) Die Kälteresistenz mediterraner Immergrüner und ihre Beeinflußbarkeit. Planta 44:607–638

Larcher W (1957) Frosttrocknis an der Waldgrenze und in der alpinen Zwergstrauchheide auf dem Patscherkofel bei Innsbruck. Ferdinandeum, Innsbruck 37:49–81

Larcher W (1961) Jahresgang des Assimilations- und Respirationsvermögens von Olea europaea L. ssp. sativa Hoff. et Link., Quercus ilex L. und Quercus pubescens Willd. aus dem nördlichen Gardaseegebiet. Planta 56:575–606

Larcher W (1963a) Zur Frage des Zusammenhanges zwischen Austrocknungsresistenz und Frosthärte bei Immergrünen. Protoplasma 57:569–587

Larcher W (1963b) Zur spätwinterlichen Erschwerung der Wasserbilanz von Holzpflanzen an der Waldgrenze. Ber Naturwiss Med Ver (Innsbruck) 53:125–137

Larcher W (1964) Winterfrostschäden in den Parks und Gärten von Arco und Riva am Gardasee. Ferdinandeum, Innsbruck 43:153–199

Larcher W (1968) Die Temperaturresistenz als Konstitutionsmerkmal der Pflanzen. Dtsch Akad Landwirtschaftswiss Tagungsber 100:7–21

Larcher W (1969a) Zunahme des Frostabhärtungsvermögens von Quercus ilex im Laufe der Individualentwicklung. Planta 88:130–135

Larcher W (1969b) Anwendung und Zuverlässigkeit der Tetrazoliummethode zur Feststellung von Schäden in pflanzlichen Geweben. Mikroskopie 25:207–218

Larcher W (1970) Kälteresistenz und Überwinterungsvermögen mediterraner Holzpflanzen. Oecol Plant 5:267–286
Larcher W (1971) Die Kälteresistenz von Obstbäumen und Ziergehölzen subtropischer Herkunft. Oecol Plant 6:1–14
Larcher W (1972) Der Wasserhaushalt immergrüner Pflanzen im Winter. Ber Dtsch Bot Ges 85: 315–327
Larcher W (1973) Gradual progress of damage due to temperature stress. Temperature resistance and survival. In: Precht H, Christophersen J, Hensel H, Larcher W (eds) Temperature and life. Springer, Berlin Heidelberg New York, pp 194–231
Larcher W (1975) Pflanzenökologische Beobachtungen in der Páramostufe der venezolanischen Anden. Anz Math.-Naturwiss Kl Österr Akad Wiss 11:1–20
Larcher W (1977) Ergebnisse des IBP-Projekts „Zwergstrauchheide Patscherkofel". Sitzungsber Österr Akad Wiss Math.-Naturwiss Kl Abt I 186:301–371
Larcher W (1978) Climate and plant life of Arco. Azienda Aut Cura Arco
Larcher W (1980a) Klimastreß im Gebirge – Adaptationstraining und Selektionsfilter für Pflanzen. Vortr Rheinisch-Westf Akad Wiss 291:49–88. Westdtsch Verlag, Leverkusen
Larcher W (1980b) Untersuchungen zur Frostresistenz von Palmen. Anz Österr Akad Wiss Math.-Naturwiss Kl 3:37–49
Larcher W (1980c) La posizione delle piante sempreverdi mediterranee nella evoluzione della resistenza al freddo. Atti Ist Veneto Sci Lett Arti 138:103–111
Larcher W (1981a) Effects of low temperature stress and frost injury on plant productivity. In: Johnson CD (ed) Physiological processes limiting plant productivity. Butterworths, London, pp 253–269
Larcher W (1981b) Resistenzphysiologische Grundlagen der evolutiven Kälteakklimatisation von Sproßpflanzen. Plant System Evol 137:145–180
Larcher W (1981c) Low temperature effects on mediterranean sclerophylls: An unconventional viewpoint. In: Margaris NS, Mooney HA (eds) Components of productivity of mediterranean region, basic and applied aspects. Junk, Den Haag, pp 259–266
Larcher W (1982) Typology of freezing phenomena among vascular plants and evolutionary trends in frost acclimation. In: Li PH, Sakai A (eds) Plant cold hardiness and freezing stress, vol II. Academic Press, London New York, pp 417–426
Larcher W (1983a) Physiological plant ecology, 2nd edn. Springer, Berlin Heidelberg New York
Larcher W (1983b) Ökophysiologische Konstitutionseigenschaften von Gebirgspflanzen. Ber Dtsch Bot Ges 96:73–85
Larcher W (1985a) Kälte und Frost. In: Sorauer P (found) Handbuch der Pflanzenkrankheiten, vol I, 7th edn. Parey, Berlin, pp 107–326
Larcher W (1985b) Winter stress in high mountains. In: Turner H, Tranquillini W (eds) Establishment and tending of subalpine forest: research and management. Eidgen Anst Forstl Versuchsw Ber 270:11–19
Larcher W (1987a) Kälteresistenz. In: Kreeb KH (ed) Methoden der Pflanzenökologie, 2nd edn. Fischer, Jena
Larcher W (1987b) Vitalitätsbestimmung. In: Kreeb KH (ed) Methoden der Pflanzenökologie, 2nd edn. Fischer, Jena
Larcher W (1987c) Streß bei Pflanzen. Naturwiss (in press)
Larcher W, Bauer H (1981) Ecological significance of resistance to low temperature. In: Lange OL, Nobel PS, Osmond CB, Ziegler H (eds) Physiological plant ecology I. Encyclopedia of plant physiology, vol 12A. Springer, Berlin Heidelberg New York, pp 403–437
Larcher W, Eggarter H (1960) Anwendung des Tripheyltetrazoliumchlorids zur Beurteilung von Frostschäden in verschiedenen Achsengeweben bei Pirus-Arten, und Jahresgang der Resistenz. Protoplasma 51:595–619
Larcher W, Mair B (1968) Das Kälteresistenzverhalten von Quercus pubescens, Ostrya carpinifolia und Fraxinus ornus auf drei thermisch unterschiedlichen Standorten. Oecol Plant 3:255–270
Larcher W, Mair B (1969) Die Temperaturresistenz als ökophysiologisches Konstitutionsmerkmal. I: Quercus ilex und andere Eichenarten des Mittelmeergebietes. Oecol Plant 4:347–376

Larcher W, Nagele M (1985) Induktionskinetik der Chlorophyllfluoreszenz unterkühlter und gefrorener Blätter von Rhododendron ferrugineum beim Übergang vom gefrierempfindlichen zum gefriertoleranten Zustand. Sitzungsber Österr Akad Wiss Math.-Naturwiss Kl Abt I 194:187–195
Larcher W, Siegwolf R (1985) Development of acute frost drought in Rhododendron ferrungineum at the alpine timberline. Oecologia (Berlin) 67:298–300
Larcher W, Wagner J (1976) Temperaturgrenzen der CO_2-Aufnahme und Temperaturresistenz der Blätter von Gebirgspflanzen im vegetationsaktiven Zustand. Oecol Plant 11:361–374
Larcher W, Wagner J (1983) Ökologischer Zeigerwert und physiologische Konstitution von Sempervivum montanum. Verh Ges Ökol 11:253–264
Larcher W, Winter A (1981) Frost susceptibility of palms: Experimental data and their interpretation. Principes 25:143–152
Larsen A (1978a) Freezing tolerance in grasses. Effect of different water contents in growth media. Medd Norg Landbrukshφgsk 57(15):1–19
Larsen A (1978b) Freezing tolerance in grasses. Methods for testing in controlled environments. Medd Norg Landbrukshφgsk 57(23):1–56
Larsen A (1979) Freezing tolerance in grasses. Variation within populations and response to selection. Medd Norg Landbrukshφgsk 58:1–28
Larsen JA (1974) Ecology of the northern continental forest border. In: Ives JD, Barry RG (eds) Arctic and alpine environments. Methuen, London, pp 341–369
Larsen JB (1978a) Die Frostresistenz der Douglasie (Pseudotsuga menziesii (Mirb.) Franco) verschiedener Herkünfte mit unterschiedlicher Höhenlage. Silv Genet 27:150–156
Larsen JB (1978b) Die Klimaresistenz der Abies grandis (Dougl.) Lindl. I: Die Frostresistenz von 23 Herkünften aus dem IUFRO-Provenienzversuch von 1974. Silv Genet 27:156–161
Larsen JB (1981) Geographic variation in winter drought resistance of Douglas-fir (Pseudotsuga menziesii Mirb. Franco). Silv Genet 30:109–114
Larsen JB, Ruetz WF (1980) Frostresistenz verschiedener Herkünfte der Douglasie (Pseudotsuga menziesii) und der Küstentanne (Abies grandis) entlang des 44. Breitengrades in Mittel Oregon. Forstw Centralbl 99:222–233
Lauscher A, Lauscher F (1980) Vom Schneeklima der Ostalpen. Jahresber Sonnblick Verein 1978 bis 1980:15–23
Lauscher F (1946) Langjährige Durchschnittswerte für Frost und Frostwechsel in Österreich. Jahrb Zentralanst Meteorol Geodyn (Wien) (Anhang 4)
Lauscher F (1960) Lufttemperatur. In: Steinhauser F (ed) Klimatographie von Österreich. Denkschr Österr Akad Wiss 3(2):138–206
Lauscher F (1976) Methoden zur Weltklimatologie der Hydrometeore. Der Anteil des festen Niederschlags am Gesamtniederschlag. Arch Meteorol Geophys Bioklimatol Ser B 24:129–176
Lauscher F (1976/77) Ergebnisse der Beobachtungen an den nordchilenischen Hochgebirgsstationen Collahuasi and Chuquicamata. 74–75. Jahresber Sonnblick Verein 1976–1977
Lauscher F (1985) Beiträge zur Wetterchronik seit dem Mittelalter. Sitzungsber Österr Akad Wiss Math-Naturwiss Kl Abt II 194:93–131
Lavagne A, Muotte P (1971) Premiéres observations chorologiques et phénologiques sur les ripisilves á Nerium oleander (Neriaies) en Provence. Ann Univ Provence, Marseille 15:135–155
Levitt J (1956) The hardiness of plants. Academic Press, London New York
Levitt J (1958) Frost, drought, and heat resistance. Protoplasmatologia VIII/6. Springer, Berlin Wien
Levitt J (1969) Growth and survival of plant at extremes of temperature. A unified concept. Symp Soc Exp Biol 23:395–448
Levitt J (1980) Responses of plants to environmental stresses. Vol I. Chilling, freezing, and high temperature stresses, 2nd edn. Academic Press, London New York
Lewis MC, Callaghan TV (1976) Tundra. In: Monteith JL (ed) Vegetation and the atmosphere, vol II. Academic Press, London New York, pp 399–433
Li PH (1984) Subzero temperature stress physiology of herbaceous plants. Hortic Rev 6:373–417
Li PH, Palta JP (1978) Frost hardening and freezing stress in tuberbearing Solanum species. In: Li PH, Sakai A (eds) Plant cold hardiness and freezing stress, vol I. Academic Press, London New York, pp 49–71

Li PH, Weiser CJ (1969) Metabolism of nucleic acids in one-year old apple twig during cold hardening and dehardening. Plant Cell Physiol 10:21–30

Li PH, Palta JP, Hawkes JG (1980) A science note: Interrelationship between frost-hardiness and elevation of genotype origin. Plant Physiol 66:414–421

Limin AE, Fowler DB (1982) The expression of cold hardiness in Triticum species amphiploids. Can J Genet Cytol 24:51–56

Lindegren RM (1933) Decay of wood and growth of some Hymenomycetes as affected by temperature. Phytopathology 23:73–81

Lindow SE (1982) Population dynamics of epiphytic ice nucleation active bacteria on frost sensitive plants and frost control by means of antagonistic bacteria. In: Li PH, Sakai A (eds) Plant cold hardiness and freezing stress, vol II. Academic Press, London New York, pp 395–416

Lindow SE (1983) The role of bacterial ice nucleation in frost injury to plants. Annu Rev Phytopathol 21:363–384

Lindow SE, Arny DC, Upper CD, Barchet WR (1978) The role of bacterial ice nuclei in frost injury to sensitive plants. In: Li PH, Sakai A (eds) Plant cold hardiness and freezing stress, vol I. Academic Press, London New York, pp 249–263

Lindstrom OM, Carter JV (1985) Injury to potato leaves exposed to subzero temperatures in the absence of freezing. Planta 164:512–516

Lipman CB (1936) The tolerance of liquid air temperatures by dry moss protonema. Bull Torrey Bot Club 63:515–518

Lockwood JG (1974) World climatology: An environmental approach. Arnold, London

Loewel EL, Karnatz H (1956) Untersuchungen über die Frostresistenz der Obstgehölze im Baumschulstadium. I. Problemstellung und Versuchsmethodik. Züchter 26:117–120

Lösch R, Kappen L (1981) The cold resistance of Macaronesian Sempervivoideae. Oecologia (Berlin) 50:98–102

Lösch R, Kappen L, Wolf A (1983) Productivity and temperature biology of two snowbed bryophytes. Polar Biol 1:243–248

Lozina-Lozinskii LK (1974) Studies in cryobiology. Wiley & Son, New York

Lucas JW (1954) Subcooling and ice nucleation in lemons. Plant Physiol 29:245–251

Ludlow MM (1980) Stress physiology of tropical pasture plants. Trop Grassl 14:136–145

Lundquist V, Pellett H (1976) Preliminary survey of cold hardiness levels of several bulbous ornamental plant species. HortSci 11:161–162

Lüning K (1985) Meeresbotanik. Thieme, Stuttgart

Luyet BJ (1937) The vitrification of organ colloids and of protoplasm. Biodynamica 29:1–15

Luyet BJ (1967) On the possible biological significance of some physical changes encountered in the cooling and the rewarming of aqueous solutions. In: Asahina E (ed) Cellular injury and resistance in freezing organisms. Inst Low Temp Sci, Hokkaido Univ, Sapporo, pp 1–20

Luyet BJ, Gehenio PM (1938) The survival of moss vitrified in liquid air and its relation to water content. Biodynamica 42:1–7

Luyet BJ, Rapatz G (1958) Patterns of ice formation in some aqueous solutions. Biodynamica 8: 1–68

Luyet BJ, Rasmussen D (1968) Study by differential thermal analysis of the temperatures of instability of rapidly cooled solutions of glycerol, ethylene glycol, sucrose and glucose. Biodynamica 10:167–191

Luyet BJ, Thoennes G (1938) The survival of plant cells immersed in liquid air. Science 23:284–285

Lydolph PE (1977) Climates of the Soviet Union. In: Landsberg HE (ed) World survey of climatology, vol 7. Amsterdam, Elsevier, pp 1–443

Lyons JM (1973) Chilling injury in plants. Annu Rev Plant Physiol 24:445–466

Lyons JM, Graham D, Raison JK (eds) (1979) Low temperature stress in crop plants. Academic Press, London New York

Lyons LM, Raison JK (1970) Oxidative activity of mitochondria isolated from plant tissues sensitive and resistant to chilling injury. Plant Physiol 45:386–389

MacHattie LB (1963) Winter injury of lodgepole pine foliage. Weather 19:301–307

Mägdefrau K (1968) Paläobiologie der Pflanzen, 4th edn. Fischer, Stuttgart

Mai DH (1981) Entwicklung und klimatische Differenzierung der Laubwaldflora Mitteleuropas im Tertiär. Flora (Jena) 171:525–582

Mair B (1967) Frostresistenz der Achsen und Knospen submediterraner und mediterraner Pflanzen. Diss, Univ Innsbruck

Mair B (1968) Frosthärtegradienten entlang der Knospenfolge auf Eschentrieben. Planta 82:164–169

Maki LR, Galyan ME, Chien ME, Caldwell DR (1974) Ice nucleation induced by Pseudomonas syringae. Appl Microbiol 28:456–459

Malek L, Bewley JD (1978) Protein synthesis related to cold temperature stress in the desiccation-tolerant moss Tortula ruralis. Physiol Plant 43:313–319

Manis RE, Knight RJ (1967) Avocado germplasm evaluation: Technique used in screening for cold tolerance. Proc Fl State Hortic Soc 80:387–391

Maotani T, Machida Y (1967) Studies on leaf water stress in fruit trees. V: Seasonal changes in leaf water potential and leaf diffusion resistance of Satsuma mandarin trees. J Jpn Soc Hortic Sci 45:261–266

Marcellos H, Single WV (1975) Temperatures in wheat during radiation frost. Aust J Exp Agric Anim Husb 15:818–822

Marcellos H, Single WV (1979) Supercooling and heterogeneous nucleation of freezing in tissues of tender plants. Cryobiology 16:74–77

Marchand PJ, Chabot BF (1978) Winter water relations of tree-line plant species on Mt. Washington, New Hampshire. Arctic Alpine Res 10:105–116

Marini HP, Boyce BR (1977) Susceptibility of crown tissues of 'Catskill' strawberry plants to low-temperature injury. J Am Soc Hortic Sci 102:515–516

Markley JL, McMillan C, Thompson GA (1982) Latitudinal differentiation in response to chilling temperatures among populations of three mangroves, Avicennia germinans, Laguncularia racemosa, and Rhizophora mangle, from the western tropical Atlantic and Pacific Panama. Can J Bot 60:2704–2715

Maronek DM, Flint ML (1974) Cold hardiness of needles of Pinus strobus L. as a function of geographic source. For Sci 20:135–141

Marro M, Deveronico M (1979) Danni occulti delle gelate tardive sul melo. Frutticoltura 41(12): 27–30

Marshall HG (1965) A technique of freezing plants crowns to determine the cold resistance of winter oats. Crop Sci 5:83–86

Marshall HG, Olien CR, Everson EH (1981) Techniques for selection of cold hardiness in cereals. In: Olien CR, Smith MN (eds) Analysis and improvement of plant cold hardiness. CRC Press, Boca Raton, pp 139–159

Martin B, Mårtensson O, Öquist G (1978) Effects of frost hardening and dehardening on photosynthetic electron transport and fluorescence properties in isolated chloroplasts of Pinus silvestris. Physiol Plant 43:297–305

Martsolf JD, Ritter CM, Hatch AH (1975) Effect of white latex paint on temperature of stone fruit tree trunks in winter. J Am Soc Hortic Sci 100:122–129

Maruta E (1983) Growth and survival of current-year seedlings of Polygonum cuspidatum at the upper distribution limit on Mt. Fuji. Oecologia (Berlin) 60:316–320

Mathys H (1974) Klimatische Aspekte zu der Frostverwitterung in der Hochgebirgsregion. Geogr Inst Univ Bern

Matile Ph (1968) Lysosomes of root tip cells in corn seedlings. Planta 79:181–196

Matile Ph (1975) The lytic compartment of plant cells. Springer, Berlin Heidelberg New York

Matile Ph, Moor H (1968) Vacuolation: Origin and development of lysosomal apparatus in root-tip cells. J Ultrastruct Res 5:193–200

Matile Ph, Winkenbach F (1971) Function of lysosomes and lysosomal enzymes in the senescing corolla of the morning glory (Ipomea purpurea). J Exp Bot 22:759–771

Matsuda T (1964) Microclimate in the community of mosses near Showa Base at East Ongul Island, Antarctica. Nankyoku Shiryo (Antarct Res) 21:12–24

Matsui T, Eguchi H, Mori K (1981) Control of dew and frost formations on leaf by radiative cooling. Environ Control Biol 19:51–57

Matuszkiewicz W (1977) Spät- und Frühfröste als standortsökologischer Faktor in den Waldgesellschaften des Bialowieza Nationalparks (Polen). In: Dierschke H (ed) Vegetation und Klima. Cramer, Vaduz, pp 195–233

Maximov NA (1914) Experimentelle und kritische Untersuchungen über das Gefrieren und Erfrieren der Pflanzen. Jahrb Wiss Bot 53:327–420

Mazur P (1963) Kinetics of water loss from cells at subzero temperatures and the likelihood of intracellular freezing. J Gen Physiol 47:347–469

Mazur P (1966) Physical and chemical basis of injury in single-celled microorganisms subjected to freezing and thawing. In: Meryman HT (ed) Cryobiology. Academic Press, London New York, pp 214–315

Mazur P (1977) The role of intracellular freezing in the death of cells cooled at supraoptimal rates. Cryobiology 14:251–272

Mazur P, Rall W, Rigopoulos N (1981) Relative contributions of the fraction of unfrozen water and of salt concentration to the survival of slowly frozen human erythrocytes. Biophys J 36: 653–675

McCurrach JC (1960) Palms of the world. Harper, New York

McMillan C (1975) Adaptive differentiation to chilling in mangrove populations. In: Walsh GE, Snedaker SC, Teas HJ (eds) International symposium for biological management of mangroves. Univ Florida Press, Gainesville, pp 62–68

McMillan JA, Los SC (1965) Vitreous ice: Irreversible transformations during warm-up. Nature (London) 206:806–807

Meeuse BJD (1975) Thermogenic respiration in aroids. Annu Rev Plant Physiol 26:117–126

Mellor M (1964) Snow and ice on the earth's surface. Cold regions sci. and energ. Monogr. II-C-1 US Army CRREL; cited in: Alford D (1974) Snow. In: Ives JD, Barry RG (eds) Arctic and alpine environments. Methuen, London, pp 85–110

Meryman HT (1968) Modified model for the mechanism of freezing injury in erythrocytes. Nature (London) 218:333–336

Meusel H (1965) Die Reliktvegetation der Kanarischen Inseln in ihren Beziehungen zur süd- und mitteleuropäischen Flora. In: Gersch M (ed) Vorträge über moderne Probleme der Abstammungslehre, vol II. Univ Jena Press, pp 117–136

Michaelis P (1934a) Ökologische Studien an der alpinen Baumgrenze. IV: Zur Kenntnis des winterlichen Wasserhaushaltes. Jahrb Wiss Bot 80:169–298

Michaelis P (1934b) Ökologische Studien an der alpinen Baumgrenze. V: Osmotischer Wert und Wassergehalt während des Winters in den verschiedenen Höhenlagen. Jahrb Wiss Bot 80:337–362

Miller JD (1976) Cold tolerance in sugar cane relatives. Sugar Azucar, vol 71, March 1976

Minckler LS (1951) Southern pines from different geographic sources show different responses to low temperatures. J For 49:915–916

Mittelstädt H (1969) Gefriervorgänge an Pflanzenteilen. DAL Tagungsber 96:149–173

Mittelstädt H, Müller-Stoll WR (1984) Veränderungen der Vakuolengröße in Rindenparenchymzellen von Gehölzen nach Frosteinwirkung. Flora (Jena) 175:231–242

Mittelstädt H, Murawski H (1975) Beiträge zur Züchtungsforschung beim Apfel. XVII: Untersuchungen zur Frostresistenz an Apfelsorten. Arch Züchtungsforsch 5:71–81

Modlibowska I (1956) Le problème des gelées printanières et la fruitière. Rapp Gen Congr Pomol Int, Namur 1956, pp 83–111

Modlibowska I (1961) Sur les mécanismes du gel et de la reprise d'une vie normale au retour a la température ordinaire. Bull Soc Fr Physiol Vég 7:123–134

Modlibowska I, Rogers WS (1955) Freezing of plant tissues under the microscope. J Exp Bot 6: 384–391

Mohn CA, Pauley S (1969) Early performance of cottonwood seed sources in Minnesota. Minn For Res Serv Notes 207:1–4

Molisch H (1897) Untersuchungen über das Erfrieren der Pflanzen. Fischer, Jena

Monange Y (1961) Le froid de l'hiver 1956 et l'anatomie du bois des quelques gymnospermes. Trav Lab For (Toulouse) 96:1–6

Mooney HA (1972) The carbon balance of plants. Annu Rev Ecol Syst 3:315–346

Mooney HA, Weisser PJ, Gulmon SL (1977) Environmental adaptations of the Atacaman desert cactus Copiapoa haseltoniana. Flora (Jena) 166:117–124

Moor H (1964) Die Gefrier-Fixation lebender Zellen und ihre Anwendung in der Elektronenmikroskopie. Z Zellforsch 62:546–580

Moor H, Mühlethaler K (1963) Fine structure in frozen-etched yeast cells. J Cell Biol 17:609–628
Moore HE Jr (1973) The major groups of palms and their distribution. Gent Herb 11(2):27–140
Moore RM, Williams JD (1976) A study of a subalpine woodland–grassland boundary. Aust J Ecol 1:145–153
Morettini A (1961) Sulla ricostituzione degli olivi danneggiati dalle basse temperature del 1956. Accad Econom Agr Georgof 8:1–40
Moser M (1958) Der Einfluß tiefer Temperaturen auf das Wachstum und die Lebenstätigkeit höherer Pilze mit spezieller Berücksichtigung von Mykorrhizapilzen. Sydowia 12:386–399
Moser W, Brzoska W, Zachhuber K, Larcher W (1977) Ergebnisse des IBP-Projekts „Hoher Nebelkogel 3184 m". Sitzungsber Österr Akad Wiss Math.-Naturwiss Kl Abt I 186:387–419
Müller-Thurgau H (1886) Über das Erfrieren der Pflanzen. I. Teil. Landwirtsch Jahrb 15:453–610
Münch E (1923) Die Knospenentfaltung der Fichte und die Spätfrostgefahr. Allg Frost Jagd Z 1923:241–265
Murawski H (1961) Beiträge zur Züchtungsforschung beim Apfel. VI: Untersuchungen über die Vererbung der Frostresistenz an Sämlingen der Sorten Glogierowka und Jonas Hannes. Züchter 31:52–57
Murawski H (1962) Probleme und Aussichten bei der Züchtung frostresistenter Obstsorten. Sitzungsber Dtsch Akad Landwirtsch Wiss (Berlin) 11:39–59
Murawski H (1968) 40 Jahre Obstzüchtung in Müncheberg. Arch Gartenbau 16:400–430
Muto J, Horiuchi T (1974) (Provenance difference in cold damage of Cryptomeria japonica). J Jpn For Soc 56:210–215
Nag KK, Street HE (1973) Carrot embryogenesis from frozen cultured cells. Nature (London) 245: 270–272
Nag KK, Street HE (1975) Freeze preservation of cultured plant cells. Physiol Plant 34:254–260
Nakagawa S, Sagisaka S (1984) Increase in enzyme activities related to ascorbate metabolism during cold acclimation in poplar twigs. Plant Cell Physiol 25:899–906
Nakayama A, Harada S (1973) Splitting injury to stems of young tea plants by artificial freezing treatment. Tea Res J 38:11–14
Nakhutsrishvili G, Gamtsemlidze ZG (1984) Zhizne rastenii v extremalnykh usloviyakh vysokogornii. Nauka, Leningrad
Nath J, Anderson JO (1975) Effect of freezing and freeze-drying on the viability and storage of Lilium longiflorum L. and Zea mays L. pollen. Cryobiology 12:81–88
Nei T, Okada J (1967) Designing of a new cryomicroscope. Low Temp Sci Ser B 25:149–153
Nelson RB, Davis DW, Palta JP, Laing DR (1983) Measurement of soil waterlogging tolerance in Phaseolus vulgaris L.: a comparison of screening techniques. Sci Hortic 20:303–313
Niki T, Sakai A (1981) Ultrastructural changes related to frost hardiness in the cortical parenchyma cells from mulberry twigs. Plant Cell Physiol 22:171–183
Niki T, Sakai A (1983) Effect of cycloheximide on the freezing tolerance and ultrastructure of cortical parenchyma cells from mulberry twigs. Can J Bot 6:2205–2221
Nilsen ET (1985) Seasonal and diurnal leaf movements of Rhododendron maximum L. in contrasting irradiance environments. Oecologia (Berlin) 65:296–302
Nilsson-Ehle H (1912) Zur Kenntnis der Erblichkeitsverhältnisse der Eigenschaft Winterfestigkeit beim Weizen. Z Pflanzenzücht 1:3–12
Nissila PC, Fuchigami LH (1978) The relationship between vegetation maturity and the first stage of cold acclimation. J Am Hortic Sci 103:710–711
Nitta R (1981) Some features of snow damage and control in Japanese forests. Proc XVII IUFRO World Congr, Div 1, pp 339–344
Nobel PS (1980) Morphology, surface temperatures, and northern limits of columnar cacti in the Sonoran desert. Ecology 61:1–7
Nobel PS (1982) Low-temperature tolerance and cold hardening of cacti. Ecology 63:1650–1656
Nobel PS (1984) Extreme temperatures and thermal tolerances for seedlings of desert succulents. Oecologia (Berlin) 62:310–317
Nobel PS (1985) Desert succulents. In: Chabot BF, Mooney HA (eds) Physiological Ecology of North American plant communities. Chapman & Hall, New York, pp 181–197
Nobel PS, Smith SD (1983) High and low temperature tolerances and their relationships to distribution of agaves. Plant Cell Environ 6:711–719

Nogami M, Koaze T, Fukuda M (1980) Perigalcial environment in Japan. Present and past. Geojournal 4:125–132
Noshiro M, Sakai A (1979) Freezing resistance of herbaceous plants. Low Temp Sci Ser B 37:11–18
Numata M (1979) The relationship of limiting factors to the distribution and growth of bamboo. In: Numata M (ed) Ecology of grasslands and bamboo lands in the world. Junk, Den Haag, pp 259–275
Oechel WC, Lawrence WT (1985) Taiga. In: Chabot BF, Mooney HA (eds) Physiological ecology of North American plant communities. Chapman & Hall, New York, pp 66–94
Ogolevets IV (1976) (Study of hardening of isolated callus tissue of trees with different frost resistance.) Fiziol Rast 23:139–145
Okada S, Mukaide H, Sakai A (1973) Genetic variation in Sakhalin fir from different areas of Hokkaido. Silv Genet 22:24–29
Okamura T (1973) Studies on the state of water in soybean seed in relation to moisture content. Bull Obihiro Zootech Univ 8:261–317
Okitsu S, Ito K (1983) Dynamic ecology of the Pinus pumila community of Mts. Taisetsu, Hokkaido, Japan. Environ Sci (Hokkaido) 6:151–184
Oldén EJ (1956) Obervationer bland stenfruktutträden vid Balsgård efter vintern, 1954–1955. Bålsgård Fruit Breed Inst 41–42:1–21
Oldén EJ (1957) Vinterhärdigheten hos plommon. Foereningen vaestfoeroedling frukt (Bålsgård) 43–45:17–39
Olien CR (1964) Freezing processes in the crown of 'Hudson' barley, Hordeum vulgare (L. emend. Lam.) Hudson. Crop Sci 4:91–95
Olien CR (1978) Analyses of freezing stresses and plant responses. In: Li PH, Sakai A (eds) Plant cold hardiness and freezing stresses, vol I. Academic Press, London New York, pp 317–348
Olien CR (1981) Analysis of midwinter freezing stress. In: Olien CR, Smith MN (eds) Analysis and improvement of plant cold hardiness. CRC Press, Boca Raton, pp 35–59
Olien CR, Marchetti BL (1976) Recovery of hardened barley from winter injuries. Crop Sci 16: 201–204
Olien CR, Smith MN (eds) (1981) Analysis and improvement of plant cold hardiness. CRC Press, Boca Raton
Omasa K, Hashimoto Y, Aiga I (1984) Image instrumentation of plants exposed to air pollutants. (4) Methods for automatic evaluation of the degree of necrotic and chlorotic visible injury. Res Rep Natl Inst Environ Stud, Ibaraki 66:99–104
O'Neil LC (1962) Some effects of artifical defoliation on the growth of Jack pine (Pinus banksiana Lamb.). Can J Bot 40:273–280
O'Neill SD, Priestley DA, Chabot BF (1981) Temperature and aging effects on leaf membranes of a cold hardy perennial, Fragaria virginiana. Plant Physiol 68:1409–1415
Ono S, Inuma M (1969) (Effects of snow quality and its depth on snow damage in heavy snow areas). Sourin 240:2–9
Oohata S (1979) Distribution of pine and its physiological characteristics. Hoppo Ringyo 31:377–382
Oohata S, Sakai A (1982) Freezing resistance and thermal indices of the genus Pinus with reference to their distributions. In: Li PH, Sakai A (eds) Plant cold hardiness and freezing stress, vol II. Academic Press, London New York, pp 437–446
Oohata S, Hasegawa Y, Sakai A (1981) Studies on freezing resistance and distribution of the genus Pinus. I. Seasonal changes in freezing resistance. Jpn J Ecol 31:79–89
Oppenheimer HR (1949) Frost effects on vegetation. Pal J Bot (Rehovot) 7:36–40
Oppenheimer HR (1963) Zur Kenntnis kritischer Wassersättigungsdefizite in Blättern und ihrer Bestimmung. Planta 60:51–69
Oppenheimer HR, Jacoby B (1961) Usefulness of autofluorescence tests as criterion of life in plant tissues. Protoplasma 53:220–226
Öquist G (1983) Effects of low temperature on photosynthesis. Plant Cell Environ 6:281–300
Öquist G, Martin B (1980) Inhibition of photosynthetic electron transport and formation of inactive chlorophyll in winter stressed Pinus silvestris. Physiol Plant 48:33–38
Orr W, Singh J, Brown DCW (1985) Induction of freezing tolerance in alfalfa suspension cultures. Plant Cell Rep 14:15–18

Orvig S (1970) Climates of the polar regions. In: Landsberg HE (ed) World survey of climatology, vol 14. Elsevier, Amsterdam New York
Otsuka K (1971) Survival of pollen cells at super-low temperatures. Low Temp Sci Ser B 29:107–111
Otsuka K (1972) The ultrastructure of mulberry cortical parenchyma cells related to the changes of the freezing resistance in spring. Low Temp Sci 30:13–44
Ouellet CE, Sherk LC (1968) Zones de rusticité pour les plantes au Canada. Ministère Agr Can, Ottawa, 1968
Palta JP, Li PH (1978) Cell membrane properties in relation to freezing injury. In: Li PH, Sakai A (eds) Plant cold hardiness and freezing stress, vol 1. Academic Press, New York, pp 93–115
Palta JP, Li PH (1979) Frost-hardiness in relation to leaf anatomy and natural distribution of several Solanum species. Crop Sci 19:665–671
Palta JP, Levitt J, Stadelmann EJ (1978) Plant viability assay. Cryobiology 15:249–255
Palta JP, Jensen KG, Li PH (1982) Cell membrane alterations following a slow freeze thaw cycle: Ion leakage, injury, and recovery. In: Li PH, Sakai A (eds) Plant cold hardiness and freezing stress, vol II. Academic Press, London New York, pp 221–242
Paquin R, Mehuys GR (1980) Influence of soil moisture on cold tolerance of alfalfa. Can J Plant Sci 60:139–147
Paquin R, Pelletier H (1980) Influence de l'environnement sur l'acclimatation au froid de la luzerne (Medicago media Pers.) et sa résistance au gel. Can J Plant Sci 60:1351–1366
Parker J (1953) Some applications and limitations of tetrazolium chloride. Science 118:77–79
Parker J (1959) Seasonal variations in sugars of conifers with some observations on cold resistance. Forest Sci 5:56–63
Parker J (1960a) Survival of woody plants at extremely low temperatures. Nature (London) 187: 1133
Parker J (1960b) Seasonal changes in cold-hardiness of Fucus vesiculosus. Biol Bull 119:474–478
Parker J (1962) Seasonal changes in cold resistance and free sugars of some hardwood tree barks. Forest Sci 8:255–262
Parker J (1963) Cold resistance in woody plants. Bot Rev 29:123–201
Paton DM (1972) Frost resistance in Eucalyptus: A new method for assessment of frost injury in altitudinal provenances of E. viminalis. Aust J Bot 20:127–139
Paton DM (1978) Eucalyptus physiology. I. Photoperiodic responses. Aust J Bot 26:633–642
Paton DM (1982) A mechanism for frost resistance in Eucalyptus. In: Li PH, Sakai A (eds) Cold hardiness and freezing stress, vol II. Academic Press, London New York, pp 77–92
Paton DM, Slattery HD, Willing RR (1979) Low root temperature delays dehardening of frost resistance Eucalyptus shoots. Ann Bot (London) 43:123–124
Patterson BD, Murata T, Graham D (1976) Electrolyte leakage induced by chilling in Passiflora species tolerant to different climates. Aust J Plant Physiol 3:435–442
Patterson BD, Paull R, Smillie RM (1978) Chilling resistance in Lycopersicon hirsutum Humb. & Bompl., a wild tomato with a wide altitudinal distribution. Aust J Plant Physiol 5:609–617
Pauley S, Perry TO (1954) Ecotypic variation of the photoperiodic responses in poplars. J Arnold Arbor 35:167–188
Pearce RS (1985) A freeze-fracture study of the cell membranes of wheat adapted to extracellular freezing and to growth at low temperatures. J Exp Bot 36:369–381
Pearce RS, Beckett A (1985) Water droplets in intercellular spaces of barley leaves examined by low temperature scanning electron microscopy. Planta 166:335–340
Pearce RS, Willison JHM (1985a) Wheat tissues freeze-etched during exposure to extracellular freezing: distribution of ice. Planta 163:295–303
Pearce RS, Willison JHM (1985b) A freeze-etch study of the effects of extracellular freezing on cellular membranes of wheat. Planta 163:304–316
Pellett H (1971) Comparison of cold hardiness levels of root and stem tissue. Can J Plant Sci 51: 193–195
Pellett H, Gearhart M, Dirr M (1981) Cold hardiness capability of woody ornamental plant taxa. J Am Soc Hortic Sci 106:239–243
Perry TO (1971) Dormancy of trees in winter. Science 171:29–36
Petty JA, Worrell R (1981) Stability of coniferous tree stems in relation to damage by snow. Forestry 54:115–128

Philippis de A (1937) Classificazioni ed indici del clima, un rapporto alla vegetazione forestale italiana. NG Bot Ital 44:1–169

Pierquet P, Stushnoff C, Burke MJ (1977) Low temperature exotherms in stem and bud tissues of Vitis riparia Michx. J Am Soc Hortic Sci 102:54–55

Pignatti S (1978) Evolutionary trends in mediterranean flora and vegetation. Vegetatio 37:175–185

Pisek A (1958) Versuche zur Frostresistenzprüfung von Rinde, Winterknospen und Blüten einiger Arten von Obstgehölzen. Gartenbauwissenschaft 23:54–74

Pisek A, Eggarter H (1959) Beobachtungen zur Überwindung von Frostschäden an Zweigen und Laubknospen von Apfel und Birne. Gartenbauwissenschaft 24:446–456

Pisek A, Larcher W (1954) Zusammenhang zwischen Austrocknungsresistenz und Frosthärte bei Immergrünen. Protoplasma 44:30–46

Pisek A, Schiessl R (1947) Die Temperaturbeeinflußbarkeit der Frosthärte von Nadelhölzern und Zwergsträuchern an der alpinen Waldgrenze. Ber Naturwiss Med Verein (Innsbruck) 47:33–52

Pisek A, Larcher W, Unterholzner R (1967) Kardinale Temperaturbereiche der Photosynthese und Grenztemperaturen des Lebens der Blätter verschiedener Spermatophyten. I: Temperaturminimum der Netto-Assimilation, Gefrier- und Frostschadensbereiche der Blätter. Flora (Jena) Abt B 157:239–264

Podbielkowska M, Kacperska-Palacz A (1971) Effects of phosfon-D and of low temperature on the morphology of cell protoplasts. Protoplasma 73:469–474

Pogosyan KS, Sakai A (1969) Freezing resistance in grape vines. Low Temp Sci Ser B 27:125–142

Pomeroy MK, Andrews CJ (1979) Metabolic and ultrastructural changes associated with flooding at low temperature in winter wheat and barley. Plant Physiol 64:635–639

Pomeroy MK, Siminovitch D (1971) Seasonal cytological changes in secondary phloem parenchyma cells in Robinia pseudoacacia in relation to cold hardiness. Can J Bot 49:787–795

Pomeroy MK, Pihakaski SJ, Andrews CJ (1983) Membrane properties of isolated winter wheat cells in relation to icing stress. Plant Physiol 72:535–539

Popovich S (1984) Frost crack injuries on some hybrid poplar trees planted on different sites in Quebec. Int Poplar Commiss FAO, 17th Sess, Ottawa

Prieur R, Cousin R (1978) Contribution a la mise au point d'une technique de sélection pour la résistance au froid des Pis d'hiver. Ann Amelior Plantes 28:157–163

Prillieux E (1869) Sur la formation de glacons à l'intérieur des plantes. Ann Sci Nat Bot 12:125–134

Proebsting EL (1959) Cold hardiness of "Elberta" peach fruit buds during winters. Proc Am Soc Hortic Sci 74:144–153

Proebsting EL (1963) The role of air temperatures and bud development in determining hardiness of dormant Elberta peach fruit buds. Am Soc Hortic Sci 83:259–269

Proebsting EL, Mills HH (1978) Low temperature resistance of developing flower buds of six deciduous fruit species. J Am Soc Hortic Sci 103:191–198

Proebsting EL, Sakai A (1979) Determining T_{50} of peach flower buds with exotherm analysis. HortSci 14:597–598

Proebsting EL, Ahmedullah M, Brummund VP (1980) Seasonal changes in low temperatures resistance of grape buds. Am J Enol Vitic 31:329–336

Pryor LD (1956) Variation in snow gum (Eucalyptus pauciflora Sieb.). Proc Linn Soc NSW 81: 299–311

Puth G, Lüttge U (1973) Sulfitwirkung auf die Membranpermeabilität von Pflanzenzellen: SO_3-Hemmung des n-Butanol-induzierten Betacyaninefflux aus dem Gewebe roter Rüben. Biochem Physiol Pflanzen 164:195–198

Quamme HA (1976) Relationship of the low temperature exotherm to apple and pear production in North America. Can J Plant Sci 56:493–500

Quamme HA (1978) Mechanism of supercooling in overwintering peach flower buds. J Am Soc Hortic Sci 103:57–61

Quamme H, Stushnoff C, Weiser CJ (1972) The relationship of exotherms to cold injury in apple stem tissues. J Am Soc Hortic Sci 97:608–613

Quamme H, Stushnoff C, Weiser CJ (1973) The mechanism of freezing injury in xylem of winter apple twigs. Plant Physiol 51:273–277

Quamme HA, Layne REC, Jackson HO, Spearman GA (1975) An improved exotherm method for measuring cold hardiness of peach flower buds. HortSci 10:521–523

Quinn PJ (1985) A lipid-phase separation model of low-temperature damage to biological membranes. Cryobiology 22:128–146

Rada F, Goldstein G, Azocar A, Meinzer F (1985a) Daily and seasonal osmotic changes in a tropical treeline species. J Exp Bot 36:989–1000

Rada F, Goldstein G, Azocar A, Meinzer F (1985b) Freezing avoidance in Andean giant rosette plants. Plant Cell Environ 8:501–507

Raese JI, Williams MV, Billingstey HD (1977) Sorbitol and other carbohydrates in dormant apple shoots as influenced by controlled temperatures. Cryobiology 14:373–378

Raison JK, Lyons JM, Thomson WW (1971a) The influence of membranes on the temperature-induced changes in the kinetics of some respiratory enzymes of mitochondria. Arch Biochem Biophys 142:83–90

Raison JK, Lyons JM, Keith AD (1971b) Temperature-induced phase changes in mitochondria membranes detected by spin labeling. J Biol Chem 246:4036–4040

Rajashekar C, Burke MJ (1978) The occurence of deep undercooling in the genera Pyrus, Prunus and Rosa: A preliminary report. In: Li PH, Sakai A (eds) Plant cold hardiness and freezing stresses. Academic Press, London New York, pp 213–225

Rajashekar C, Burke MJ (1982) Liquid water during slow freezing based on cell water relations and limited experimental testing. In: Li PH, Sakai A (eds) Plant cold hardiness and freezing stress, vol II. Academic Press, London New York, pp 211–220

Rajashekar C, Gusta LV, Burke MJ (1979) Membrane structure transition: probable relation to frost damage in hardy herbaceous species. In: Lyon JM, Graham D, Raison JK (eds) Low temperature stress in crop plants: The role of membrane. Academic Press, London New York, pp 255–274

Rajashekar C, Tao D, Li PH (1983) Freezing resistance and cold acclimation in turfgrasses. HortSci 18:91–93

Rakitina ZG (1965) O pronicaemosti dlja O_2 i CO_2 v svjazi s izučeniem pričin gibeli ozimych zlakov pod ledjanoj korkoj. (The permeability of ice with respect to CO_2 and O_2 in connection with a study of the cause of death of winter cereals under an ice crust). Fiziol Rast 12:909–919

Rakitina ZG (1967) Vlijanie gazovogo sostava atmosfery na morozostojkost rastenii ozimoj pšenicy. (Effect of the gas composition of the atmosphere on the frost resistance of winter wheat plants). Fiziol Rast 14:328–336

Rakitina ZG (1968) O vlijanie aeracii na morozoustojčivost pobegov drevesnych rastenii. (On the effect of aeration on the frost resistance of woody plant shoots). Fiziol Rast 15:235–245

Rakitina ZG (1970a) Vlijanie ledjanoj korki na gazovyi sostav vnutrennej atmosfery ozimoj pšenicy. (Effect of ice crust on gas composition of the interior atmosphere on winter wheat). Fiziol Rast 17:907–912

Rakitina ZG (1970b) Vlijanie aeratsii na morozostojkost korenvych sistem drevesnych rastenii. (Effect of aeration on the root frost-resistance in woody plants). Fiziol Rast 17:808–818

Rasmussen DH, MacKenzie AP (1972) Effect of solute on ice-solution interfacial free energy; calculation from measured homogeneous nucleation temperatures. In: Jellinek HHG (ed) Water structure at the water-polymer interface. Plenum, New York, pp 126–145

Raunkiaer C (1910) Statistik der Lebensformen als Grundlage für die biologische Pflanzengeographie. Beih Bot Centralbl 27II:171–206

Reader RJ (1979) Flower cold hardiness: a potential determinant of the flowering sequence exhibited by bog ericads. Can J Bot 57:997–999

Rees ap T, Fuller WA, Green JH (1981) Extremely high activities of phosphoenolpyruvate carboxylase in thermogenic tissues of Araceae. Planta 152:79–86

Rehfeldt GE (1977) Growth and cold hardiness of intervarietal hybrids of Douglas-fir. Theor Appl Genet 50:3–15

Rehfeldt GE (1979) Variation in cold hardiness among populations of Pseudotsuga menziesii var. glauca. USDA For Serv Res Pap INT-233 Ogden, Utah, USA, pp 1–11

Rehfeldt GE (1980) Cold acclimation in populations of Pinus contorta from the northern Rocky Mountains. Bot Gaz 141:458–463

Reid WS, Harris RE, McKenzie JS (1976) A portable freezing cabinet for cold-stressing plants growing in the field. Can J Plant Sci 56:623–625

Reuther G (1971) Die Dynamik des Kohlenhydratmetabolismus als Kriterium der Frostresistenz von Obstgehölzen in Abhängigkeit von der Winterruhe. Ber Dtsch Bot Ges 84:571–583

Richards JH, Bliss LC (1986) Winter water relations of a deciduous timberline conifer, Larix lyallii Parl. Oecologia (Berlin) 69:16–24

Richardson StG, Salisbury FB (1977) Plant responses to the light penetrating snow. Ecology 58: 1152–1158

Riedmüller-Schölm HE (1974) The temperature resistance of Alaskan plants from the continental boreal zone. Flora (Jena) 163:230–250

Rikin A, Blumenfeld A, Richmond AE (1976) Chilling resistance as affected by stressing environments and abscisic acid. Bot Gaz 137:307–312

Roberts DWA (1969a) A comparison of the peroxidase isoenzymes of wheat plants grown at 6 °C and 20 °C. Can J Bot 47:263–265

Roberts DWA (1969b) Some possible roles for isoenzyme substitutions during cold hardening in plants. Int Rev Cytol 26:303–328

Roberts DWA (1975) The inverse complement of cold-hardy and cold sensitive wheat leaves. Can J Bot 53:1333–1337

Roberts DWA, Grant MM (1968) Changes in cold hardiness accompanying development in winter wheat. Can J Plant Sci 48:369–376

Robins JK, Sunset JP (1974) Red Belt in Alberta. Inf Rep NOR-X-99, North For Cent, Edmonton

Rochat E, Therrien HP (1967a) Metabolism of ribonucleic acids of winter wheat Triticum aestivum L. during hardening at low temperatures. Nat Can 103:451–456

Rochat E, Therrien HP (1976b) Effects of antimetabolites and of some exogenous substances on hardening under cold conditions of winter wheat Triticum aestivum. Nat Can 103:451–456

Roche de la IA, Andrews CJ, Pomeroy MK, Weinberger P, Kates M (1972) Lipid changes in winter wheat seedlings (Triticum aestivum) at temperature inducing cold hardiness. Can J Bot 50:2401–2409

Roche de la IA, Pomeroy MK, Andrews CJ (1975) Changes in fatty acid composition in wheat cultivars of contrasting hardiness. Cryobiology 12:506–512

Rodinov VS, Nyuppieva KA, Zakharova LS (1973) Effect of low temperature on concentration of galacto and phospholipids in potato leaves. Fiziol Rast 20:523–531

Rogers RA, Dunn JH, Nelson CJ (1977) Photosynthesis and cold hardening in Zoysia and Bermudagrass. Crop Sci 17:727–732

Rohde CR, Pulham CF (1960) Heritability estimates of winter hardiness in winter barley determined by the standard unit method of resistance analysis. Agron J 52:584–586

Roos IMM, Hattingh MJ (1983) Scanning electron microscopy of Pseudomonas syringae cv. morsprunorum on sweet cherry leaves. Phytopathol Z 108:18–25

Rottenburg W (1972) Gefriervorgänge in lebendem Pflanzengewebe: Der Beginn des intrazellulären Gefrierens. Plant Cell Physiol 13:563–573

Rowley JA (1976) Development of freezing tolerance in leaves of C_4-grasses. Aust J Plant Physiol 3:597–603

Rowley JA, Tunnicliffe CG, Taylor AO (1975) Freezing sensitivity of leaf tissue of C_4 grasses. Aust J Plant Physiol 2:447–451

Rudolf TD, Nienstaedt H (1962) Polygenic inheritance of resistance to winter injury in jack pine lodgepole pine hybrids. J For 60:138–139

Rudorf W (1938) Keimstimmung und Photoperiode in ihrer Bedeutung für die Kälteresistenz. Züchter 10:238–246

Running SW, Reid CP (1980) Soil temperature influences on root resistance of Pinus contorta seedlings. Plant Physiol 65:635–640

Ruthsatz B (1978) Las plantas es cojin de los semi-desiertos andinos del Nordeste Argentino. Darwiniana 21:492–539

Rychnovská-Soudková M (1963) Study of the reversibility of the water saturation deficit as one of the methods of causal phytogeography. Biol Plant 5:175–180

Sachs J (1860) Krystallbildung bei dem Gefrieren und Veränderung der Zellhäute bei dem Aufthauen saftiger Pflanzenteile. Verh K Sächs Ges Wiss Math Phys Kl 12:1–50

Sadakane H, Hatano S (1982) Isoenzymes of glucose-6-phosphate dehydrogenase in relation to frost hardiness of Chlorella ellipsoidea. In: Li PH, Sakai A (eds) Plant cold hardiness and freezing stress, vol II. Academic Press, London New York, pp 157–167

Sadakane H, Kabata K, Ishibashi K, Watanabe T, Hatano S (1980) Studies on frost hardiness in Chlorella ellipsoidea. V. The role of glucose and related compounds. Environ Exp Bot 20:297–305

Sagisaka S (1969) Studies on cryobiochemistry in plants. III. Regulatory steps of sugarphosphate flow in winter bark of Populus gelrica. Low Temp Sci Ser B 27:81–88

Sagisaka S (1972) Decrease of glucose 6-phosphate and 6-phosphogluconate dehydrogenase activities in the xylem of Populus gelrica on budding. Plant Physiol 50:750–755

Sagisaka S (1974) Transition of metabolisms in living poplar bark from growing to wintering stage and vice versa. Plant Physiol 54:544–549

Sagisaka S (1982) Comparative studies on the metabolic function of differentiated xylem and living bark of wintering perennials. Plant Cell Physiol 23:1337–1346

Sagisaka S, Araki T (1983) Amino acid pools in perennial plants at the wintering stage and at the beginning of growth. Plant Cell Physiol 24:479–494

Sagisaka S, Asada M (1983) Coordinate and non-coordinate changes in enzyme activities in pentose phosphate cycle in poplar: A control of enzyme activities in differentiated xylem. Plant Cell Physiol 22:1459–1468

Sakai A (1956) Survival of plant tissue at super-low temperatures. Low Temp Sci Ser B 14:17–23

Sakai A (1958) The frost-hardening process of woody plant II. Relation of increase of carbohydrate and water soluble protein to frost-resistance. Low Temp Sci B 15:17–19

Sakai A (1959) The frost-hardening process of woody plant. V. The relationship between developmental stage and frost-hardening. Low Temp Sci B 17:43–49

Sakai A (1960a) Survival of the twigs of woody plants at −196 °C. Nature (London) 185:393–394

Sakai A (1960b) The frost hardiness of bulbs and tubers. J Hortic Assoc Jpn 29:233–238

Sakai A (1961) Effect of polyhydric alcohols to frost hardiness in plants. Nature (London) 189: 416–417

Sakai A (1962a) Studies on the frost hardiness of woody plants. 1. The causal relation between sugar content and frost hardiness. Contr Inst Low Temp Sci B11:1–40

Sakai A (1962b) Survival of woody plants in liquid helium. Low Temp Sci Ser B 20:121–122

Sakai A (1965) Survival of plant tissue at super-low temperatures. III. Relation between effective prefreezing temperatures and degree of frost hardiness. Plant Physiol 40:882–887

Sakai A (1966a) Studies of frost hardiness in woody plants. II: Effect of temperature on hardening. Plant Physiol 41:353–359

Sakai A (1966b) Survival of plant tissue at super-low temperature. IV: Cell survival with rapid cooling and rewarming. Plant Physiol 41:1050–1054

Sakai A (1966c) Temperature fluctuation in wintering trees. Physiol Plant 19:105–114

Sakai A (1966d) Seasonal variations in the amounts of polyhydric alcohol and sugar in fruit trees. J Hortic Sci 41:207–213

Sakai A (1968a) Frost damage on basal stems in young trees. Contrib Low Temp Sci Ser B 15:1–14

Sakai A (1968b) Mechanism of desiccation damage of forest trees in winter. Contrib Inst Low Temp Sci Ser B 15:15–35

Sakai A (1970a) Freezing resistance in willows from different climates. Ecology 51:485–491

Sakai A (1970b) Mechanism of desiccation damage of conifers wintering in soil-frozen areas. Ecology 51:657–664

Sakai A (1971a) Some factors contributing to the survival of rapidly cooled plant cells. Cryobiology 8:225–234

Sakai A (1971b) Freezing resistance of relicts from the arctotertiary flora. New Phytol 70:1199–1205

Sakai A (1972) Freezing resistance of evergreen and broad-leaf trees indigenous to Japan. J Jpn For Soc 54:333–339

Sakai A (1973) Characteristics of winter hardiness in extremely hardy twigs of woody plants. Plant Cell Physiol 14:1–9

Sakai A (1976) Adaptation of plants to deposited snow. Low Temp Ser B 34:47–76

Sakai A (1978a) Freezing tolerance of evergreen and deciduous broadleaved trees in Japan with reference to tree regions. Low Temp Sci Ser B 36:1–19
Sakai A (1978b) Low temperature exotherm of winter buds of hardy conifers. Plant Cell Physiol 19:1439–1446
Sakai A (1978c) Freezing tolerance of primitive willows ranging at subtropics and tropics. Low Temp Sci Ser B 36:21–29
Sakai A (1979a) Freezing avoidance mechanism of primordial shoots of conifer buds. Plant Cell Physiol 20:1381–1390
Sakai A (1979b) Deep supercooling of winter flower buds of Cornus florida L. Hort Sci 14:69–70
Sakai A (1980) Freezing resistance of broad-leaved evergreen trees in the warm-temperate zone. Low Temp Sci Ser B 38:1–14
Sakai A (1982a) Extraorgan freezing of primordial shoots of winter buds of conifer. In: Li PH, Sakai A (eds) Plant cold hardiness and freezing stress, vol II. Academic Press, London New York, pp 199–209
Sakai A (1982b) Freezing tolerance of shoot and flower primordia of coniferous buds by extraorgan freezing. Plant Cell Physiol 23:1219–1227
Sakai A (1982c) Freezing resistance of ornamental trees and shrubs. J Am Soc Hort 107:572–581
Sakai A (1983) Comparative study on freezing resistance of conifers with special reference to cold adaptation and its evolutive aspects. Can J Bot 9:2323–2332
Sakai A, Hakoda N (1979) Cold hardiness of the genus Camellia. Am Soc Hortic Sci 104:53–57
Sakai A, Horiuchi T (1972) Freezing resistance of tree trunks. J Jpn For Soc 54:379–382
Sakai A, Malla SB (1981) Winter hardiness of tree species at high altitudes in the East Himalaya, Nepal. Ecology 62:1288–1298
Sakai A, Miwa S (1979) Frost hardiness of Ericoideae. J Am Soc Hortic Sci 104:26–28
Sakai A, Okada S (1971) Freezing resistance of conifers. Silv Genet 20:91–97
Sakai A, Otsuka K, Yoshida S (1968) Mechanism of survival in plant cells at super-low temperatures by rapid cooling and rewarming. Cryobiology 4:165–173
Sakai A, Otsuka K (1970) Freezing resistance of alpine plants. Ecology 51:665–671
Sakai A, Wardle P (1978) Freezing resistance of New Zealand trees and shrubs. N Z J Ecol 1:51–61
Sakai A, Weiser CJ (1973) Freezing resistance of trees in North America with reference to tree regions. Ecology 54:118–126
Sakai A, Yoshida S (1967) Survival of plant tissue at superlow temperature. VI. Effects of cooling and rewarming rates on survival. Plant Physiol 42:1695–1701
Sakai A, Yoshida S (1968) The role of sugar and related compounds in variations of freezing resistance. Cryobiology 5:160–174
Sakai A, Yoshie F (1984) Freezing tolerance of ornamental bulbs and corms. J Jpn Soc Hortic Sci 52:445–449
Sakai A, Yoshida S, Saito M, Zoltai SC (1979) Growth rate of spruce related to the thickness of permafrost active layer near Inuvik, northwestern Canada. Low Temp Sci 37:19–32
Sakai A, Paton DM, Wardle P (1981) Freezing resistance of trees of the south temperate zone, especially subalpine species of Australasia. Ecology 62:563–570
Sakai A, Fuchigami L, Weiser CJ (1986) Cold hardiness in the genus Rhododendron. J Am Soc Hortic 111:273–280
Salt RW, Kaku S (1967) Ice nucleation and propagation in spruce needles. Can J Bot 45:1335–1346
Samygin GA (1974) Pričiny vymerzanija rastenii. Nauka, Moscow
Samygin GA, Volkova LA, Popov AS (1985) Sravnenie rasnykh metodov dlya otsenki zhiznesposobnosti kletok suspensionnykh i kallusnykh kultur (Comparison of various methods for assessing cell viability in suspension and callus cultures). Fiziol Rast 32:813–818
Santarius KA (1982) The mechanism of cryoprotection of biomembrane systems by carbohydrates. In: Li PH, Sakai A (eds) Plant cold hardiness and freezing stress, vol II. Academic Press, London New York, pp 475–486
Sato T (1982) Phenology and wintering capacity of sporophytes and gametophytes of ferns native to northern Japan. Oecologia (Berlin) 55:53–61
Sato T (1983) Freezing resistance of warm temperate ferns as related to their alternation of generations. Jpn J Ecol 33:27–35

Sato T, Sakai A (1980a) Freezing resistance of gametophytes of the temperate fern, Polystichum retroso-paleaceum. Can J Bot 58:1144–1148

Sato T, Sakai A (1980b) Phenological study of the leaf of Pterophyta in Hokkaido. Jpn J Ecol 30: 369–375

Sato T, Sakai A (1981a) Cold tolerance of gametophytes and sporophytes of some cool temperate ferns native to Hokkaido. Can J Bot 59:604–608

Sato T, Sakai A (1981b) Observation on the spore-dispersal period of Pterophyta in Hokkaido. Jpn J Ecol 31:91–97

Saure MC (1985) Dormancy release in deciduous fruit trees. Hortic Rev 7:239–300

Scheumann W (1965) Möglichkeiten und Ergebnisse der Frostresistenzprüfung in der Douglasien- und Lärchenzüchtung. Dtsch Akad Landwirtsch Wiss Berlin, Tagungsber 69:189–199

Scheumann W (1968) Die Dynamik der Frostresistenz und ihre Bestimmung an Gehölzen im Massentest. Dtsch Akad Landwirtsch Wiss, Berlin, Tagungsber 100:45–54

Scheumann W, Hoffmann K (1967) Die serienmäßige Prüfung der Frostresistenz einjähriger Fichtensämlinge. Arch Forstw 16:701–705

Scheumann W, Schönbach H (1968) Die Prüfung der Frostresistenz von 25 Larix leptolepis-Herkünften eines internationalen Provenienzversuches mit Hilfe von Labor-Prüfverfahren. Arch Forstw 17:597–611

Schmidt M (1942) Beiträge zur Züchtung frostwiderstandsfähiger Obstsorten. Züchter 14:1–19

Schmitt JM, Schramm MJ, Pflanz H, Coughlan S, Heber U (1985) Damage to chloroplast membranes during dehydration and freezing. Cryobiology 22:93–104

Schnell RC, Valli G (1972) Atmospheric ice nuclei from decomposing vegetation. Nature (London) 236:163–165

Schnelle F (ed) (1963) Frostschutz im Pflanzenbau, vol I: Die meteorologischen und biologischen Grundlagen der Frostschadensverhütung. Bayerischer Landwirtschaftsverlag, München

Scholander PF, Flagg W, Hock RF, Irving L (1953) Studies on the physiology of frozen plants and animals in the arctic. J Cell Comp Physiol 42 Suppl 1:1–56

Schölm HE (1968) Untersuchungen zur Hitze- und Frostresistenz einheimischer Süßwasseralgen. Protoplasma 65:97–118

Schönbach H, Bellmann E (1967) Frostresistenz der Nachkommenschaften von Kreuzungen „grüner“ und „blauer“ Formen der Douglasie (Pseudotsuga menziesii (Mirb.) Franco). Arch Forstw 16:707–711

Schüepp M (1967) Klimatologie der Schweiz. Lufttemperatur, pt 3, 4, C63–C106. Swiss Inst Meteor, Zürich

Schüepp M (1968) Klimatologie der Schweiz. Lufttemperatur, pt 5–8, C107–C153. Swiss Inst Meteor, Zürich

Schulze E-D, Beck E, Scheibe R, Ziegler P (1985) Carbon dioxide assimilation and stomatal response of afroalpine giant rosette plants. Oecologia (Berlin) 65:207–213

Schütt P (1981) Erste Ansätze zur experimentellen Klärung des Tannensterbens. Schweiz Z Forstw 132:443–452

Schwartz GJ, Diller KR (1982) Design and fabrication of a simple, versatile cryomicroscopy stage. Cryobiology 19:529–538

Schwarz W (1970) Der Einfluß der Photoperiode auf das Austreiben, die Frosthärte und die Hitzeresistenz von Zirben und Alpenrosen. Flora (Jena) 159:258–285

Schwarzbach E (1972) The inheritance of frost hardening ability in crosses between spring and winter barley. In: Rajki S (ed) Proceedings of a colloquium on the winter hardiness of cereals. Agric Res Inst Hung Acad Sci, Martonvásár, pp 83–91

Schwintzer CR (1971) Energy budgets and temperatures of nyctinastic leaves on freezing nights. Plant Physiol 48:203–207

Scott KR, Spangelo LPS (1964) Portable low temperature chamber for winter hardiness testing of fruit trees. Proc Am Soc Hortic Sci 84:131–136

Senser M (1982) Frost resistance in spruce (Picea abies (L.) Karst.): III. Seasonal changes in the phospho- and galactolipids of spruce needles. Z Pflanzenphysiol 105:229–239

Senser M, Beck E (1979) Kälteresistenz der Fichte. II. Einfluß von Photoperiode und Temperature auf die Struktur und photochemische Reaktionen von Chloroplasten. Ber Dtsch Bot Ges 92: 243–259

Senser M, Beck E (1982a) Frost resistance in spruce (Picea abies (L.) Karst.): IV. The lipid composition of frost resistant and frost sensitive spruce chloroplasts. Z Pflanzenphysiol 105:241–253
Senser M, Beck E (1982b) Frost resistance in spruce (Picea abies (L.) Karst.): V. Influence of photoperiod and temperature on the membrane lipids of the needles. Z Pflanzenphysiol 108:71–85
Senser M, Beck E (1984) Correlation of chloroplast ultrastructure and membrane lipid composition to the different degrees of frost resistance achieved in leaves of spinach, ivy, and spruce. J Plant Physiol 117:41–55
Senser M, Schötz F, Beck E (1975) Seasonal changes in structure and function of spruce chloroplasts. Planta 126:1–10
Shigo AL (1972) Ring and ray shakes associated with wounds in trees. Holzforschung 26:60–62
Shinozaki J (1954) The velocity of crystallization of ice from the blood of prepupae of slug moth. Low Temp Sci Ser B 11:1–11
Shitei T (1954) Studies on the damages to forest trees by snow pressure. Bull Gov For Exp Stn 73: 1–89
Shitei T (1974) Climatic and distribution of vegetation zones. In: Numata M (ed) The flora and vegetation of Japan. Kodansha & Elsevier, Tokyo Amsterdam, pp 20–21
Shmueli E (1960) Chilling and frost damage in banana leaves. Bull Res Counc Isr 8D:225–288
Shomer-Ilan, Waisel Y (1975) Cold hardiness of plants: correlation with changes in electrophoretic mobility, composition of amino acids and average hydrophobicity of fraction 1 protein. Physiol Plant 34:90–96
Shreve F (1911) The influence of low temperatures on the distribution of the giant cactus. Plant World 14:136–146
Shreve F (1914) The role of winter temperatures in determining the distribution of plants. Am J Bot 1:194–202
Siegel SM, Speitel T, Stoecker R (1969) Life in earth extreme environments: a study of cryobiotic potentialities. Cryobiology 6:160–181
Sigafoos RS (1952) Frost action as a primary physical factor in tundra plant communities. Ecology 33:480–487
Sikorska E, Kacperska-Palacz A (1979) Phospholipid involvement in frost tolerance. Physiol Plant 47:144–150
Sikorska E, Kacperska A (1982) Freezing-induced membrane alterations: Injury or adaption? In: Li P, Sakai A (eds) Plant cold hardiness and freezing stress. Academic Press, London New York, pp 261–272
Silberbauer-Gottsberger I, Morawetz W, Gottsberger G (1977) Frost damage of Cerrado plants in Botucatu, Brazil, as related to the geographical distribution of the species. Biotropica 9:253–261
Simatos D, Faure M, Bonjour E, Couach M (1975) The physical state of water at low temperatures in plasma with different water contents as studies by differential thermal analysis and differential scanning calorimetry. Cryobiology 12:202–208
Siminovitch D (1981) Common and disparate elements in the processes of adaptation of herbaceous and woody plants to freezing – a perspective. Cryobiology 18:166–185
Siminovitch D, Briggs DR (1953) Studies on the chemistry of the living bark of the black locust in relation to its hardiness. III. The validity of plasmolysis and desiccation tests for determining the frost hardiness of bark tissues. Plant Physiol 28:15–34
Siminovitch D, Therrien H, Wilner J, Gfeller F (1962) The release of amino acids and other ninhydrinreacting substances from plant cells after injury by freezing; a sensitive criterion for the estimation of frost injury in plant tissues. Can J Bot 40:1267–1269
Siminovitch D, Rheaume B, Pomeroy K, Lepage M (1968) Phospholipid, protein, and nucleic acid increases in protoplasm and membrane structures associated with development of extreme freezing resistance in black locust tree cells. Cryobiology 5:202–225
Siminovitch D, Singh J, Roche IA (1975) Studies on membranes in plant cells resistant to extreme freezing. I: Augmentation of phospholipids and membrane substance without changes in unsaturation of fatty acids during hardening of black locust bark. Cryobiology 12:144–153
Siminovitch D, Singh J, Roche AI (1978) Freezing behavior of protoplast of winter rye. Cryobiology 15:205–213

Simura T (1957) Breeding of polyploid varieties of the tea plant with special reference to their cold resistance. Cytol Proc Int Genet Symp 1956, pp 321–324

Singh J (1979) Ultrastructural alterations in cells of hardened and non-hardened winter rye during hyperosmotic and extracellular freezing stress. Protoplasma 98:329–341

Singh J, Miller RW (1980) Spin-label studies of membranes in rye protoplasts during extracellular freezing. Plant Physiol 66:349–352

Singh J, Miller RW (1982) Spin-probe studies during freezing of cells isolated from cold-hardened and non-hardened winter rye. Molecular mechanism of membrane freezing injury. Plant Physiol 69:1423–1428

Smith AP (1974) Bud temperature in relation to nyctinastic leaf movements in an Andean giant rosette plant. Biotropica 6:263–266

Smith D (1968a) Carbohydrates in grasses. IV. Influence of temperature on the sugar and fructosan composition of Timothy (Phleum pratense) plant parts at anthesis. Crop Sci 8:331–334

Smith D (1968b) Varietal chemical differences associated with freezing resistance in forage plants. Cryobiology 5:148–159

Smith MN, Olien CR (1978) Pathological factors affecting survival of winter barley following controlled freeze tests. Phytopathology 68:773–777

Smith MN, Olien CR (1981) Recovery from winter injury. In: Olien CR, Smith MN (eds) Analysis and improvement of plant cold hardiness. CRC Press, Boca Raton, pp 117–138

Smith WH (1970) Tree pathology. Academic Press, London New York

Smith WK (1985a) Western montane forests. In: Chabot BF, Mooney HA (eds) Physiological ecology of North American plant communities. Chapman & Hall, New York, pp 95–126

Smith WK (1985b) Environmental limitations on leaf conductance in Central Rocky Mountain conifers, USA. In: Turner H, Tranquillini W (eds) Establishment and tending of subalpine forest: research and management. Ber Eidgen Anst Forstl Versuchw 270:95–101

Smithberg MH, Weiser CJ (1968) Patterns of variation among climatic races of red-osier dogwood. Ecology 49:495–505

Smolenska G, Kuiper PJC (1977) Effect of low temperature upon lipid and fatty acid composition of roots and leaves of winter rape plants. Physiol Plant 41:29–35

Sobczyk EA, Kacperska-Palacz A (1978) Adenine nucleotide changes during cold acclimation of winter rape plants. Plant Physiol 62:875–878

Sobczyk EA, Kacperska-Palacz (1980) Changes in some enzyme activities during cold acclimation of winter rape plants. Acta Physiol Plant 2:123–131

Sobczyk EA, Shcherbakova A, Kacperska A (1980) Effect of cold acclimation on the stability of some enzymes in winter rape plants. Z Pflanzenphysiol 100:113–119

Sobczyk EA, Rybka Z, Kacperska A (1984) Modification of pyruvate kinase activity in cold-sensitive and colc-resistant leaf tissues. Z Pflanzenphysiol 114:285–293

Sobczyke E, Marszalek A, Kacperska A (1985) ATP involvement in plant tissue responses to low temperature. Physiol Plant 63:399–405

Soeder C, Stengel E (1974) Physico-chemical factors affecting metabolism and growth rate. In: Stewart WDP (ed) Algal physiology and biochemistry. Biol Monogr, vol X. Blackwell, Oxford, pp 714–740

Sosińska A, Kacperska-Palacz A (1979) Ribulosediphospate and phosphoenolpyruvate carboxylase activities in winter rape as related to cold acclimation. Z Pflanzenphysiol 92:455–458

Soule J (1985) Glossary for horticultural crops. Wiley & Son, New York

Souzu H (1973) The phospholipid degradation and cellular death caused by freeze-thawing or freeze-drying of yeast. Cryobiology 10:427–431

Sparks D, Payne JA (1977) Freeze injury susceptibility of non-juvenile trunks in pecan. Hort Sci 12:497–498

Sparks D, Payne A, Horton BD (1976) Effect of sub-freezing temperatures on bud break of pecan. HortSci 11:415–416

Spomer GG (1979) Prospects of hormonal mechanisms in cold adaptations of arctic plants. In: Comparative mechanisms of cold adaptations. Academic Press, London New York, pp 311–321

Srivastava LM, O'Brien TP (1966a) On the ultrastructure of cambium and its vascular derivaties. I. Cambium of Pinus strobus L. Protoplasma 61:257–276

Srivastava LM, O'Brien TP (1966b) Secondary phloem of Pinus strobus. Protoplasma 61:277–293

Stebbins GL (1952) Aridity as a stimulus to evolution. Am Nat 86:33–44

Stebbins L (1965) The probable growth habit of the earliest flowering plants. Ann Miss Bot Garden 52:457–468

Stefansson E, Sinko M (1967) Experiments with provenances of Scots pine with special regard to high-lying forests in Northern Sweden. Stud For Suec 47:1–108

Steponkus PL (1981) Responses to extreme temperatures. Cellular and sub-cellular bases. In: Lange OL et al. (eds) Physiological plant ecology I. In: Lange OL, Nobel PS, Osmond CB, Ziegler H (eds) Encyclopedia of plant physiology, vol 12A. Springer, Berlin Heidelberg New York, pp 371–402

Steponkus PL (1984) Role of the plasma membrane in freezing injury and cold acclimation. Annu Rev Plant Physiol 35:543–584

Steponkus PL, Lanphear FO (1967) Refinement of the triphenyl tetrazolium chloride method of determining cold injury. Plant Physiol 42:1423–1426

Steponkus PL, Wiest SC (1979) Freeze-thaw induced lesions in the plasma membrane. In: Lyons J, Graham D, Raison J (eds) Low temperature stress in crop plants. Academic Press, London New York, pp 231–254

Steponkus PL, Garber MP, Myers SP, Unebergen RD (1977) Effects of cold acclimation and freezing on structure and function of chloroplast thylakoids. Cryobiology 14:303–321

Steponkus PL, Dowgert MF, Evans RY, Gordon-Kamm W (1982) Cryobiology of isolated protoplasts. In: Li PH, Sakai A (eds) Plant cold hardiness and freezing stress, vol II. Academic Press, London New York, pp 459–474

Steponkus PL, Dowgert MF, Gordon-Kamm WJ (1983) Destabilization of plasma membrane of isolated plant protoplasts during a freeze-thaw cycle: The influence of cold acclimation. Cryobiology 20:448–465

Stergios BG, Howell GS (1977) Effects of defoliation, trellis height, and cropping stress on the cold hardiness of 'Concord' grapevines. Am J Enol Vitic 28:34–42

Stern K (1974) Beiträge zum geographischen Variationsmuster der Douglasie. Silv Genet 23:53–58

Steubing L, Alberdi M, Wenzel H (1983) Seasonal changes of cold resistance of Proteaceae of the South Chilean laurel forest. Vegetatio 52:35–44

Stiles W (1927) The exosmosis of dissolved substance from storage tissue into water. Protoplasma 2:577–601

Stout DG (1979) Plant plasma membrane water permeability and slow freezing injury. Plant Cell Environ 2:273–275

Stout DG, Majak W, Reaney M (1980) In vivo detection of membrane injury at freezing temperatures. Plant Physiol 66:74–77

Strand M, Öquist G (1985) Inhibition of photosynthesis by freezing temperatures and high light levels in cold-acclimated seedlings of Scots pine (Pinus sylvestris). II. Effects on chlorophyll fluorescence at room temperature and 77 K. Physiol Plant 65:117–123

Sturm H, Rangel O (1985) Ecologia de los Paramos Andinos. Mus Hist Nat Bibl Jose Jeronimo Triana 9 (Bogotá)

Stushnoff C (1972) Breeding and selection methods for cold hardiness in deciduous fruit crops. HortSci 7:10–13

Stushnoff C, Junttila O (1978) Resistance to low temperature injury in hydrated lettuce seed by supercooling. In: Li PH, Sakai A (eds) Plant hardiness and freezing stress. Academic Press, London New York, pp 241–248

Sugawara Y, Sakai A (1974) Survival of suspension cultured sycamore cells cooled to the temperature of liquid nitrogen. Plant Physiol 54:722–724

Sugawara Y, Sakai A (1978) Cold acclimation of callus cultures of Jerusalem artichoke. In: Li PH, Sakai A (eds) Plant cold hardiness and freezing stress, vol I. Academic Press, London New York, pp 197–210

Sugiyama N, Simura T (1967) Studies on the varietal differentiation of frost resistance of the tea plant. IV. The effect of sugar level combined with protein in chloroplasts on the frost resistance. Jap J Breed 17:292–296

Sugiyama N, Simura T (1968) Studies on the varietal differentiation of frost resistance of the tea plant. VI. The distribution of sugars in cells during frost hardening in the tea plant with special reference to histochemical observation by using ^{14}C sucrose. Jpn J Breed 18:27–32

Suneson CA, Peltier G (1934) Cold resistance adjustments of field hardened winter wheats as determined by artificial freezing. J Am Soc Agron 26:50–58

Sutka J (1981) Genetic studies of frost resistance in wheat. Theor Appl Genet 59:145–152

Sutton-Jones B, Street E (1968) Studies on the growth in culture of plant cells. II. Changes in fine structure during growth of Acer pseudoplatanus L. cells in suspension culture. J Exp Bot 19: 114–118

Svejda F (1979) Inheritance of winter-hardiness in roses. Euphytica 28:309–314

Swiss Meteorological Institute (1930) Annalen der schweizerischen meteorologischen Zentralanstalt, vol 67. Zürich

Swiss Meteorological Institute (1955–1975) Annalen der schweizerischen meteorologischen Zentralanstalt, vol 92–112. Zürich

Takahashi K (1960) (Plant distribution and snow cover). Ritchi Shinrin 2:19–24

Takhtajan AL (1970) Proiskholzhdenie i rasselenie tsvetkovykh rastenii. Nauka, Leningrad

Tanai T (1967) Miocene floras and climate in East Asia. Abb Zent Geol Inst 10:195–205

Tanai T (1972) Tertiary history of vegetation in Japan. In: Graham A (ed) Floristics and paleofloristics of Asia and Eastern North America. Elsevier, Amsterdam, pp 235–255

Tanino KK, McKersie BD (1985) Injury within the crown of winter wheat seedlings after freezing and icing stress. Can J Bot 63:432–436

Tao D, Li PH, Carter JV (1983) Role of cell wall in freezing tolerance of cultured potato cells and their protoplasts. Physiol Plant 58:527–532

Tappeiner U (1985) Bestandesstruktur, Mikroklima und Energiehaushalt einer naturnahen Almweide und einer begrünten Schipistenplanierung im Gasteiner Tal (Hohe Tauern). Diss, Univ Innsbruck

Terumoto I (1959) Frost and drought injury in lake balls. Low Temp Sci Ser B 17:1–7

Terumoto I (1960) Ice masses in root tissue of table beet. Low Temp Sci Ser B 18:39–42

Terumoto I (1964) Frost resistance in some marine algae from the winter intertidal zone. Low Temp Sci Ser B 22:19–28

Terumoto I (1965) Freezing and drying in a red marine alga, Porphyra yezoensis Ueda. Low Temp Sci Ser B 23:11–20

Thomas DA, Barker HN (1976) Studies on leaf characteristics of a line of Eucalyptus urnigera from Mount Wellington, Tasmania: 1. Water repellency and freezing leaves. Aust J Bot 22:501–512

Tieszen LL (ed) (1978) Vegetation and production ecology of an Alaskan arctic tundra. Ecol Stud 29. Springer, Berlin Heidelberg New York

Tikhomirov BA (1970) Forest limits as the most important geographical boundary in the North. Ecology of the subarctic regions. UNESCO Proc Helsinki Symp, Ecol Conserv 1:35–40

Tikhomirov BA (1971) Osobennosti biosfery Kraynego severa. Priroda 1971 (11):30–42

Till O (1956) Über die Frosthärte von Pflanzen sommergrüner Laubwälder. Flora (Jena) 143:499–542

Timmis R (1977) Critical frost temperature for Douglas-fir cone buds. Can J For Res 7:19–22

Towill LE, Mazur P (1976) Osmotic shrinkage as a factor in freezing injury in plant tissue cultures. Plant Physiol 57:290–296

Toyao T (1982) Inheritance of cold hardiness of tea plants in crosses between var. sinensis and var. assamica. In: Li PH, Sakai A (eds) Plant cold hardiness and freezing stress, vol II. Academic Press, London New York, pp 591–603

Tranquillini W (1957) Standortsklima, Wasserbilanz und CO_2-Gaswechsel junger Zirben (Pinus cembra L.) an der alpinen Waldgrenze. Planta 40:612–661

Tranquillini W (1958) Die Frosthärte der Zirbe unter besonderer Berücksichtigung autochthoner und aus Frostgärten stammender Jungpflanzen. Forstw Centralbl 77:65–128

Tranquillini W (1976) Water relations and alpine timberline. In: Lange OL, Kappen L, Schulze E-D (eds) Water and plant life. Ecol Stud 19. Springer, Berlin Heidelberg New York, pp 473–491

Tranquillini W (1979a) Über die Frostgefährdung von Fichten in verschiedener Höhenlage. Wiss Mitt Meteorol Inst München 35:51–57

Tranquillini W (1979b) Physiological ecology of the alpine timberline. Ecol Stud 31. Springer, Berlin Heidelberg New York
Tranquillini W (1982) Frost-drought and its ecological significance. In: Lange OL, Nobel PS, Osmond CB, Ziegler H (eds) Encyclopedia of plant physiology, vol 12B. Springer, Berlin Heidelberg New York, pp 379–400
Tranquillini W, Holzer K (1958) Über das Gefrieren und Auftauen von Coniferennadeln. Ber Dtsch Bot Ges 71:143–156
Tranquillini W, Platter W (1983) Der winterliche Wasserhaushalt der Lärche (Larix decidua Mill.) an der alpinen Waldgrenze. Verh Ges Ökol 11:433–443
Troll C (1960) The relationship between the climates, ecology and plant geography of the southern cold temperate zone and of the tropical high mountains. Proc R Soc London B 152:529–532
Troll C (1961) Klima und Pflanzenkleid der Erde in dreidimensionaler Sicht. Naturwissenschaften 48:332–348
Troll C (1964) Karte der Jahreszeitklimate der Erde. Erdkunde 18:5–28
Tronsmo AM (1984) Predisposing effects of low temperature on resistance to winter stress factor in grasses. Acta Agric Scand 34:210–220
Trunova II (1982) Mechanism of winter wheat hardening at low temperature. In: Li PH, Sakai A (eds) Plant cold hardiness and freezing stress, vol II. Academic Press, London New York, pp 41–54
Trunova TI, Zvereva CN (1974) (Protective effect of sugars during freezing of chloroplast suspensions). Fiziol Rast 21:1000–1006
Tseng MJ, Li PH (1986) Cold acclimation and ABA induced changes in the population of translatable messenger RNA in tuberbearing Solanum potato. 3rd Plant Cold Hardiness Seminar, Abstr 28, Shanghai
Tuhkanen S (1980) Climatic parameters and indices in plant geography. Acta Phytogeogr 67:5–110
Tukhanen S (1984) A circumboreal system of climatic-phytogeographical regions. Acta Bot Fenn 127:1–50
Tumanov II (1931) Das Abhärten winterannueller Pflanzen gegen niedrige Temperaturen. Phytopathol Z 3:231–339
Tumanov II (1967) The frost hardening process in plants. In: Troshin AS (ed) The cell and environmental temperature. Pergamon Press, Oxford, pp 6–14
Tumanov II (1979) Fiziologija zakalivanija i morozostoikosti rastenii. Nauka, Moscow
Tumanov II, Khvalin NN (1967) Prichiny slaboi morozostoikosti kornei u plodovykh derevev. (Causes of the weak frost resistance of fruit tree roots). Fiziol Rast 14:908–918
Tumanov II, Krasvatsev OA (1959) Zakalivanie severnych drevesnych rastenii otricatelvnymi temperaturami. (Hardening of northern woody plants by negative temperature treatment). Fiziol Rast 6:654–667
Tumanov II, Krasavtsev OA (1975) Razvitie morozostoikosti rastenii. (Development of resistance to frost in plants). XIIth Int Bot Congr, Leningrad, Abstr II, p 488
Tumanov II, Trunova TI (1963) (First phase of frost hardiness of winter plants in the dark on sugar solution). Fiziol Rast 10:140–149
Tumanov II, Krasavtsev OA, Khvalin II (1959) (Promotion of resistance of birch and black current to −253 °C temperature by frost hardening). Dokl Akad Nauk SSSR 127:1301–1303
Tumanov II, Kuzina GV, Karnikova LD (1965) Vliyanie fotoperiodov na morozostoikost abrikosa i chernoi smorodiny. (Effect of photoperiod on the frost resistance of apricots and black currants). Fiziol Rast 12:665–682
Tumanov II, Kuzina GV, Karnikova LD (1972) Vliyanie prodolzhitelnosti vegetattsii u drevesnykh rastenii na nakoplenie zapesnykh uglevodov i kharakter fotoperiodicheskoi reaktsii. (Effect of the duration of vegetation in trees on the accumulation of reserve carbohydrates and the character of photoperiodic reaction). Fiziol Rast 19:1122–1131
Tumanov II, Kuzina GV, Karnikova LD (1973) Period pokoya i sposobiost drevesnykh rastenii zakalivatsya nizhkimi temperatuami. (Dormancy period and ability of trees for hardening with low temperatures). Fiziol Rast 20:5–16
Turner H (1980) Types of microclimate at high elevations. NZFS FRI Technol Pap 70:21–26
Turner H, Tranquillini W (1985) Establishment and tending of subalpine forest: research and management. Ber Eidgen Anst Forstl Versuchsw Birmensdorf 270

Tuyama T (1968) Differentiation of the wild Camellia japonica. In: Tuyama T (ed) Camellias of Japan. Takeda Sci Found, Osaka, pp 3–13
Tyler NJ, Fowler DB, Gusta LV (1981) The effect of salt stress on the cold hardiness of winter wheat. Can J Plant Sci 61:543–548
Tyurina MM (1957) Issledovanie morozostoikosti rastenii v usloviyakh vysokogorii pamira. Nauk Tadzhik, Stalinabad
Tyurina MM, Gogoleva GA, Jegurasdova AS, Bulatova TG (1978) Interaction between development of frost resistance and dormancy in plants. Acta Hortic 81:51–60
Uemura M, Yoshida S (1983) Isolation and identification of plasma membrane from light-grown winter rye seedlings (Secale cereale L. cv. Puma). Plant Physiol 73:586–597
Uemura M, Yoshida S (1984) Involvement of plasma membrane alterations in cold acclimation of winter rye seedlings (Secale cereale L. cv. Puma). Plant Physiol 75:818–826
Ullrich H, Mäde A (1940) Studien über die Ursachen der Frostresistenz. II. Untersuchungen über den Temperaturverlauf beim Gefrieren von Blättern und Vergleichsobjekten. Planta 31:251–263
Ulmer W (1937) Über den Jahresgang der Frosthärte einiger immergrüner Arten der alpinen Stufe, sowie der Zirbe und der Fichte. Jahrb Wiss Bot 84:553–592
Usui H (1961) Phytoecological revision of the dominant species of Sasatype undergrowth. Silvicultural application of the research on Japanese forest vegetation. Spec Bull, Coll Agric, Utsunomiya Univ 11:1–43
Valk van der GGM (1970) Frost injury to flowerbulb crops. Acta Hortic 23:345–349
Vareschi V (1970) Flora de los Páramos de Venezuela. Mérida Edic Univ de los Andes
Vasilyev IM (1934) Yarovization of winter wheat varieties and frost resistance. Dokl Akad Nauk SSSR 4:158–161
Vasilyev IM (1961) Wintering of plants. Am Inst Biol Sci (Washington)
Veblen TT, Aston DH, Schlegel FM, Veblen AT (1977) Plant succession in a timberline depressed by volcanism in south-central Chile. J Biogeogr 4:275–294
Vegis A (1973) Dependence of the growth processes on temperature. In: Precht H, Christophersen J, Hensel H, Larcher W (eds) Temperature and life. Springer, Berlin Heidelberg New York, pp 145–170
Verkleij AJ, Ververgaert PHJT (1978) Freeze-fracture morphology of biological membranes. Biochim Biophys Acta 515:303–327
Vieweg GH, Ziegler H (1969) Zur Physiologie von Myrothamnus flabellifolia. Ber Dtsch Bot Ges 82:29–36
Vigh L, Horváth I, Horváth LI, Dudits D, Farkas T (1979a) Protoplast plasmalemma fluidity of hardened wheats correlates with frost resistance. FEBS Lett 107:291–294
Vigh L, Horváth I, Farkas T, Horváth LI, Belea A (1979b) Adaptation of membrane fluidity of rye and wheat seedlings according to temperature. Phytochemistry 18:787–789
Villiers TA (1975) Dormancy and survival of plants. Arnold, London
Vintejoux C, Dereuddre J (1981) Étude de quelques aspects de l'alternance saisonnière chez les végétaux. Bull Soc Bot Fr 128:7–21
Vojnikov V, Korzun A, Pobezhimova T, Varakina N (1984) Effect of cold shock on mitochondrial activity and on the temperature on winter wheat seedlings. Biochem Physiol Pflanzenk 179: 327–330
Wagner HWJ, Sharp JA (1963) A remarkably reduced vascular plant in the United States. Science 142:1483–1484
Walter H (1976) Die ökologischen Systeme der Kontinente (Biogeosphäre). Fischer, Stuttgart
Walter H (1984) Vegetation und Klimazonen, 5th edn. Ulmer, Stuttgart
Walter H, Lieth H (1967) Klimadiagramm – Weltatlas. Fischer, Jena
Walter H, Medina E (1969) Die Bodentemperatur als ausschlaggebender Faktor für die Gliederung der subalpinen und alpinen Stufe in den Anden Venezuelas. Ber Dtsch Bot Ges 82:275–281
Wardle P (1963) Evolution and distribution of the New Zealand flora, as affected by Quaternary climates. N Z J Bot 1:3–17
Wardle P (1971) An explanation for alpine timberline. N Z J Bot 9:371–402
Wardle P (1974) Alpine timberlines. In: Ives JD, Barry RG (eds) Arctic and alpine environments. Methuen, London, pp 371–402

Wardle P (1981) Winter desiccation of conifer needles simulated by artificial freezing. Arctic Alpine Res 13:419–423

Wardle P (1985) New Zealand timberlines. 3: A synthesis. N Z J Bot 23:263–271

Wardle P, Campbell AD (1976a) Seasonal cycle of tolerance to low temperatures in three native woody plants, in relation to their ecology and post-glacial history. Proc N Z Ecol Soc 23:85–91

Wardle P, Campbell AD (1976b) Winter dormancy in seedlings of mountain beech (Nothofagus solandri var. cliffortioides) near timberline. N Z J Bot 14:183–186

Wareing PF, Phillips IDJ (1978) The control of growth and differentiation in plants. Pergamon Press, Oxford New York

Wareing PF, Saunders PF (1971) Hormons and dormancy. Annu Rev Plant Physiol 22:261–288

Warrington IJ, Rook DA (1980) Evaluation of techniques used in determining frost tolerance of forest planting stock: A review. N Z J For Sci 10:116–132

Watkins R, Spangelo LPS (1970) Components of genetic variance for plant survival and vigor of apple trees. Theor Appl Genet 40:195–203

Weiser CJ (1970) Cold resistance and injury in woody plants. Science 169:1269–1278

Whightman F (1979) Modern chromatographic methods for the identification and quantification of plant growth regulators and their application to studies of the changes in hormonal substances in winter wheat during acclimation to cold stress conditions. In: Scott TK (ed) Plant regulation and world agriculture. Plenum Press, New York London, pp 327–377

White WC, Weiser CJ (1964) The relation of tissue desiccation, extreme cold, and rapid temperature fluctuations to winter injury of American arbovitae. Proc Am Soc Hortic Sci 85:554–563

Whittaker RH (1975) Communities and ecosystems, 2nd edn. McMillan, New York

Widholm JM (1972) The use of fluorescein diacetate and phenosafranine for determining viability of cultured plant cells. Stain Technol 47:189–194

Wiest SC, Steponkus PL (1978) Freeze thaw injury to isolated spinach protoplasts and its simulation at above-freezing temperatures. Plant Physiol 62:599–605

Wiest SC, Good GL, Steponkus PL (1976) Evaluation of root viability following freezing by the release of ninhydrin-rective compounds. HortSci 11:197–199

Wilhelm F (1975) Lehrbuch der allgemeinen Geographie III, 3: Schnee- und Gletscherkunde. De Gruyter, Berlin

Willemot C (1977) Simultaneous inhibition of linolenic acid synthesis in winter wheat roots and frost hardening by BASF 13–338, a derivative of pyridazinone. Plant Physiol 60:1–4

Williams BJ, Pellet NE, Klein RM (1972) Phytochrome control of growth cessation and initiation of cold adaptation in selected woody plants. Plant Physiol 50:262–265

Williams MW (1984) The cold hardiness adaptive response of green ash to geoclimate gradients. PhD Thesis, Penn State Univ

Wilner J (1960) Relative and absolute electrolytic conductance tests for frost hardiness of apple varieties. Can J Plant Sci 40:630–637

Wilner J, Brach EJ (1979) Utilization of bioelectric tests in biological research. Agric Can, Ottawa, Publ 6289-I-139

Wilson RF, Rinne RW (1976) Effect of freezing and cold storage on phospholipids in developing soybean cotyledons. Plant Physiol 57:270–273

Wiltbank WJ, Rouse RE (1973) A portable freezing unit for determining leaf freezing points of Citrus leaves. HortSci 8:510–511

Winkler E, Moser W (1967) Die Vegetationszeit in zentralalpinen Lagen Tirols in Abhängigkeit von den Temperatur- und Niederschlagsverhältnissen. Ferdinandeum Innsbruck 47:121–147

Withers LA (1978) The freeze-preservation of synchronously dividing cultured cells of Acer pseudoplatanus L. Cryobiology 15:87–92

Withers LA, King PJ (1979) A noted cryoprotectant for the freeze preservation of cultured cells of Zea mays L. Plant Physiol 64:675–678

Withers LA, Street HE (1977) Freeze preservation of cultured plant cells. III. The pregrowth phase. Physiol Plant 39:171–178

Wright JW, Kung FH, Read RA, Lemien WA, Bright JN (1971) Genetic variation in Rocky Mountain Douglas-fir. Silv Genet 20:54–60

Yamaki S, Uritani I (1973) Mechanism of chilling injury in sweet potato. Plant Physiol 51:883–898

Yamaki S, Uritani I (1974) Mechanism of chilling injury in sweet potato. XI. Irreversibility of physiological deterioration. Plant Cell Physiol 15:385–388

Yelenosky G (1975) Cold hardening in Citrus stems. Plant Physiol 56:540–543

Yelenosky G (1977) The potential of Citrus to survive freezes. Proc Int Soc Citric 1:199–203

Yelenosky G (1978) Cold hardening 'Valencia' orange trees to tolerate –6.7 °C without injury. J Am Soc Hortic Sci 103:449–452

Yelenosky G (1979) Water-stress-induced cold hardening of young Citrus trees. J Am Soc Hortic Sci 104:270–273

Yelenosky G (1985) Cold hardiness in Citrus. Hort Reviews 7:201–238

Yelenosky G, Horanic G (1969) Subcooling in wood of citrus seedlings. Cryobiology 5:281–283

Yelenosky G, Barrett H, Young R (1978) Cold hardiness of young hybrid trees of Eremocitrus glauca (Lindl.) Swing. HortScience 13:257–258

Yoe de DR, Brown GN (1979) Glycerolipid and fatty acid changes in eastern white pine chloroplast lamellae during the onset of winter. Plant Physiol 64:924–929

Yoshida S (1974) Studies on lipid changes associated with frost hardiness in cortex in woody plants. Contrib Inst Low Temp Sci Ser B 18:1–43

Yoshida S (1982) Fluorescence polarization studies on plasma membrane isolated from mulberry bark tissues. In: Li PH, Sakai A (eds) Plant cold hardiness and freezing stress, vol II. Academic Press, London New York, pp 273–284

Yoshida S (1984a) Studies on freezing injury of plant cells. I. Relation between thermotropic properties of isolated plasma membrane vesicles and freezing injury. Plant Physiol 75:38–42

Yoshida S (1984b) Chemical and biophysical changes in the plasma membrane during cold acclimation of mulberry bark cells (Morus bombycis Koidz. cv. Goroji). Plant Physiol 76:257–265

Yoshida S, Sakai A (1968) The effect of thawing rate on freezing injury in plants. Low Temp Sci, Hokkaido Univ Ser B 26:23–31

Yoshida S, Sakai A (1973) Phospholipid changes associated with the cold hardiness of cortical cells from poplar stem. Plant Cell Physiol 14:353–359

Yoshida S, Sakai A (1974) Phospholipid degradation in frozen plant cells associated with freezing injury. Plant Physiol 53:509–511

Yoshida S, Uemura M (1984) Protein and lipid compositions of isolated plasma membranes from orchard grass (Dactylis glomerata L.) and changes during cold acclimation. Plant Physiol 75: 31–37

Yoshida S, Uemura M, Niki T, Sakai A, Gusta L (1983) Partition of membrane particles in aqueous two-polymer phase system and its practical use for purification of plasma membrane from plants. Plant Physiol 72:105–114

Yoshie F, Sakai A (1981a) Freezing resistance of plants in relation to their life form and habitat. Jpn J Ecol 31:395–404

Yoshie F, Sakai A (1981b) Water status and freezing resistance of arctic plants. In: Kinosita S (ed) Joint studies on physical and biological environments in the permafrost zone, North Canada. Inst Low Temp Sci, Sapporo, Contrib 2427:95–110

Yoshie F, Sakai A (1982) Freezing resistance of temperate deciduous forest plants in relation to their life form and microhabitat. In: Li PH, Sakai A (eds) Plant cold hardiness and freezing stress, vol II. Academic Press, London New York, pp 427–436

Yoshimura F (1967) Studies on the cold injury of citrus trees. Mem Fac Agric Kochi Univ 18:79–134

Yurugi Y, Nagata H (1981) Studies on dormancy in woody plants. II. Growth patterns and distribution of evergreen broadleaf trees. Bull Fac Agric Mie Univ 63:199–203

Ziegler P, Kandler O (1980) Tonoplast stability as a critical factor in frost injury and hardening of spruce (Picea abies L.) needles. Z Pflanzenphysiol 99:393–410

Zimmermann M (1964) Effect of low temperature on ascent of sap in trees. Plant Physiol 39: 568–572

Zimmermann MH (1983) Xylem structure and the ascent of sap. Springer, Berlin Heidelberg New York

Zwintscher M (1957) Beiträge zur Vererbung des Frostverhaltens der Obstgehölze. Gartenbauwissenschaft 22:50–70

Terminology and Definitions

Freezing Parameters

Frost: Environmental temperatures below 0 °C.

Freezing: The procedure of ice formation.

Freezing point: The temperature at which a liquid exists in equilibrium with its solid phase (Encyclopaedia Britannica 1963).

Freezing point depression: The lowering of the freezing point of a solution compared to the pure solvent; the depression is proportional to the active mass of the solute in a given amount of solvent (Lapedes 1978).

Threshold freezing temperature: The highest subzero temperature below which ice nucleation in living cells (tissues) is detectable (Larcher 1985a).

Supercooling: The cooling of a liquid below its freezing point without freezing taking place; this results in a metastable state, i.e. an excited stationary energy state whose lifetime is unusually long (Lapedes 1978).

Supercooling temperature: The lowest subfreezing temperature attained before ice formation (Levitt 1980).

Transient supercooling: Short-term labile supercooling of plant cells (tissues) before heterogeneous nucleation occurs (Modlibowska 1956; Larcher 1985a).

Persistent supercooling: Supercooling of living cells (tissues) which are capable of avoiding ice nucleation for long periods (Modlibowska 1956; Larcher 1985a).

Deep supercooling: Persistent supercooling to the homogeneous nucleation point (adapted from Burke et al. 1976).

Ice nucleation: Initiation of freezing at the limit of supercooling (Burke et al. 1976).

Homogeneous nucleation point: The temperature at which spontaneous ice nucleation occurs in the absence of nucleating substances (Rasmussen and MacKenzie 1972).

Heterogeneous nucleation point: The temperature at which ice formation is initiated by nonaqueous catalysts (e.g. impurities, microbes or shock).

Exotherm: Temperature increase as a result of the heat of crystallization released during freezing.

Freezing Patterns

Extracellular (extraplasmatic) freezing: Ice formation on the surface of the cell or between the protoplast and the cell wall.

Intracellular (intraplasmatic) freezing: Ice formation within the protoplast (protoplasm, vacuole).

Extraorgan freezing: Ice segregation from a supercooled organ to a specific space outside, resulting in dehydration of the organ (Sakai 1979b).

Translocated freezing: Migration of water from supercooled tissues to nucleation centres in adjacent spaces (Larcher 1985a).

Equilibrium freezing: Liquid-solid phase transition outside the protoplast at small displacements of temperature from the balanced state associated with low crystallization energy (Olien 1981).

Nonequilibrium freezing: Ice formation after a large displacement from thermodynamic equilibrium due to supercooling and the generation of sufficient free energy to be disruptive (Olien 1981).

Injury and Survival

Injury: Irreversible damage of cells, tissues or the whole plant.

Chilling injury: Low temperature damage in the absence of freezing (Molisch 1897).

Freezing injury: Damage connected with ice formation in plant tissues.

Frost resistance (frost hardiness): The ability to survive temperature below 0 °C.

Freezing resistance (freeze hardiness): The ability of cells to survive frost by means of avoidance or tolerance of ice formation in plant tissues (Levitt 1958, 1980).

Freezing avoidance: The ability of living cells to prevent (avoid) the establishment of thermodynamic equilibrium with the frost stress (Levitt 1980).

Freezing tolerance: The ability of living cells to resist the internal freezing stress without suffering injury (Levitt 1980).

Frost hardening: see acclimation.

Frost survival: The ability of a plant (individual, population) to survive frost events by prevention of freezing, by freezing resistance and by recovery after partial damage (see p. 59).

Frost drought (winter desiccation): Damage due to evaporative water loss whilst water supply is interrupted to soil or stem frost.

Survival capacity: The ability of a plant to withstand the complex environmental stress effects during adverse weather conditions and in extreme habitats (Larcher 1973).

Adaptive Events

Adaptation: Heritable changes of structure and functions that increase the probability of an organism's surviving and reproducing in a particular environment (Kramer 1980); change brought about in a population as a result of exposure to particular environmental conditions (Soule 1985, p. 531).

Acclimation: Nonheritable modification of structure and functions as a response to a given climatic constraint which minimizes damage and improves the fitness of an individual plant (adapted from Kramer 1980 and Harper 1982). Acclimation includes also the process of seasonal adjustment to oncoming unfavourable weather conditions (Giertych 1974).

Dormancy: A state of rest in which growth is temporarily suspended (Wareing and Phillips 1978); a state of reduced activity or development without visible growth of any structure containing a meristem (Lang et al. 1985).

Imposed dormancy: A dormant condition caused by external conditions unfavourable for growth, the release from which is brought about as soon as the conditions improve (Giertych 1974).

True dormancy: Endogenous inhibition of growth which can be broken only after specific requirements, e.g. chilling, changes in light supply and length of photoperiod, are satisfied; there is no true domancy if plants can resume growth when environmental conditions become favourable for metabolic activity (adapted from Giertych 1974 and Saure 1985).

Taxonomic Index

Subject Index

For plant genera and species see Taxonomic Index

MIX
Papier aus verantwortungsvollen Quellen
Paper from responsible sources
FSC® C105338

If you have any concerns about our products,
you can contact us on
ProductSafety@springernature.com

In case Publisher is established outside the EU,
the EU authorized representative is:
Springer Nature Customer Service Center GmbH
Europaplatz 3, 69115 Heidelberg, Germany

Printed by Libri Plureos GmbH
in Hamburg, Germany